Tech Manual to Accompany

AUTOMOTIVE TECHNOLOGY

A SYSTEMS APPROACH

6th Edition

Rob Thompson

Jack Erjavec

CENGAGE

Australia • Brazil • Mexico • Singapore • United Kingdom • United States

Tech Manual to Accompany Automotive Technology: A Systems Approach, 6e

Rob Thompson & Jack Erjavec

VP, General Manager, Skills and Planning: Dawn Gerrain

Director, Development, Global Product Management, Skills: Marah Bellegarde

Product Manager: Erin Brennan

Senior Product Development Manager: Larry Main

Senior Content Developer: Meaghan Tomaso

Product Assistant: Scott Royael

Marketing Manager: Linda Kuper

Market Development Manager: Jonathon Sheehan

Senior Production Director: Wendy Troeger

Production Manager: Mark Bernard

Senior Content Project Manager: Cheri Plasse

Art Director: Cheryl Pearl

Media Developer: Debbie Bordeaux

Cover image(s): ©denverdave/Shutterstock

Library of Congress Control Number: 2013940142

ISBN-13: 978-1-133-93373-1

ISBN-10: 1-133-93373-4

Cengage

20 Channel Center Street
Boston, MA 02210
USA

Cengage is a leading provider of customized learning solutions with office locations around the globe, including Singapore, the United Kingdom, Australia, Mexico, Brazil, and Japan. Locate your local office at: **www.cengage.com/global**

Cengage products are represented in Canada by Nelson Education, Ltd.

To learn more about Cengage platforms and services, register or access your online learning solution, or purchase materials for your course, visit **www.cengage.com**

Notice to the Reader

AUTOMOTIVE TECHNOLOGY

CONTENTS

CHAPTER 46 Suspension Systems   594

CHAPTER 47 Steering Systems   608

CHAPTER 48 Restraint Systems: Theory, Diagnosis, and Service   624

CHAPTER 49 Wheel Alignment   634

CHAPTER 55 Air-Conditioning Diagnosis and Service   747

PREFACE

Alone, raw knowledge of an automobile's components and systems does not make someone a technician. Today's automotive technician must be able to think logically about each of the vehicle's systems and its function and performance within the whole of the automobile. To comprehend the fundamental concepts that define how the modern vehicle operates, this *Tech Manual* focuses on developing an understanding of the foundational principles and basic skills necessary for diagnosing and servicing today's cars and light trucks.

Throughout this educational package, professionalism is a constant theme. This *Tech Manual* continues this theme by offering hands-on shop experience with an emphasis on practical, real-life skills and logical troubleshooting. The manual follows the text's section and chapter breakdown, and offers these features to help you build your professional skills:

- *Concept Activities* are assignments that develop the fundamental understanding of principles and concepts related to the automobile. Many of these activities require supplemental teaching aids to provide visual, hands-on, and demonstrative experience.

- *Job Sheets* are lab activities that focus on diagnostic, service, and repair procedures. Most *Job Sheets* focus on basic skill development and service procedures meant to reinforce the *Concept Activities* and the textbook.

- *Review Questions* test your understanding and skills.

- *ASE Prep Tests* are to help prepare you to pass the ASE automobile tests.

ASSUMPTIONS

This *Tech Manual* includes several assumptions regarding the delivery of its content and use of the *Job Sheets*.

- To be used effectively, many of the *Concept Activities* will require some preparation by the instructor. Some of the activities specify common lab items; some use items commonly used in a school; and others may require preplanning to use as outlined.

- The *Tech Manual* assumes that students have access to adequate service information, whether in the form of manuals or electronic media. Service information is no longer listed in the tools and equipment necessary in the *Job Sheets*.

- The *Tech Manual* assumes all appropriate safety instruction has taken place before the lab work begins. Because proper protective equipment is mandated by most or all training programs, the reiteration of required safety glasses, shoes, and other safety apparel has been removed.

- Because professionalism is a constant theme, the *Tech Manual* assumes fender covers, floor mats, and other protective means are being taught and used when servicing vehicles.

The author hopes the concepts and activities included in the *Tech Manual* will provide you with new and effective ways of teaching automotive technology.

CAREERS IN THE AUTOMOTIVE INDUSTRY

CHAPTER

1

OVERVIEW

In Chapter 1 of Automotive Technology: A Systems Approach, *many aspects of the automotive industry were looked at. One point from your reading should have stood out—there is and will continue to be a need for qualified automotive service technicians. This need plus many other benefits make being a technician a worthwhile career.*

THE IMPORTANCE OF AUTOMOTIVE TECHNICIANS

Technology is rapidly changing in the automotive industry, and so are the people working in the industry. Electronics plays a major role in the operation of automobiles today, controlling most of the functions that allow the vehicle to operate. The changes in technology and the extensive use of electronics have changed the requirements for the technician needed to service the automobile of today and tomorrow.

CAREER OPPORTUNITIES

There are many opportunities for those who would like to enter the automotive service industry in new car dealerships, independent service shops, franchise repair shops, store-associated shops, and fleet service and maintenance.

Concept Activity 1-1
Explore the Automotive Industry

Use an *Occupation Outlook Handbook* or go to the online version at http://www.bls.gov/OOH/ to gain additional information about automotive service technicians. Automotive technicians are listed under the heading of Installation, Maintenance, and Repair occupations.

1. Describe some of the work performed by automotive service technicians.

_____

_____

_____

2. Describe the job outlook for automotive technicians.

_____

_____

3. What is the income range for technicians?

_____

_____

_____

■

JOB CLASSIFICATIONS

A variety of job classifications exists in the automotive industry. Jobs related to the service technician are service advisor, service manager, and parts manager. Other related areas of employment are parts distribution, marketing and sales, body and paint repair, and trainers for various manufacturers or instructors for automotive programs.

Concept Activity 1-2
Explore Job Opportunities

Search for automotive technician job openings in your area. It is common for dealerships and larger shops to place their job openings on their websites. In addition, many local newspaper and television station websites will have job postings. You can also look through the help-wanted section of the Sunday edition of your local newspaper. Create a list of three of these jobs, and describe what skills, experience, and certifications are required for each.

Job 1: _____

_____

Job 2: _____

_____

Job 3: _____

_____

Do you see yourself, in the next year or two, in one of the jobs you found? Consider these jobs, and think about your personal education and career goals. What will you need to do to accomplish these goals?

_____

_____

_____

■

TRAINING FOR A CAREER IN AUTOMOTIVE SERVICE

Extensive training is available for those wanting to enter or update themselves in the automotive industry. Training is available through manufacturers, trade schools, vocational schools, and community colleges. Manufacturers offer entry-level programs such as Chrysler CAPS, Ford ASSET, General Motors ASEP, and Toyota T-TEN, which are programs dedicated to the manufacturers' specific products. Trade schools, vocational schools, and community colleges offer automotive programs under several formats: mentoring programs, cooperative programs, apprenticeship programs, and postgraduate education.

Due to the ever changing technology used in today's vehicles, automotive technicians must constantly update their knowledge base. Training programs are offered by a variety of sources such as OEMs, aftermarket part suppliers, and independent training groups. Other informal sources of information regarding new products and new vehicle systems include trade magazines, vehicle magazines, and current automotive textbooks.

Concept Activity 1-3
Explore Training and Education Options

1. If you are a high school student, make a list of colleges or other postsecondary schools in your area that offer automotive training.

2. If you are a postsecondary student, list what type of automotive training programs are available at the school you attend.

3. Determine what education and training opportunities are available in your area from sources other than schools. For example, list any local automotive parts suppliers that offer training for technicians.

■

ASE CERTIFICATION

ASE certification is recognized by the automotive industry as proof of minimum competency. Many service facilities require that their technicians be ASE certified. As a student, you may take the ASE Student Certification tests, which are valid for two years. More information about this certification can be found at www.asestudentcertification.com/.

You can also take the professional certification exams while in school. This certification lasts for five years. If you pass the professional certification exam(s), you will not become certified until you have had two years of experience. But you will not need to take the test again to become certified.

Concept Activity 1-4
Explore ASE Certification

Visit the ASE website at http://www.asecert.org/ and explore certification.

1. What are the requirements to become an ASE certified technician?

_____

_____

2. Which areas of ASE certification interest you the most and why?

_____

_____

_____

■

REVIEW QUESTIONS

Review Chapter 1 of the textbook to answer these questions:

1. What types of jobs are available for those wanting to work in the automotive industry?

   _____

   _____

   _____

2. What is a cooperative education program?

   _____

   _____

   _____

3. Explain what a new car warranty is and two different types of warranties provided with a new car or truck.

   _____

   _____

   _____

4. List three things that service technicians must be able to do on a regular basis.

   _____

   _____

   _____

5. What are the requirements for obtaining ASE certification?

   _____

   _____

   _____

   _____

   _____

OVERVIEW

Your ability to develop career goals, write cover letters, create résumés, prepare for interviews, and finally accept the offered position of employment and its pay structure were addressed in Chapter 2 of Automotive Technology: A Systems Approach. *Once you obtain employment, the opportunity to demonstrate other workplace skills presents itself. These include creating good employer–employee relationships, being a good communicator, and having the ability to solve problems using critical thinking. Demonstrating a professional approach to your job in dealing with customers and fellow workers is expected. Both verbal and written communication skills are important when your position requires writing repair work orders and estimates. Your ability to summarize the customer concerns and write these concerns precisely on the repair order is critical to communicating with the technician for the repair of the vehicle. Accurate and detailed notations regarding your diagnostic and repair activities help ensure that you are properly compensated for your work.*

SEEKING AND APPLYING FOR EMPLOYMENT

Begin by creating a criteria list for the type of employment you want. The list would include some or all of the following: location (e.g., city or suburbs); technician or other service department positions; expected wages; work schedule; and dealership or aftermarket shop.

Employment Plan

Your employment plan is a blueprint of what your career goals are for the next 6 months, 5 years, and 10 years. Career goals allow you to judge your career progress by what your expectations are for the given time period. Your career goals may become a topic in a job interview. For example, employers are interested in knowing if the new hire will become a long-term employee. An employment plan usually serves as the basis for your résumé and cover letter.

Concept Activity 2-1
Create an Employment Plan

Create an employment plan by honestly appraising yourself and your goals.

Employment goals: _____

_____

Time frame to meet employment goals: _____

_____

Skills, both automotive and others: _____

_____

List of potential employers: _____

_____

Personal strengths related to goals: _____

_____

Areas requiring improvement: _____

_____

■

Preparing Your Résumé

Your résumé is a summary of your life experience in the areas of education, volunteer work, and work experience. If you have little or no experience in preparing a résumé, look up the topic at the library and on the Internet. Once you have created a first draft, have a family member or relative read it and provide constructive criticism. It may take several drafts before it is ready to present to a potential employer.

 Concept Activity 2-2
Preparing a Résumé

Prepare a résumé including the following elements: contact information, career objectives, skills and/or accomplishments, work experience, education, and a statement about references. To help with this, résumé templates are available in Microsoft Word and on the Internet. ■

Preparing Your Cover Letter

The cover letter is a tool for showing how your experience, education, communication skills, ability to be a team player, ability to work independently, leadership skills, and task-oriented skills apply to a job description. It is your opportunity to sell yourself to the prospective employer. The goal of the cover letter is to obtain an interview for the position.

Contacting Potential Employers

The job advertisement or posting will give directions on how to contact the potential employer. For example, the advertisement may state that interested applicants should respond by e-mail, or it may say "no phone calls" or "do not contact us regarding the progress of the applicant selection." The information may state that only those selected for an interview will be contacted. The job posting most often has a closing date and time. It is important that you follow the stated directions; otherwise, your application may end up in the wastepaper basket.

If you are canvassing certain areas of the city to give out résumés, with the hope of talking to the person doing the hiring, make a phone call to the place of business first. The phone call gives you an opportunity to identify the contact person, when that person would be available to receive your résumé, and if there is an opportunity to complete a job application form. Equally important, it gives the prospective employer some familiarity with the name attached to the résumé. It is also important to note that some employers frown on people dropping in unannounced, which creates a negative impression.

Applications

Most businesses use online applications. In some cases, the application is completed at the place of business, whereas other applications can be completed from any computer. Some places still use a traditional paper application. Regardless of the type of application, the job application form is a legal document that allows you, as the job applicant, to enter accurate and precise information regarding who you are. There is considerable information to list on the document, so it is important that you are prepared with the information you expect to give. A small notebook of facts, figures, and work history, as well as a pocket dictionary, are helpful when entering information.

Preparing for the Interview

Many questions may be asked during an interview, which may last 45 to 60 minutes. Remember that all you can do is answer the questions to the best of your ability. Most people agree that an interview is a stressful process, but preparation can reduce the stress. Although you can never anticipate all of the questions that may be asked during an interview, having ready answers for the following possible interview questions can help you to be prepared:

- How would you describe yourself?
- What do you consider your greatest strength?
- How would you calm an angry customer?
- Why did you leave your previous employer?
- Why did you apply for this position?
- What was your most embarrassing customer relations experience?
- What are your career goals?
- Do you have any questions about the job or company?

The Interview

An interview is a process that allows the employer to have face-to-face contact with the job applicant. The employer may conduct the interview personally or may be represented by a selection panel consisting of a number of people responsible for determining the best candidate for the position. Each panel member will ask one or more questions of the candidate. During an interview, your ability to answer the questions and your body language (nonverbal communication) will be observed.

After the Interview

When the interview is complete, reflect on the experience. Making a list of the pros and cons about the interview allows you to learn from the experience. You will become more confident as your awareness grows about the types of questions asked and what is expected.

ACCEPTING EMPLOYMENT

If you are the successful candidate for the position, an offer of employment will be made to you. The conditions of employment and employee expectations will be discussed at this time, such as hours of work, supervisor, dress code, fringe benefits, and so on. One of the more important employment conditions is payment. The company will offer a salary or hourly wage, and the person offering the position will indicate what your starting wage will be, what the increments are, and when you will rise to the next level. During the discussion, a starting date will be agreed on. The starting date will depend on whether you are currently employed and must give notice to your present employer, among other possible circumstances.

WORKING AS AN AUTOMOTIVE TECHNICIAN

Being the successful candidate for an automotive position launches your career as an automotive technician. How long you have the position depends totally on you. How well you interact with supervisors and colleagues and perform your duties as a technician, your punctuality, and your positive attitude will determine your longevity in the position.

A successful automotive technician participates in lifelong learning as a means of keeping up to date with technological change in the workplace. Good writing skills to communicate information, as well as reading skills to interpret service manual information, are a must.

Technicians need to have basic computer and keyboarding skills in the workplace. Computerization is present in so many areas of our daily lives that it is critical to have the basic skills. In the automotive environment, the computer controls many of the vehicle operating systems, aids in diagnosing vehicle faults, allows for tracking customers and for record keeping, and acts as a source of information. For example, technical information may be sourced by computer programs such as Mitchell1® and ALLDATA®.

Employer–Employee Relationships

As an employee you exchange time and job skills for money. A good employee has many attributes besides job skills. Other things must be considered as part of the employee–employer interaction.

The employer has certain responsibilities: instruction and supervision; providing a clean, safe, and harassment-free work environment; paying for services rendered; providing the fringe benefits as agreed at the point of hiring; and providing the opportunity to succeed, with the possibility of career advancement without prejudice or nepotism.

Employees have certain responsibilities to the employer. Employees are expected to show up for work on time and have regular attendance; be responsible for their actions; give an honest day's work; and demonstrate loyalty to the company.

COMMUNICATIONS

Good communication involves interaction between one or more persons who demonstrate the skills of speaking, listening, reading, and writing. These skills allow us to convey our thoughts and ideas about a given subject, and our ability to totally understand or comprehend the intent of the message depends on its clarity. In face-to-face speaking and listening, nonverbal communication plays an important role.

Nonverbal Communication

Nonverbal communication is a complex part of any interaction. It is the facial expression, voice tone, body expression, and body position that both the speaker and listener project while a conversation is taking place. These visual signs depict happiness, sorrow, confusion, or anger, to name a few emotions, and are interpreted by both participants. In reading and writing, the lack of nonverbal communication means the writer must write into the communication the expression of feeling by very carefully choosing the right words to convey the mood.

SOLVING PROBLEMS AND CRITICAL THINKING

Because of the varying levels of complexity in repairing system faults in a vehicle, the technician must be able to convert theoretical knowledge and interpret the service manual information to create a logical and methodical approach to the practical repair of the vehicle. The ability of the technician to solve problems using critical thinking is the skill needed to be able to repair the vehicle right the first time.

Diagnosis

Diagnosing a customer's vehicle begins with the complete understanding of the customer's concern. The next step is to verify whether the concern is normal operation or a vehicle fault. When it has been determined that there is a fault, the goal is to repair the vehicle right the first time. To accomplish this, you must have a methodical and systematic approach to pinpoint the cause of the concern. It is important to consider all the necessary resources available to you, such as diagnostic tables and repair flowcharts, which help in narrowing down possibilities and in finally determining the cause of the fault.

PROFESSIONALISM

Others notice how you conduct yourself in dealing with and treating them. Verbal and nonverbal communications play a major role in your projection of professionalism. Respect for yourself and other people is of the utmost importance. Interacting with people in a mature, respectful way is a major part of professionalism, which also involves demonstrating a positive attitude, displaying the appropriate behavior, and accepting responsibility for your actions. As a professional demonstrating these and other positive qualities, you will earn respect in all aspects of your life.

Coping with Change

Change is a personal dilemma, one that may not be easy to accept. Change is unavoidable, whether it is at home or work. Change creates stress in most people. When your environment is stable and your feeling is one of contentment and comfort, you feel safe and your stress levels are presumed to be low. When your comfort zone is interrupted by change, an adjustment period invariably takes place.

People either embrace or resist change to varying degrees, but resisting change causes greater levels of stress. Typical changes experienced by an automotive technician include a change in supervisors, technological change, or personal change.

INTERPERSONAL RELATIONSHIPS

As an employee of a company, you are expected to act in a way that demonstrates mutual respect toward your fellow workers. Most companies stress teamwork, where you work with colleagues in a cooperative, supportive, cohesive unit toward the common goals of the company.

Customer Relations

Customer relations are a very important aspect of doing business. Customers like to feel that their patronage is important to you. The interaction you have with the customer, creating an honest and trusting relationship, allows your company to have repeat business.

The customer expects you to repair the vehicle right the first time and to complete the repair procedure at the designated scheduled time. The service department's interaction with the customer influences the customer's decision to purchase another vehicle from that dealer.

REPAIR ORDERS

Information written on the work order describing the customer's concern with the vehicle must be accurate and precise. This information is critical for the technician to have the best opportunity to diagnose and repair the vehicle fault. Information concerning the vehicle particulars such as make, model, VIN, correct odometer reading, and the customer's signature on the repair order are extremely important. A repair order is a legal document that may be required if litigation arises because of unforeseen circumstances.

 REVIEW QUESTIONS

Review Chapter 2 of the textbook to answer these questions:

1. Explain the purpose of a cover letter.

   _____

   _____

   _____

2. Describe the different ways automotive technicians are compensated.

   _____

   _____

   _____

3. List four responsibilities of the employee in the employee–employer relationship.

_____

_____

_____

4. Describe the difference between verbal and nonverbal communication.

_____

_____

_____

5. Explain several characteristics of positive interpersonal relationships.

_____

_____

_____

ASIC THEORIES ND MATH

CHAPTER

3

OVERVIEW

This chapter provides an introduction to or review of a number of basic theories and expands on the definitions of terms needed to understand the theories. Some of the theories discussed include mathematical formulas for calculating values applicable to an automotive context. The major topics of the chapter are matter, heat, energy, volume, force, motion, friction, simple machines, waves, hydraulics, gases, chemical properties, and applications of electromagnetism.

Suggested Materials: To perform the activities and job sheets in this chapter, it is recommended to have the following items:

Thermometer

Beakers

Hot plate

Vacuum gauges

Syringes

Plastic water bottle

Balloons

Baking soda

Carbonated soda drink

Compass

Magnets

MATTER AND HEAT

All matter exists in one of three forms, as a solid, a liquid, or a gas. Some types of matter, such as water, can exist in all three forms, though not without an outside factor, such as heat, causing the change in state. Water is the only substance on earth that we experience as a solid, liquid, and a gas. For water to change state from a solid to a liquid, to a gas, heat transfer must take place. Heat can be detected down to absolute zero, about –459°F (–273.15°C), at which all molecular activity stops.

Water will freeze, becoming a solid, when enough heat is removed. If heat is added to water, enough for it to reach 212°F (100°C), it boils, turning it into a gas. Once the water vapor cools (or as the heat is removed), it condenses back into liquid water.

**Concept Activity 3-1
Heat**

1. Place some cold tap water into a glass or similar container. Place a thermometer into the water and record its temperature. _____ °F (°C)

2. Next, place some ice into the water, and note the water temperature every few minutes until you achieve the lowest steady reading.

Time _____ Temperature °F (°C) _____

Time _____ Temperature °F (°C) _____

Time _____ Temperature °F (°C) _____

Time _____ Temperature °F (°C) _____

Time _____ Temperature °F (°C) _____

3. What is the lowest temperature the water reached? _____

4. Why does the water not get any colder? _____

5. What would need to take place to cause the entire quantity of water to freeze? _____

When water is heated to its boiling point, 212°F (100°C), the heat causes the water to turn to a vapor. Ask permission to perform this activity before continuing. Boiling water can cause severe burns and serious injury.

6. Measure the water temperature, and then begin to heat a pan or beaker of water while observing the temperature with an appropriate thermometer.

Time _____ Temperature °F (°C) _____

Time _____ Temperature °F (°C) _____

Time _____ Temperature °F (°C) _____

Time _____ Temperature °F (°C) _____

Time _____ Temperature °F (°C) _____

7. What is the highest temperature the water reached? _____

8. Why does the water not get any hotter? _____

9. What happens to the quantity of water as it boils? _____

■

Every substance needs a specific amount of heat, either applied or removed, to cause it to change state. Changing from a liquid to a gas is called evaporation. Changing from a gas into a liquid is called condensation. This heat transfer is the basis of air-conditioning systems.

ENERGY

Energy is neither created nor destroyed; it is simply converted from one form into another. When a spring is compressed, it stores kinetic energy, which it can release to perform work. When a substance is burned, it releases heat energy. Automobiles use many forms of energy to accomplish their various functions.

 Concept Activity 3-2
Energy Conversion

Match the following types of energy conversion with its corresponding components.

Chemical to thermal   Headlight

Chemical to electrical   Window motor

Electrical to mechanical   Air/fuel burning

Thermal to mechanical   Brake pads

Electrical to radiant   Battery

Mechanical to thermal   Combustion

Mechanical to electrical   Generator

■

CONVERSION TABLES AND COMMONLY USED FORMULAS

TABLE 3-1 Conversion Tables and Commonly Used Formulas.

Into Metric

If you know	Multiply by	To Get
Length		
inches	2.54	centimeters
foot	30	centimeters
yards	0.91	meters
miles	1.6	kilometers
Area		
sq. inches	6.5	sq. centimeters
sq. feet	0.09	sq. meters
sq. yards	0.8	sq. meters
sq. miles	2.6	sq. kilometers
acres	0.4	hectares
Mass (Weight)		
ounces	28	grams
pounds	0.45	kilograms
short ton	0.9	metric ton
Volume		
teaspoons	5	milliliters
tablespoons	15	milliliters
fluid ounces	30	milliliters
cups	0.24	liters
pints	0.47	liters
quarts	0.95	liters
gallons	3.8	liters
cubic feet	0.03	cubic meters
cubic yards	0.76	cubic meters
Temperature		
Fahrenheit	Subtract 32, then multiply by 5/9ths	Celsius

Out of Metric

If you know	Multiply by	To Get
Length		
millimeters	0.04	inches
centimeters	0.4	inches
meters	3.3	feet
kilometers	0.62	miles
Area		
sq. centimeters	0.16	sq. inches
sq. meters	1.2	sq. yards
sq. kilometers	0.4	sq. miles
hectares	2.47	acres
Mass (Weight)		
grams	0.035	ounces
kilograms	2.2	pounds
metric tons	1.1	short tons
Volume		
milliliters	0.03	fluid ounces
liters	2.1	pints
liters	1.06	quarts
liters	0.26	gallons
cubic meters	35	cubic feet
cubic meters	1.3	cubic yards
Temperature		
Celsius	Multiply by 9/5ths, then add 32	Fahrenheit

TABLE 3-2 Common Conversions.

Shapes	Formula
	Rectangle Area = Length × Width A = lw Perimeter = 2 × Lengths + 2 × Widths P = 2l + 2w
	Triangle Area = 1/2 of the base × the height a = 1/2 bh Perimeter = a + b + c (add the length of the three sides)
	Circle The distance around the circle is the circumference. The distance across the circle is the diameter (d). The radius (r) is the distance from the center to a point on the circle. (Pi = 3.14) d = 2r c = πd = 2πr A = πr² (π = 3.14)
	Rectangular Solid Volume = Length × Width × Height V = lwh Surface = 2lw + 2lh + 2wh
	Cylinder Volume = πr² × height V = πr² h Surface = 2π radius × height S = 2πrh + 2πr²

© Cengage Learning 2015

Proportions and ratios: If the cylinder is half full, the proportions are equal.

© Cengage Learning

_____ Full
_____ 1/2 full

If this cylinder is 10 inches high and 6 inches across, we can find the volume by using
$V = \pi r^2 h$. $V = 3.1416 \times 3^2 \times 10$, so the volume equals 282.74 cubic inches.

If a 10-cubic-inch cylinder is full of air and then the air is compressed into 1/5 of its previous volume, the ratio of compression is 5:1.

VEHICLE DESIGN INFLUENCES

Cars and trucks have changed greatly over the 100 plus years of their existence. Many of the changes are due to the demands of the consumer.

 Concept Activity 3-3
Vehicle Design

1. List three important factors that have influenced modern vehicle design:

   a. _____

   b. _____

   c. _____

2. Describe what features you think may be found on cars and trucks 10 to 20 years from today.

   _____

   _____

   ■

NEWTON'S LAWS OF MOTION

Over three hundred years ago, Isaac Newton formulated three laws of motion, all of which apply to the automobile. Newton's first law, sometimes referred to as the law of inertia, says that an object will stay at rest or in motion unless acted upon by an outside force. The first part of the law, an object at rest will remain at rest, is easily understandable because desks, chairs, and automobiles do not just move about on their own. The second part, though, requires some thought. What Newton is saying is that if an object were moving in a straight line, and no outside forces such as gravity or air resistance were acting on the object, it would continue to move in a straight line forever. This is not easy to experiment with, because we are all subject to gravity and various resistances to movement. However, moving objects have inertia, and anyone who has ever rollerbladed or skateboarded into a piece of gravel knows that skating along and encountering an obstacle can abruptly send you flying forward due to your momentum.

Newton's second law states that an object's momentum is a proportional result of the force acting on the object. Another way to view this law is that the acceleration of an object is proportional to the force applied and inversely proportional to its mass. If you have ever had to push a car or truck that has run out of gas, you have learned this law. One person trying to push a 2-ton vehicle on level ground will not be moving very fast. If that person were to try to push a 4-ton vehicle, he or she would only be able to push it half as fast as the 2-ton vehicle.

Newton's third law states simply that for every action there is an equal and opposite reaction. If you have ever stepped from a small boat to a dock, you have probably noticed that the force of your legs pushing you away from the boat also pushes the boat away from you. The force moving you and the force against the boat are equal and moving in opposite directions. In automotive applications, most disc brake systems rely on Newton's third law to apply pressure to the two brake pads equally. Hydraulic pressure pushes a piston out against a pad, which forces the caliper to move backward, pushing the other pad against the disc with equal force **(Figure 3-1)**.

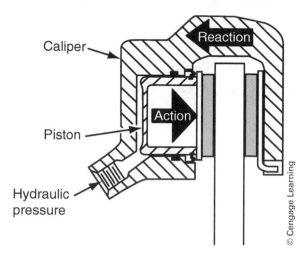

Figure 3-1 Hydraulic pressure in a sliding or floating caliper forces the piston and one pad in one direction and the caliper body and the other pad in another direction.

 Concept Activity 3-4
Newton's Laws

1. Describe another automotive situation in which Newton's first law applies.___

2. Using Newton's $f = ma$ equation, determine the amount of force needed to push a 4,000-pound car two miles per hour. ___

3. Describe another application of Newton's third law of motion. ___

FRICTION

Friction is a result of two objects in contact with each other moving against each other. Even objects that appear smooth have imperfect surfaces. These surfaces, when against other surfaces, resist movement.

 Concept Activity 3-5
Friction

1. Think about your experiences with automobiles, and list as many examples of friction as possible.

_____

The amount of friction between two objects is dependent upon the materials of which they are made and the force pushing the two surfaces together. The force needed to move the objects compared to the mass of the objects is called the coefficient of friction.

Imagine that you have a large square block of rubber in your lab that weighs 100 pounds.

2. How much force do you think would be needed to slide that block across the floor? _____

If the required force is 100 pounds and the mass is 100 pounds, then the coefficient of friction would be equal to 100/100, or 1. Now imagine that you have a 100 lb block of ice sitting on the lab floor.

3. How much force do you think would be needed to slide the ice across the floor? _____

As you can imagine, moving the ice should be easier than moving the rubber; therefore, the coefficient of friction is lower. The automotive brake system relies on the coefficient of friction to remain consistent.

4. Describe how the brakes would operate if the coefficient of friction were too low. _____

5. Describe how the brakes would operate if the coefficient of friction were too high. _____

■

THE SIMPLE MACHINES

The existence of the six simple machines dates back thousands of years. Each of the machines is designed to function needing the application of only one force. This means that each can provide mechanical advantage to make work easier. The six simple machines are as follows:

■ The inclined plane—The inclined plane, used everyday as a ramp, allows heavy loads to be carried up to a greater height more easily than those just lifted straight up. Using the plane does require the load be moved a longer distance.

■ The wheel and axle—Wheels and axles and doorknobs are everyday examples of this machine.

■ The lever—Levers are used as teeter-totters, claw hammers, pry bars, can openers, pliers, and countless other tools. The lever allows for greatly increasing a force by applying the force over a longer distance.

■ The pulley—The pulley can be used as a wheel with a rope to change the direction of a force and increase the advantage of the force over a longer length of rope. Pulleys on engines turn accessories such as the power-steering pump and air-conditioning compressor. These pulleys are different sizes than the crank-shaft pulley, to increase or decrease the component's rotational speed.

■ The wedge—The wedge is basically a portable inclined plane. The wedge is used to separate or split objects. Wedges are used when splitting wood logs into pieces.

■ The screw—The screw is another form of inclined plane, one that is wrapped into a helical shape. A screw can convert rotational force into linear force or the reverse. Screws hold engines, cars, and bridges together. A type of screw used in the vehicle's steering gearbox to transform the rotational motion of the steering wheel into the linear motion that turns the front wheels is called a worm screw.

 Concept Activity 3-6
Simple Machines

1. Match the following simple machines with their common application:

Screw	Nail
Pulley	Wheelbarrow
Wedge	Ferris wheel
Lever	Crane
Wheel and axle	Playground slide
Inclined plane	Key ring

■

TORQUE AND HORSEPOWER

Two of the most often talked about terms related to cars are torque and horsepower. Though often discussed, torque and horsepower are also often misunderstood. Torque is a force that tends to twist or rotate things, and it is measured by the force applied over the distance traveled. As an example, if you needed to tighten the lug nuts on a car to 100 foot-pounds (ft-lb), you would have to apply 100 pounds of force to a bar 1-foot long to achieve 100 ft-lb. Because we already know that levers can increase force or advantage, we can use a longer bar, so less force will need to be applied to reach the desired torque. The metric measurement of torque is the Newton-meter. One ft-lb of torque is equal to about 1.36 Newton-meters.

Concept Activity 3-7
Torque and Horsepower

1. How much force will need to be applied to a torque wrench 2 feet long to achieve a torque of 100 ft-lb?

_____

2. How much force will need to be applied to a torque wrench 0.5 m long to achieve a torque of 80 N-m?

_____

3. How much force will need to be applied to a torque wrench 6 inches long to achieve a torque of 20 ft-lb?

_____

Power is the measurement of the rate at which work is done. Power, expressed in watts, is a unit of speed combined with a unit of force. For example, if you pushed a cart with a force of one Newton at a speed of 1 meter per second, the power output would be 1 watt. The term *watt* is taken from James Watt, an English inventor who is credited with calculating horsepower. By measuring the amount of weight a horse could move over distance and time, Watt derived the calculation of horsepower. By observing that a horse could move 330 pounds 100 feet in 1 minute, he calculated 1 horsepower to 33,000 foot pounds per minute, or 550 ft-lbs per second (330 × 100 = 33,000 lb-ft per minute). Two horsepower could do the same amount of work in half the time. (330 × 100 × 2 = 66,000 lb-ft per minute. 66,000 lb-ft per minute = 33,000 lb-ft in 30 seconds.)

4. If you pushed a 3,000-pound car 11 feet in 15 seconds, how much horsepower did you produce?

_____

To find the horsepower of an engine use this formula: horsepower = torque × engine speed/5,252. The constant of 5,252 is used to covert rpm into revolutions per second. If an engine produces 250 ft-lb of torque at 4,500 rpm, it will produce 214 HP at 4,500 rpm.

5. What is the horsepower of an engine with 300 ft-lb or torque at 4,000 rpm?

_____

As you can see, torque and horsepower are related. An engine that can produce torque at higher rpm will be able to accelerate over a longer period of time than an engine that produces its maximum torque at a lower rpm. ■

VIBRATION AND SOUND

Vibrations and sounds are caused by oscillations, the back-and-forth movement of an object between two points. When a part oscillates, like a spring bouncing, it moves the air around the part, creating an airwave. The wave travels through air in a series of oscillating high and low points, called compression waves, which we can often hear as sound **(Figure 3-2)**.

Sound waves

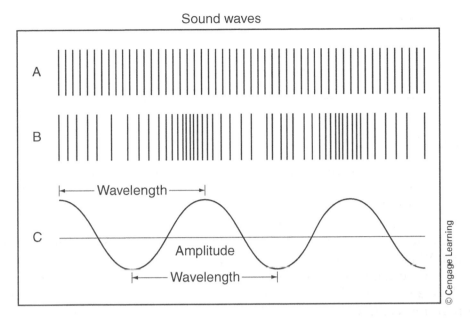

Figure 3-2   (A) Air without sound; (B) Compressions and rarefactions of sound wave; (C) Sound wave amplitude and wavelength.

Vibrations are caused by an oscillating component. If you have ever watched a running engine, you may have noticed the engine vibrating slightly. To reduce the effect of this vibration, the engine is mounted with special engine mounts, which are designed to reduce these vibrations and prevent them from transferring to the body of the vehicle, where the passengers can feel them. Vibrations can also result from a component that rotates but is not balanced correctly. If a wheel and tire assembly is not balanced correctly, as the wheel speed increases, a vibration that shakes the steering wheel or body of the vehicle can result.

Concept Activity 3-8
Vibrations and Sound

1. List three components or systems that can cause unwanted vibrations on a vehicle.

Sound is a result of vibrations, though some vibrations can occur at frequencies either below or above the human range of hearing. Frequency is the number of times a vibration occurs and is usually rated in hertz (Hz), which is the number of cycles per second. A sound with a vibration of 200 Hz will have 200 oscillations per second. A sound with a frequency of 2 KHz (kilohertz, or 2,000 Hz) will oscillate 2,000 times per second. The lower the hertz rate, the lower the frequency and the tone of the sound. The higher the frequency, the higher pitch the sound will be. Humans can usually hear sounds between 20 Hz and 20,000 Hz.

Being able to identify various sounds is an important diagnostic skill. By listening to a noise and its relation to engine speed or wheel speed, you can more easily locate the source of the problem.

2. A customer complains that there is a high-pitched rattling noise from her vehicle, which gets louder and increases in frequency when engine speed is increased. The noise occurs in park and while driving. Which of these is the most likely cause and why?

   a. Loose wheel bearings

   b. Loose drive belt

c. Loose exhaust shield

d. Loose motor mount

Why? _____

_____ ∎

PASCAL'S LAW

All modern automobiles use a hydraulically operated brake system. The same principles that define how the hydraulics work in construction equipment, airplanes, and hydraulic jacks and presses also explain automotive brake systems. Cars and light trucks have been using hydraulically operated brakes since the 1920s and will continue to use them in the foreseeable future.

French mathematician and scientist Blaise Pascal experimented with and developed the laws of hydraulic systems. What Pascal determined was that a fluid contained within a closed system could transmit force and motion. By applying pressure to the fluid via a piston, a second piston could also be moved and produce a force equal to that applied to the first piston **(Figure 3-3)**. Because the fluid can transmit motion and force, the forces can be increased or decreased by changing the sizes of the pistons. When there is a change in force, there is a change in the amount of motion or distance as well; this means the motion can also be increased or decreased **(Figure 3-4)**.

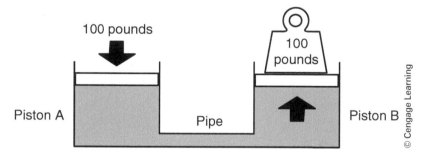

Figure 3-3   As hydraulic fluid transmits motion through a closed system, it also transmits force.

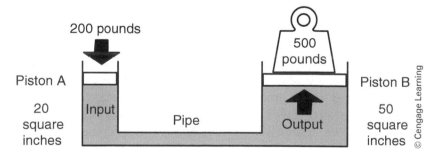

Figure 3-4   A hydraulic system also can increase force.

The automotive brake system uses the master cylinder as the hydraulic input, and the calipers and wheel cylinders as the outputs. Disc brake caliper pistons are larger than the master cylinder pistons. This allows for more force to be generated by the caliper piston. The larger caliper piston also means it will move less than the master cylinder pistons. A vehicle with rear drum brakes has a wheel cylinder to apply the brake shoes outward against the brake drum. Wheel cylinders have small pistons, smaller than the master cylinder pistons. This means the wheel cylinder pistons will apply a smaller amount of force but will be able to move a longer distance **(Figure 3-5)**.

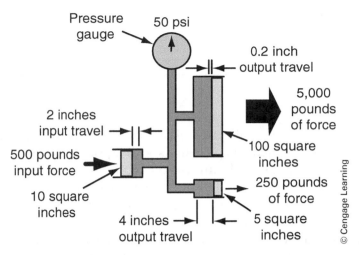

Figure 3-5 Hydraulic systems can provide an increase in force (mechanical advantage), but the output travel will decrease proportionally.

Concept Activity 3-9
Hydraulic Principles

1. Label the forces of each piston in the diagram below **(Figure 3-6)**:

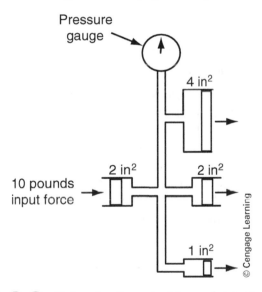

Figure 3-6 Determine the output forces for each piston.

Master cylinder output pressure: _____

Output force of 4-inch piston: _____

Output force of 2-inch piston: _____

Output force of 1-inch piston: _____

2. Why do the disc brake calipers have much larger pistons than wheel cylinders?

_____

3. Why are the front caliper pistons larger than the rear caliper pistons on a four-wheel disc brake vehicle?

_____

4. What other automotive systems use hydraulics to operate?

_____

5. List three pieces of equipment in your automotive lab that utilize hydraulics.

_____

■

GASES

Gases, as a state of matter, consist of a collection of particles without any specific shape. A gas will conform to and fill any container in which it is placed. As with liquids, gases move from higher pressure to lower pressure. When a piston in a cylinder moves down, the pressure in the cylinder decreases. The higher atmospheric pressure forces the air into the lower pressure cylinder.

Unlike fluids, gases are compressible. Once the piston begins its upward movement, the piston compresses the air in the cylinder. This compression heats the air and increases pressure in the cylinder. This allows for a much greater extraction of power when the compressed mixture of air and fuel is ignited.

Boyle's Law and Charles's Law describe the relationship between pressure, volume, and temperature of gases. The following activities will demonstrate these laws.

Concept Activity 3-10
Vacuum and Pressure

Attach a vacuum/pressure gauge to a 100 cc syringe (or similar). With the syringe plunger approximately half retracted and the gauge attached to the small opening, pull the plunger out farther, and note the reading on the gauge.

1. Gauge reading: _____

2. Next, push the plunger in, and note the gauge reading. _____

The change in pressure, both a decrease and an increase, are a result of changing the volume inside the syringe. By increasing the volume, the pressure decreased, and by decreasing the volume, the pressure increased. This is an example of Boyle's Law.

To test the Pressure Law, that the volume of a gas depends on its temperature, you will need a balloon or vacuum/pressure gauge, bottled water, and a heating element.

Place approximately 4 ounces of water in a 16-ounce water bottle and attach either a balloon or vacuum/pressure gauge to the opening of the container. Place the bottle of water into a separate container, and place on the heating element. Heat the water until it reaches boiling, and note the pressure reading on the gauge or note the balloon.

3. Pressure reading: _____ or

   Balloon diameter: _____

4. Why did the pressure in the balloon increase? _____

5. What would happen to the pressure in the balloon if the water were cooled to the freezing point?

_____

■

CHEMICAL (OR PHYSICAL) PROPERTIES

A substance's chemical properties describe how one type of matter reacts with another during a chemical reaction. Some materials, such as iron and steel, react with oxygen to form rust. When oxygen bonds with hydrogen, water is formed.

There are many different chemical reactions associated with an automobile. The battery, through a chemical reaction between electrolyte and the cells is able to store and release electrical energy.

The components of gasoline and diesel fuel, when mixed with air and compressed and ignited, release a great amount of stored chemical energy during combustion.

Concept Activity 3-11
Chemical Reactions

Obtain three rusted bolts from your instructor. Place one bolt in a container with tap water; place another bolt in a container with a carbonated soft drink; and place the third bolt in a container with tap water and baking soda. Monitor the condition of the bolts over several days, and note their appearance.

Day 1 Bolt in water: _____

Bolt in soft drink: _____

Bolt in baking soda and water: _____

Day 2 Bolt in water: _____

Bolt in soft drink: _____

Bolt in baking soda and water: _____

Day 3 Bolt in water: _____

Bolt in soft drink: _____

Bolt in baking soda and water: _____

1. Describe your findings on the bolts after several days of soaking in the three containers.

_____

2. Why do you think the bolts reacted the way they did?

_____

■

OXIDATION AND REDUCTION (REDOX)

Oxidation and reduction refer to when a substance combines with oxygen (oxidation) or gains hydrogen atoms or electrons (reduction). Every day we witness signs of oxidation and reduction. When iron atoms combine with oxygen, rust is formed. In automotive systems, the catalytic converter converts NO and CO into N_2 and CO_2 (reduction) and converts HC, CO, and O_2 into H_2O and CO_2 (oxidation).

ELECTROMAGNETISM

Electromagnetism refers to the interrelated principles of electricity and magnetism. The modern world we know would not exist without electromagnetism. Whenever a current is carried by a conductor, a magnetic field forms around the conductor. When a conductor is moved at a right angle through a magnetic field, a current is induced into the conductor (**Figure 3-7**).

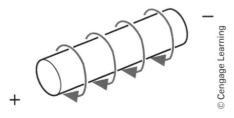

© Cengage Learning

Figure 3-7 A comparison of English and metric systems.

Concept Activity 3-12
Electromagnetism

Obtain a magnetic compass or short finder from your instructor. Place the compass along a vehicle's battery cable, and turn on the headlights.

1. What was the reaction of the compass?

_____

2. Crank the engine over, and note the compass while cranking. Was the response the same as with the headlights?

_____

3. How does increased current flow affect the compass?

_____

Obtain a digital multimeter that can measure AC volts; a length of wire; and a magnet from your instructor. Position the wire and magnet so that you can attach the meter leads to the ends of the wire. The magnet must be able to pass very close to the wire so that the magnetic lines of force move perpendicular to the wire. Rapidly move the magnet back and forth near the wire, and note the reading on the meter **(Figure 3-8)**.

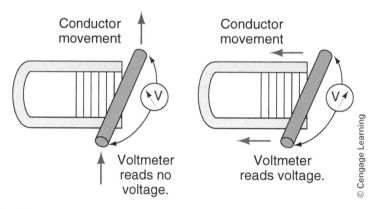

Figure 3-8   Electricity can be produced by moving a magnetic field over a conductor.

4. How much AC voltage was produced?

_____

5. Move the magnet at a slower rate than before. How much AC voltage was produced?

_____

6. Move the magnet quickly past the wire but at a great distance. How much AC voltage was produced?

_____

   If possible, add additional magnets and repeat.

7. Did the amount of AC voltage change with more magnets?

_____

8. What conclusions can you make about AC voltage induction based on this experiment?

_____

As you complete the activities in this chapter, you should have furthered your understanding of some of the basic principles on which the automobile operates.

■

☐ JOB SHEET 3–1

Using the Bore and Stroke of an Engine to Calculate Displacement

Name _____ Station _____ Date _____

Objective

Upon completion of this job sheet, you will have demonstrated the ability to calculate the displacement of an engine in both English and metric terms. Before beginning, review the material on the topic in Chapter 3 of *Automotive Technology*.

Tools and Materials

Service information

Calculator

Description of Vehicle

Year _____ Make _____ Model _____

Engine Type and Size _____

Source of information _____

PROCEDURE

1. Using your designated source of information, find the specifications for this engine.

2. What is the specified stroke for this engine? _____ in. _____ mm

3. What is the specified bore of the cylinders? _____ in. _____ mm

4. How many cylinders does this engine have? _____

5. Using the formula displacement = $\pi \times R^3 \times L \times N$, determine the English and metric displacement of the engine. Show your math here:

Displacement is: _____ CID and _____ cc.

6. Do your calculations agree with the advertised size of the engine? If they do not, explain why.

_____

_____

_____

7. If the bores in this engine were increased by 0.030 inch to correct problems on the walls of the cylinders, what would the new displacement of the engine be? _____ CID and _____ cc.

Problems Encountered

_____

_____

Instructor's Comments

_____

_____

☐ JOB SHEET 3–2

Calculating Gear Ratios

Name _____ Station _____ Date _____

Objective

Upon completion of this job sheet, you will have demonstrated the ability to calculate gear ratios of a gearset and torque changes through a drive train. Before beginning, review the material on the topic in Chapter 3 of *Automotive Technology.*

Tools and Materials

Service manual
Calculator

PROCEDURE

1. Using the formula

$$\frac{\text{driven (A)}}{\text{drive (A)}} \times \frac{\text{driven (B)}}{\text{drive (B)}} = \text{Gear ratio}$$

To calculate the gear ratios for the following gearsets. Remember power moves through a transmission via four gears (that is two sets of two gears).

A. Input gear has 20 teeth; the "A" gear on the cluster gear has 40 teeth; the "B" cluster gear has 15 teeth; and the output gear has 35 teeth. The gear ratio of this set of gears is: _____

B. Input gear has 20 teeth; the "A" gear on the cluster gear has 40 teeth; the "B" cluster gear has 30 teeth; and the output gear has 20 teeth. The gear ratio of this set of gears is: _____

C. Input gear has 25 teeth; the "A" gear on the cluster gear has 40 teeth; the "B" cluster gear has 15 teeth; and the output gear has 30 teeth. The gear ratio of this set of gears is: _____

D. Input gear has 25 teeth; the "A" gear on the cluster gear has 50 teeth; the "B" cluster gear has 25 teeth; and the output gear has 15 teeth. The gear ratio of this set of gears is: _____

2. Select a vehicle to research. This vehicle should have a manual transmission and be one where specifications are readily available.

Year _____ Make _____ Model _____

Engine Size and Type _____

RWD or FWD? _____ Number of Transmission Gears _____

Source(s) of information: _____

3. What is the torque rating of this engine? _____

4. What is the ratio of the final drive gears? _____

5. What is the gear ratio of first gear? _____

6. What is the gear ratio of second gear? _____

7. What is the gear ratio of third gear? _____

8. What is the gear ratio of fourth gear? _____

9. What is the gear ratio of fifth gear? _____

10. What is the gear ratio of reverse gear? _____

11. What is the overall gear ratio of the drive train in first gear? _____

12. What is the overall gear ratio of the drive train in second gear? _____

13. What is the overall gear ratio of the drive train in third gear? _____

14. What is the overall gear ratio of the drive train in fourth gear? _____

15. What is the overall gear ratio of the drive train in fifth gear? _____

16. What is the overall gear ratio of the drive train in reverse gear? _____

17. Using peak engine torque and the speed at which it occurs, calculate the torque and speed that is delivered to the drive axles for each of the following speed gears. For example, if an engine produces 251 lb-ft @ 4,900 rpm and has a first gear ratio of 4.21:1 and a final drive gear ratio of 3.15:1, the torque applied to the drive axles would be approximately 3,330 lb-ft @ 370 rpm.

A. The torque and speed applied to the drive axles in first gear is:

_____

_____

B. The torque and speed applied to the drive axles in second gear is:

_____

_____

C. The torque and speed applied to the drive axles in third gear is:

_____

_____

D. The torque and speed applied to the drive axles in fourth gear is:

_____

_____

E. The torque and speed applied to the drive axles in fifth gear is:

_____

_____

F. The torque and speed applied to the drive axles in reverse gear is:

_____

_____

Problems Encountered

_____

_____

Instructor's Comments

_____

_____

☐ JOB SHEET 3–3

Applying Pascal's Law

Name _____ Station _____ Date _____

Objective

Upon completion of this job sheet, you will have demonstrated the ability to use the basics of Pascal's Law to understand and predict how a hydraulic circuit will behave when a force is applied to the fluid. Before beginning, review the material on the topic in Chapter 3 of *Automotive Technology*.

Tools and Materials

Calculator

PROCEDURE

1. According to Blaise Pascal, pressure applied to a liquid-filled container will be transmitted . . .

_____

_____

2. According to Pascal's Law, what is the formula for calculating force?

_____

_____

3. If a hydraulic pump provides 100 psi, there will be _____ pounds of pressure on every square inch of the system. If the system includes a piston with an area of 2 square inches, each square inch receives _____ pounds of pressure, and _____ pounds of force are applied to that piston.

4. If a hydraulic pump provides 40 psi, there will be _____ pounds of pressure on every square inch of the system. If the system includes a piston with an area of 5 square inches, each square inch receives _____ pounds of pressure and _____ pounds of force are applied to that piston.

5. If a hydraulic pump provides 10 psi, there will be _____ pounds of pressure on every square inch of the system. If the system includes a piston with an area of 50 square inches, each square inch receives _____ pounds of pressure and _____ pounds of force are applied to that piston.

6. If a hydraulic system has two cylinders, one with a 1-inch piston and the other with a 2-inch, why will the larger piston move less than the smaller one?

_____

_____

Problems Encountered

_____

_____

Instructor's Comments

_____

_____

 **REVIEW QUESTIONS**

Review Chapter 3 of the textbook to answer these questions:

1. Define the three states of matter.

   _____

   _____

   _____

2. Explain Newton's three laws of motion as they apply to the automobile.

   _____

   _____

   _____

   _____

3. Define the coefficient of friction and how it relates to modern brake system design.

   _____

   _____

   _____

4. List the six simple machines, and site an example of each.

   _____

   _____

   _____

   _____

   _____

5. Explain work, torque, and horsepower.

   _____

   _____

   _____

   _____

UTOMOTIVE SYSTEMS

OVERVIEW

Chapter 4 gives an overview of the major design influences in the development of the automobile. Vehicle construction and manufacturing processes are discussed as well as the major systems that make up the modern automobile.

Today's cars and trucks use gasoline- and diesel-powered engines to drive two or more wheels. Hybrids use an electric motor to either assist the internal combustion engine or actually drive the wheels. Electronics are being used more in the suspension, steering, and brake systems than ever before. Increased demand for passenger comfort, safety, and convenience has driven the development of adaptive suspensions, ABS with traction control, and voice-activated communication and entertainment systems.

Concept Activity 4-1
Automotive Design Influences

Explain the three major influences in the design evolution of modern vehicles.

_____

_____

_____

_____

_____

THE BASIC ENGINE

The engine provides the power to drive the wheels and operate the accessories of the vehicle. An engine is composed of many different parts bolted together to make a complete assembly. These components include the cylinder block, cylinder head, intake manifold, and exhaust manifold.

Even though today's engines are much more reliable, powerful, and cleaner than those of the past, engine operation has not changed much in the last 100 years. The burning of the air-fuel mixture in the cylinder creates pressure, which forces the piston to turn a crankshaft.

Concept Activity 4-2
Engine Design and Construction

Using a vehicle in lab, describe the following:

Year _____ Make _____ Model _____

a.   Engine design (inline, opposed, V, W): _____

b.   Number of cylinders: _____

c.   Fuel type (gasoline, diesel, E85, CNG): _____

d.   Cylinder block construction material: _____

    e.   Cylinder head construction material: _____

    f.   Valvetrain design (OHV, OHC, DOHC): _____

    g.   Intake manifold construction material: _____

_____

■

ENGINE SYSTEMS

The engine has a number of supporting systems that enable it to function properly: lubrication, cooling, fuel and air, emission control, and electrical. The electrical system consists of the ignition system, starting system, charging system, and various electronic control systems. When properly maintained and functioning, these systems become a highly efficient power plant.

Concept Activity 4-3
Identify Vehicle Fluids and Emission Devices

Using a shop vehicle, locate the following information: Some of this information can be found on the vehicle emission control information (VECI) decal, shown in **Figure 4-1**.

Year _____ Make _____ Model _____

    a.   Recommended engine oil: _____

    b.   Recommended engine coolant: _____

    c.   Emission devices: _____

Vehicle Emission Control Information - XXXXXXX
Engine Family XXXXXXXXX
Evaporative Family XXXX
Displacement XXXXcm³

CATALYST

Tune-up specifications

 Tune-up conditions:

   Engine at normal operating temperature, all accessories turned off, cooling fan off, transmission in neutral

Adjustments to be made in accordance with indications given in shop manual

Idle speed	5 speed transmission	800 ± 50 rpm
	Automatic transmission	800 ± 50 rpm
Ignition timing at idle		15 ± 2° BTDC
Valve lash	Setting points between camshaft and rocker arm	In. 0.17 ± 0.02mm cold
		Ex. 0.19 ± 0.02mm cold
No other adjustments needed		

© Cengage Learning 2015

Figure 4-1   This image shows emissions equipment and family as well as emission system calibration information.

    d.   A/C refrigerant type and charge amount: _____

_____

■

The drivetrain is made up of the components that transfer power from the engine to the driving wheels. Vehicle drivetrains can be either front-wheel-drive (FWD), rear-wheel-drive (RWD), four-wheel-drive (4WD), or all-wheel-drive (AWD). Most cars are FWD. Some luxury and performance cars are RWD. Most trucks are either RWD or 4WD. Sport-utility vehicles (SUV) can be FWD, RWD, 4WD, or AWD. Transmissions are either automatic or manually shifted. Some newer performance vehicles have automatic transmissions that can be shift like a manual transmission but without using a clutch to change gears. A 4WD vehicle will have a transfer case to split power to the front and rear differentials.

Concept Activity 4-4
Identify Drivetrain

Using a shop vehicle, locate the following information:

Year _____ Make _____ Model _____ _____

a. Transmission type: _____

b. Drivetrain layout (FWD, RWD, 4WD, AWD): _____

c. Recommended transmission fluid: _____

_____

■

The running gear includes the parts used to control the vehicle, which include the wheels and tires and the suspension, steering, and brake systems.

The suspension carries the weight of the vehicle and allows for movement of the wheels and tires when going over bumps and dips in the road. The steering system works with the suspension to provide precise driver control over the turning of the vehicle. The brake system provides for safe and controlled slowing and stopping.

Concept Activity 4-5
Identify Suspension and Brake System

Using a shop vehicle, locate the following information:

Year _____ Make _____ Model _____

a. Front-suspension type (strut, SLA, I-beam, multilink): _____

_____

b. Power-steering type (hydraulic or electric): _____

_____

c. Front and rear brake type: _____

_____

d. ABS type and number of channels: _____

_____

e. Recommended brake fluid type: _____

_____

 REVIEW QUESTIONS

Review Chapter 4 of the textbook to answer these questions:

1. Explain how an engine produces the power needed to move a vehicle.

_____

_____

_____

2. What is preventive maintenance and why is it so important to the life of a vehicle?

_____

_____

_____

3. Where should you look to find the proper tire inflation pressures?

_____

_____

_____

4. Describe three major types of engine cylinder configurations.

_____

_____

_____

5. Compare and contrast FWD and RWD drivetrain configurations.

_____

_____

_____

OVERVIEW

The tools of the automotive trade include many different things, such as measuring tools, fasteners, tap and dies, hand tools, power tools, shop equipment, and service information or reference materials. This chapter describes proper use of hand tools, portable power tools, pneumatic power tools, and stationary grinders, and provides general safety requirements associated with them. Remember that more accidents are caused by hand tools than by power tools. Knowing the proper use of tools will help eliminate many accidents. The Job Sheets and Concept Activities provide a method of demonstrating proper use of these tools.

Without the proper tools, it is nearly impossible for a technician to do quality work in a reasonable amount of time. Tools are a major investment for a technician. To protect that investment, you should take proper care of them and use them only for the purposes they were designed for.

Most tools are sized to fit certain size fasteners. Any discussion on tools must be preceded by a discussion of measuring systems and fasteners.

MEASURING SYSTEMS

Measuring systems are used to define the size of something in an accurate way. Basically there are two different measuring systems used when working on today's vehicles. One is based on the meter, and the other on the inch. Inch and meter measurements define the length, width, height, or diameter of something. Measuring systems also include ways to define the weight, area, volume, pressure, temperature, and torque of an object. Technicians need to be familiar with both metric and inch-based measuring systems, especially the systems for measuring length, width, height, and diameter **(Figure 5-1)**.

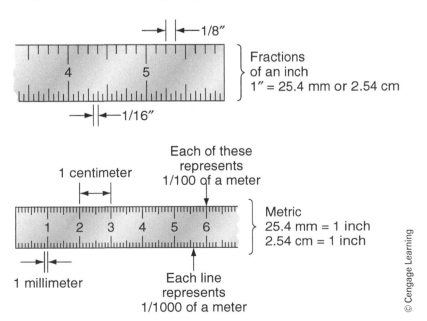

Figure 5-1 A comparison of English and metric systems.

Concept Activity 5-1
Measurement Conversion

Convert the measurements in the table below using the formulas given on pages 59–60 in the textbook.

English	Metric
10.5 inches	mm
inches	201.4 mm
0.035 inch	mm
inches	19 mm
80 ft-lb	N-m
ft-lb	64 N-m
224°F	°C
°F	39°C

© Cengage Learning 2015

■

FASTENERS

Simply put, a car is held together with fasteners, and when something needs to be replaced, some fasteners need to be loosened, the part removed, and the fasteners reinstalled and tightened. Many different types of fasteners are used today, each with its own purpose. The purpose of all fasteners is to hold things together or to fasten one part to another. The weight or pressure working against it determines the size of the fastener selected for use. The size of a fastener can be based on the metric or inch measuring system.

Concept Activity 5-2
Bolt Measurement

Get an assortment of fasteners from your instructor. Identify each using the information in your Tech Manual and textbook. List the dimensions and grades below. **Figure 5-2** provides an example of how bolts are measured.

Bolt 1:

Length _____ Diameter _____ Tread Pitch _____ Grade _____

Bolt 2:

Length _____ Diameter _____ Tread Pitch _____ Grade _____

Bolt 3:

Length _____ Diameter _____ Tread Pitch _____ Grade _____

Bolt 4:

Length _____ Diameter _____ Tread Pitch _____ Grade _____

Bolt 5:

Length _____ Diameter _____ Tread Pitch _____ Grade _____

Bolt 6:

Length _____ Diameter _____ Tread Pitch _____ Grade _____

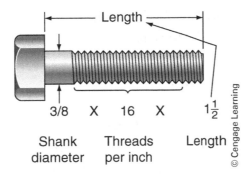

Figure 5-2 The shank diameter, the number of threads per inch, and the length are all used to determine the bolt size.

MEASURING TOOLS

Vehicles are built to exact specifications and with precise tolerances. Many parts must be installed with clearances as small as a human hair. If the installed clearance is not correct, excessive wear will result. To ensure that the part is installed correctly, technicians use precise measuring instruments.

One of the most commonly used precision measuring tools is the micrometer. A micrometer is used to measure the length, width, height, diameter, and/or depth of something. Often it is also used to determine the amount of wear or distortion of an object. Another precision measuring instrument commonly used is a dial indicator.

 Concept Activity 5-3
Using a Micrometer

Using a micrometer, list the micrometer reading of the 10 drill shanks supplied by your instructor. The masking tape on the drill shank is covering up the actual drill size; do not remove the tape. The number on the masking tape is used as a reference in the recording of the drill shank sizes.

 Concept Activity 5-4
Using a Dial Indicator

With the guidance of your instructor, raise the front of a vehicle off the ground, and set it on safety (jack) stands. Secure a dial indicator with the proper holding mechanisms so that the indicator's plunger barely contacts the outside rim of one of the front wheels. With the plunger contacting the rim, set the dial to zero. Slowly rotate the wheel and watch the indicator. Any indicator needle movement indicates that there is some wheel distortion. ■

HAND TOOLS

Hand tools are all of those tools you power with your hands and arms. The most commonly used hand tools are wrenches and screwdrivers. It is important that you thoroughly understand the purpose of each hand tool and that you know how to select the correct size for a particular fastener.

 Concept Activity 5-5
Tools

Take complete inventory of the tools in your toolbox or one assigned to you. Cross-reference the supplied tool list to the tools on hand.

- General tips for safe hand tool use include the following:
- Select the proper size and type of tool for the job.
- Use tools only for the purpose for which they are designed.
- Keep hand tools in safe working condition.
- Store hand tools safely when not in use.
- Report any breakage or malfunctions to your instructor.
- Sharp or pointed tools shall not be carried in clothing pockets unless in a sheath.
- Make sure that cutting tools are properly sharpened and in good condition.
- Secure small or sharp work with a vise or clamp.
- Do not use tools with loose or cracked handles.
- Remember, hand tool safety depends mainly on the person who uses the tool. Knowing what a tool is designed to do and how to use it properly is the key.
- Think safety and avoid surprises. Use the safety equipment that is appropriate and required.

JACKS AND LIFTS

The following are some general safety rules for using jacks, lifts, and hoists:

- Do not allow anyone to remain in the vehicle when it is being raised.
- Make sure you know how to operate the equipment and that you are aware of its limitations.
- Never overload a lift, hoist, or jack.
- Do not use any lift, hoist, or jack that you believe is defective or not operating properly.
- Make sure there is nothing or nobody in the space the vehicle will occupy as it is raised or lowered.
- Always support the vehicle with safety stands after it has been raised by a jack.
- Always engage the safety locks on the hoist or lift after the vehicle has been raised.

SERVICE INFORMATION

The automotive technician, more so today than at any other time, requires accurate, up-to-date service information to correct a vehicle fault. Service information is presented in several different formats, from printed hard copy to electronic media. The information is critical to the repair of the vehicle.

Perhaps the most important tools you will use are service manuals or service information systems. Because of the number of different type vehicles and systems, it is impossible for anyone to remember the correct procedures for doing something on all vehicles. Also, there is no reason for trying to remember them all. But it is important to know how to locate the correct service information.

The appropriate service information for a particular vehicle should be referred to before doing any maintenance, diagnostic, or repair work. Doing this will prepare you for the work you are about to do and will alert you to any precautions you need to take while performing the procedure. The service information will also give you the wear limits of parts and the correct installation tolerances.

Service information also contains diagnostic charts. These charts are extremely helpful in leading you to the cause of a problem. When diagnosing electronically controlled systems, following the diagnostic charts is a must.

Additional sources of useful information can be found in factory service bulletins (frequently called technical service bulletins, or TSBs) and in various trade magazines and publications. There are also several websites that will help you learn particular systems and components. For more information on these, go to http://www.autoed.com.

Note: *Repair procedures on different makes, models, and years of vehicles may seem similar; however, they may actually differ dramatically. Service manuals should be referenced for every repair.*

☐ JOB SHEET 5–1

Repairing and/or Replacing Damaged Threads

Name _____ Station _____ Date _____

Objective

Upon completion of this job sheet, you will have demonstrated the ability to properly use a tap and die set to repair damaged threads and a heli-coil repair kit to replace damaged threads. This task applies to ASE Education Foundation Tasks MLR I.A.6 (P-1), and AST/MAST I.A.7 (P-1).

Tools and Materials

Tap and die set	Drill
Heli-coil set	Drill bits
Center punch	

PROCEDURE TO REPAIR DAMAGED THREADS

1. Using a thread gauge, determine the proper thread pattern of the damaged threads.

   Thread Size and Pitch _____

2. Thoroughly clean the threads to be repaired.

3. Select the proper size tap to repair internal threads or die to repair external threads (**Figure 5-3** and **Figure 5-4**).

   Size and Type _____

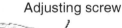

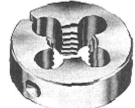

Adjustable round   Adjustable round

Solid hexagon   Solid square

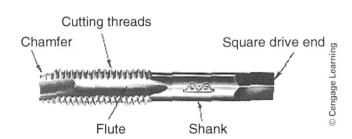

Figure 5-3 Parts of a hand tap.

Figure 5-4 Common die shapes.

4. Align the repair tool with the damaged threads, and slowly rotate the tool in the direction of the threads. This tool will clean and straighten the damaged threads.

   CAUTION: *Use a light oil to enable the taps and dies to cut easily and cleanly.*

5. When you have reached the end of the damaged threads, remove the tool by reversing its rotation.

6. Inspect the threads. Test with the properly sized bolt. Note results.

PROCEDURE TO REPLACE DAMAGED THREADS

When internal threads in an object are damaged and cannot be repaired, they are sometimes replaced with a heli-coil **(Figure 5-5)**.

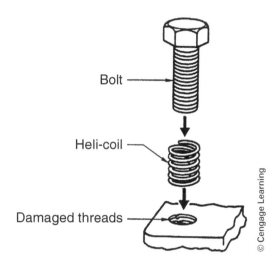

Figure 5-5   Replacing damaged threads with a heli-coil or thread insert.

1. Explain how to select the proper size of heli-coil kit. List the contents of the heli-coil kit.

   _____

   _____

   _____

2. Describe the importance of using the drill bit supplied in the heli-coil kit to drill out the hole.

   _____

   _____

   _____

3. Use the tap furnished in the heli-coil kit to thread the drilled hole. Why was the threaded hole size increased beyond its original size?

   _____

   _____

   _____

4. Use the tool provided in the heli-coil kit to install the heli-coil insert. Describe the installation process. What is the last step in the installation of the heli-coil?

_____

_____

_____

Problems Encountered

_____

_____

_____

Instructor's Comments

_____

_____

_____

☐ JOB SHEET 5–2

Cutting with a Chisel

Name _____ Station _____ Date _____

Objective

Upon completion of this job sheet, you will have demonstrated the ability to use a chisel for cutting off a rivet head properly. Before beginning, review Chapter 5 of *Automotive Technology*.

Tools and Materials

Assorted chisels   Riveted metal block

Assorted hammers   Vise

PROCEDURE

1. Secure the block with the rivet in a vise. Place the work in the vise so that cutting will be done toward the fixed jaw.

    Instructor's Check _____

    WARNING: *Never use a chisel on hardened metal.*

2. Select the proper size of chisel. The blade should be at least as large as the desired cut to be made. Select the proper size and type of hammer.

    Chisel Type _____

3. Hold the chisel firmly enough to guide it, but lightly enough to ease the shock of the hammer blows. Hold the chisel just below the head to prevent pinching your hand in the event that you miss striking the chisel.

4. Place the chisel at an angle to the rivet head that will allow it to cut just below the rivet's head and not into the metal block. Grip the end of the hammer's handle and strike the chisel head. Strike with enough force to cut the rivet head. Check your work every two or three blows. Correct the angle of the chisel as needed to cut the rivet head off without damaging the block.

    Instructor's Check _____

Problems Encountered

_____

_____

_____

Instructor's Comments

_____

_____

_____

☐ JOB SHEET 5–3

Proper Use of a Vehicle Hoist

Name _____ Station _____ Date _____

Objective

Upon completion of this job sheet, you will have properly demonstrated the ability to lift a vehicle using a frame contact hoist. Before beginning, review Chapter 5 of *Automotive Technology*.

    Lift type (in ground, aboveground) _____

    Lift capacity rating _____

    Operational controls _____

    Description of safety mechanism _____

    Inspection of service arms/contacts _____

PROCEDURE

1. Using the appropriate service information, locate the proper vehicle lift points. Describe or illustrate the vehicle lift point locations used in performing this task.

   _____

   _____

2. Position the vehicle on the hoist. Consider the vehicle's center of gravity when positioning the vehicle on the hoist. What conditions on the vehicle could upset the center of gravity?

   _____

   _____

3. Position the hoist contact pads under the lift points on the vehicle. Operate the hoist and lift the vehicle until the vehicle's wheels are two or three inches off the floor. Are the four hoist contact pads in contact with the lift points of the vehicle?

   _____

4. Shake the vehicle and observe any movement or unusual noise. If movement and noise are detected, lower the vehicle and reposition the hoist contact pads.   OK ☐   Not OK ☐

5. If repositioning of contact pads was required, repeat step 4.

6. Raise the vehicle to the desired working height. Keeps your hands on the hoist controls when raising or lowering the vehicle. Do not proceed under the vehicle until the hoist locks have been secured. What is the safety inspection interval for the vehicle hoist?

   _____

   _____

Instructor's Check _____

7. To lower the vehicle, unlock the hoist safety locks. Verify that there are no obstructions or objects under the vehicle. Lower the vehicle to the ground. Push the contact pads off the path of the tires.

Instructor's Check _____

Problems Encountered

_____

_____

_____

Instructor's Comments

_____

_____

_____

☐ JOB SHEET 5–4

Proper Use of a Hydraulic Floor Jack and Jack Stands

Name _____ Station _____ Date _____

Objective

Upon completion of this job sheet, you will have demonstrated the ability to use a hydraulic jack properly. Before beginning, review Chapter 5 of *Automotive Technology*.

Tools and Materials

Floor jack

Jack stands

Wheel blocks

Wood blocks

PROCEDURE

1. Make sure the floor jack being used is sufficiently rated to lift and sustain the load.

    Load Rating _____ _____

    Jack Stand Capacity Rating _____

2. Test the operation of the jack before attempting to lift the load. If the jack does not operate properly, tag it for "out of service" and report it to your instructor immediately.

3. Place the transmission in gear or PARK, and set the parking brake. Block the wheels that are to remain on the ground.

    CAUTION: *If both ends of the vehicle are to be lifted, jack the front end first. Place safety stands under the front before lifting the rear end of the vehicle.*

4. Position the jack under the load at a point strong enough to carry the weight. The lift should be flat and level with the floor. Position the jack so the movement of its lift will be straight up and down.

    Jack Placed _____

    CAUTION: *If one side of the vehicle is to be lifted, do not lift one side so high that it is in danger of slipping off the jack saddle.*

    Instructor's Check _____

5. Once the load has started to lift, recheck the position of the jack. If the load or jack starts to lean, lower the jack and reset it.

6. Lift the load to the required height. Do not lift higher than necessary.

    WARNING: *A floor jack must roll slightly when being jacked up, especially if the parking brake is on and the wheels are chocked.*

    WARNING: *If the floor jack does not lift the vehicle high enough, select a larger capacity floor jack that will give the desired height. Do not under any circumstances place blocks of wood between the floor jack lifting pad and the vehicle.*

7. Do not get under a vehicle that is supported only by a hydraulic jack. Place safety stands under the vehicle in positions in which they will not lean or slip and will support the load. Lower the vehicle onto the stands.

Instructor's Check _____

Problems Encountered

_____

_____

_____

Instructor's Comments

_____

_____

_____

☐ JOB SHEET 5–5

Proper Use of a Stationary Grinder

Name _____ Station _____ Date _____

Objective

Upon completion of this job sheet, you will have demonstrated the ability to use a stationary grinder to redress a chisel properly. Before beginning, review Chapter 5 of *Automotive Technology*.

Tools and Materials

Dull chisels

PROCEDURE

1. Check the grinder's electrical cord for good condition. Check the guards and guides for proper position **(Figure 5-6)**. The work rest should be adjusted to within 1/8″ (3.17 mm) of the wheel's surface. Describe overall condition. _____

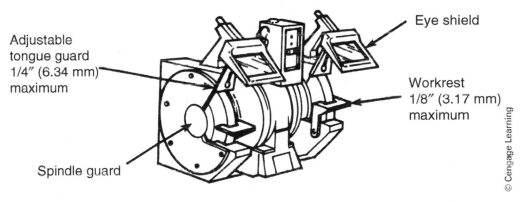

Adjustable
tongue guard
1/4″ (6.34 mm)
maximum

Eye shield

Workrest
1/8″ (3.17 mm)
maximum

Spindle guard

© Cengage Learning

Figure 5-6 Keep the grinder guards and guides in place and properly adjusted.

2. Make sure the grinding wheel has the correct grid and is in good condition. Check the spindle nut for tightness. Does the grinding wheel need to be dressed? _____

   CAUTION: *Never remove the paper from the sides of the grinding wheel. Do not use a grinding wheel that is worn to less than 1/2″ (13 mm) of its original diameter.*

3. If the chisel is too small to hold in your hands safely, clamp it into a clamping device. Eye protection or an approved face shield *must* be worn. Ensure that you are not wearing any loose clothing or jewelry.

4. Connect the electrical terminal to a grounded sprocket. Turn on the grinder and allow it to reach operating speed. Why is it important to allow the grinder to reach its operating speed before starting to grind something with the stone?

   _____

   _____

   WARNING: *Stand to one side of the wheel when turning power on. The wheel may be cracked. The speed of rotation may cause it to fly apart.*

5. Hold the chisel securely and press the face against the face of the grinding wheel. Hold the chisel at the proper angle **(Figure 5-7)**. The cutting edge should be ground convex.

   WARNING: *If the grinding wheel is cold, warm it by first applying light pressure on the chisel against the wheel.*

   Instructor's Check _____

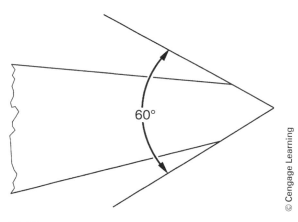

Figure 5-7   The correct angle for chisel sharpening is 60 degrees.

6. Move the chisel back and forth across the wheel face. Do not keep it in one place. As grinding proceeds, cool the chisel in water. List two reasons for cooling a chisel in water while redressing it.

   _____

   _____

   WARNING: *Never use the side of the wheel.*

7. If the wheel cuts slowly or vibrates, use a wheel dresser to expose a new cutting surface.

   WARNING: *Never use a loaded or glazed grinding wheel.*

8. Grind both edges of the chisel to obtain a 60-degree included angle. Turn off the grinder and stay there until the wheel comes to a complete stop. Unplug the electrical cord.

   Instructor's Check _____

Problems Encountered

_____

_____

_____

Instructor's Comments

_____

_____

_____

☐ JOB SHEET 5–6

Proper Use of Electronic Media

Name _____ Station _____ Date _____

Objective

Upon completion of this job sheet, you will have demonstrated the ability to properly use electronic media to look up vehicle information. This task applies to ASE Education Foundation Tasks MLR 1.A.1 – 8.A.1 (P-1), and AST/MAST 1.A.2 – 3.A.2 (P-1), 4.A.1 – 6.A.1 (P-1), 7.A.2 & 8.A.2 (P-1).

REPORT SHEET FOR USE OF ELECTRONIC SERVICE INFORMATION
1. Vehicle information
Date of manufacture _____
Manufacturer _____
Model _____
VIN _____
Engine _____
2. Specifications
Wheel and/or tire run out _____
Wheel fastener torque _____
Cylinder head bolt torque _____
Coolant refill capacity _____
3. Parts and Labor
Labor time to replace front wheel bearing(s) _____
Fuel injector part cost _____
Labor time to R&R A/C generator _____
Radiator part cost _____
4. Maintenance
Recommended oil weight and type _____
Services recommended at 30,000 miles (48,600 kilometers)
_____
_____
5. Technical Service Bulletins (TSBs)
Powertrain related TSB number _____
Electrical system TSB number _____

6. **Recall information**

   Any recall notice number _____

   Reason for recall _____

Problems Encountered

_____

_____

_____

Instructor's Comments

_____

_____

_____

REVIEW QUESTIONS

Review Chapter 5 of the textbook to answer these questions:

1. What is the main advantage of using a box-end wrench instead of an open-end wrench?

   _____

2. Technician A says that a flare nut wrench (line wrench) is used to loosen exhaust flare nuts. Technician B says that a flare nut wrench (line wrench) is used on nuts on fuel lines, brake lines, and transmission cooling lines. Who is correct?

   a. Technician A   c. Both A and B
   b. Technician B   d. Neither A nor B

3. Technician A says that when grinding a small item on a bench grinder the item should be held in your hand. Technician B says when using a bench grinder you must only wear your safety goggles if the machine doesn't have a safety shield. Who is correct?

   a. Technician A   c. Both A and B
   b. Technician B   d. Neither A nor B

4. What precision measuring tool should be used to check end play or out-of-roundness?

   _____

5. How do you determine where the proper lift points for a vehicle are?

   _____

OVERVIEW

To service and repair today's cars and trucks, a wide range of tools are necessary. Most technicians have an extensive collection of hand tools as well as a wide variety of diagnostic tools and other pieces of specialized equipment. Many shops purchase special tools for all their technicians to use, saving the technicians some of the expense.

Concept Activity 6-1
General Hand Tools and Lab Equipment

1. Does your lab provide a basic hand tool kit for students?   _____ YES   _____ NO

2. If yes, provide a brief description of the contents of the kit.

   _____

   _____

   _____

3. If no, describe where and how student tools are kept and maintained.

   _____

   _____

   _____

4. Does your lab have a tool room or rooms for tool storage?   _____ YES   _____ NO

5. If yes, describe what tools and equipment are kept in the tool room and how tools are checked in and out.

   _____

   _____

   _____

6. If no, describe how specialty tools and pieces of equipment are kept and maintained.

   _____

   _____

7. Locate and record the information for the lab equipment listed below:

   a. Hydraulic press make and model: _____

       Rated output pressure: _____

       Adjustable table height?   _____ YES   _____ NO

b. Drill press make and model: _____

Chuck size capacity: _____

Number of speeds: _____

c. Floor jack lifting capacity: _____

d. List of available scan tools: _____

_____

_____

■

Concept Activity 6-2
Engine Diagnostic and Service Tools

1. Determine if your lab has the following engine diagnostic and service tools.

   a. Compressions tester _____ YES _____ NO

   b. Cylinder leak tester _____ YES _____ NO

   c. Cooling system pressure tester _____ YES _____ NO

   d. Combustion leak detector _____ YES _____ NO

   e. Valve spring compressors _____ YES _____ NO

   f. Cylinder ridge reamers _____ YES _____ NO

   g. Dial indicators _____ YES _____ NO

   h. Piston ring installers _____ YES _____ NO

   i. Cam bearing remover/installer _____ YES _____ NO

2. Which of the tools from the list above are diagnostic tools? _____

_____

3. Which of the tools from the list above are used for engine repair? _____

_____

■

Concept Activity 6-3
Electrical Test Equipment

1. Determine whether your lab has the following electrical diagnostic and service tools:

   a. Digital multimeters _____ YES _____ NO

   b. Starter/Generator tester _____ YES _____ NO

   c. Battery tester _____ YES _____ NO

   d. Test lights _____ YES _____ NO

   e. Remote starter switch _____ YES _____ NO

   f. Terminal tool kit _____ YES _____ NO

g. Headlight aiming tools   _____ YES   _____ NO

h. Fused jumper wires   _____ YES   _____ NO

i. Battery chargers   _____ YES   _____ NO

2. List the brand(s) of digital multimeters available for student use.

_____

3. Identify the electrical safety categories of the digital multimeters.

_____

4. Why is it important to use a digital multimeter with the correct electrical safety category rating?

_____

_____

5. Why is it important for jumper wires to be fused?

_____

6. Examine a battery charger and record the charging rates listed on the charger.

_____

_____

7. List the steps to properly connect a battery charger to a battery.

_____

8. List the safety information found on the battery charger.

_____

_____

9. Why is it important to operate a battery charger by following all safety and charging rate guidelines?

_____

_____

_____

■

Concept Activity 6-4
Drivetrain Service Tools

1. Determine whether your lab has the following drivetrain service tools:

a. Transmission/Transaxle jack   _____ YES   _____ NO

b. Axle pullers   _____ YES   _____ NO

c. Drive shaft angle gauge   _____ YES   _____ NO

d. Coil spring compressor   _____ YES   _____ NO

e. Wheel alignment equipment   _____ YES   _____ NO

f. Wheel-bearing service tools   _____ YES   _____ NO

2. Explain why a transmission/transaxle jack should always be used when removing and installing a transmission/transaxle. _____

_____

3. Describe why a drive shaft angle gauge is used.

_____

4. Does your lab have internal and external coil spring compressors? _____ YES _____ NO

5. What type of suspensions requires internal coil spring compressors?

_____

6. List the manufacturer and model of your alignment equipment.

_____

7. Describe the method used by the aligner to determine wheel position (example: laser, infrared).

_____

_____

■

Concept Activity 6-5
Brake System Tools

1. Determine whether your lab has the following brake system service tools:

   a. Brake hydraulic pressure gauges   _____ YES _____ NO

   b. Drum brake spring tools   _____ YES _____ NO

   c. Asbestos reclamation equipment   _____ YES _____ NO

   d. On-car brake lathe   _____ YES _____ NO

   e. Off-car brake lathe   _____ YES _____ NO

   f. Pressure brake bleeder   _____ YES _____ NO

   g. Brake fluid testing equipment   _____ YES _____ NO

2. Describe the type of asbestos reclamation equipment available in your lab and how it is used.

_____

_____

3. Explain why asbestos reclamation equipment is used when servicing brakes and clutches.

_____

_____

4. Describe why an on-car brake lathe is used instead of an off-car lathe.

_____

5. Explain what other methods of brake bleeding are used instead of using a pressure bleeder.

_____

_____

_____

■

Concept Activity 6-6
HVAC Tools and Equipment

1. Determine whether your lab has the following HVAC tools and equipment:

   a. Refrigerant leak detector _____ YES _____ NO

   b. R134a reclaiming station _____ YES _____ NO

   c. Thermometer _____ YES _____ NO

   d. Belt tension gauges _____ YES _____ NO

   e. Serpentine belt tools _____ YES _____ NO

2. Describe what type of refrigerant leak detector(s) are available in your lab.

_____

_____

3. Explain why refrigerant reclamation and recycling equipment must be used for air-conditioning service.

_____

_____

4. Describe the type of belt tension tools available in your lab. _____

_____

5. Explain the importance of accessory drive belts being properly tensioned. _____

_____

_____

_____

■

Concept Activity 6-7
Conclusions

1. Based on your previous experiences with automotive repair facilities, how well do you think your lab is equipped compared to full-service repair shops?

_____

_____

2. How well do you think your lab is equipped to handle the types of services and repairs you will be performing during your training?

_____

_____

_____

REVIEW QUESTIONS

Review Chapter 6 of the textbook to answer these questions:

1. List five engine diagnostic and repair tools.

_____

_____

_____

_____

_____

2. Which of the following is *not* an electrical system diagnostic tool?

   a. Test light
   b. Volt meter

   c. Conductance tester
   d. Compression tester

3. Explain the purpose of a lab scope.

_____

_____

_____

4. Explain the difference between an internal and an external spring compressor.

_____

_____

_____

5. Describe the differences between an on-car and an off-car brake lathe.

_____

_____

_____

WORKING SAFELY IN THE SHOP

CHAPTER 7

OVERVIEW

Shop safety must be a primary concern for all technicians, managers, and shop owners. Safety rules and regulations must be followed to prevent injuries to yourself, fellow employees, and your customers. This chapter includes some basic safety guidelines that should be followed whenever you are in a service facility.

PERSONAL SAFETY

As the worker, it is your responsibility to take the necessary precautions to ensure your personal safety. Such precautions include wearing approved eye protection and approved footwear, dressing for safety, and using hand tools and operating shop equipment safely. Protect yourself from serious burns by wearing protective clothing while working around hot objects such as cooling system components, exhaust manifolds, catalytic converters, and exhaust mufflers. In addition, be aware that one of the greatest concerns to personal safety is the effect of long-term exposure to hazardous materials, particularly solvents, cleaning agents, and paint products used in auto body repairs.

Personal Safety Precautions

1. Always wear eye protection while working in the auto lab.
2. Additional eye protection, such as a face shield, should also be worn whenever there is a chance of liquids, dirt, dust, or metal shavings getting into your eyes.
3. Eye and face protection should be worn whenever you are working with or around a battery.
4. Never wear loose clothing that might get caught in moving parts or machinery.
5. Always wear good foot protection. Hard-soled, nonslip, steel-toed safety boots or shoes are often required.
6. Long hair should always be tied back or tucked under a hat.
7. Jewelry should not be worn.
8. Clean up all spills immediately after they occur.
9. Do not play around in the shop.
10. Do not operate a piece of equipment unless you have been trained on it.

Concept Activity 7-1
Shop Safety Map

Make a simple line drawing of the shop. Then walk around the shop and identify all safety equipment. Mark the locations of the equipment on your drawing. ∎

Concept Activity 7-2
Shop Safety Inspection

Perform an inspection of your lab checking for safety concerns such as missing shields, damaged electrical cords, or other issues. ∎

Lifting and Carrying

Incorrect lifting and carrying procedures cause back injuries to many automotive technicians. Follow the OSHA-approved method of using your leg muscles, not your back, to lift an object. Do not take a chance on injuring your back; ask for help. It is important when carrying heavy objects to keep the object close to your body to avoid back injury.

PROFESSIONAL BEHAVIOR

Many states have passed laws banning smoking in public buildings. Regardless of local laws, it is unsafe to smoke in an auto repair facility. Do not smoke while working on a customer's vehicle, and *never* smoke inside a customer's vehicle. Additionally, do not change radio stations or adjust the audio controls. Try to leave the seats and mirrors as close to where the customer left them as possible. Make every attempt to keep the vehicle as clean as possible; use steering wheel, seat, and floor covers to protect the interior.

Keep your work area of the shop neat and free of dirt, and keep oil spills wiped up. Your hand tools should be clean and kept in an organized manner.

Observe all safety precautions when using the shop equipment, and report any unsafe equipment conditions to your instructor and/or supervisor. Keep a clean appearance by changing your coveralls or uniform on a regular basis. It is important to instill confidence in your customer by presenting a professional attitude and appearance.

TOOL AND EQUIPMENT SAFETY

It is everyone's responsibility to demonstrate good safety procedures when using hand tools and shop equipment.

1. Always use the proper tool for a specific job. Keep hand tools clean and in good condition.
2. Always check electrically powered tools before using them. Make sure the cord and plug are in good condition. Never use a grounded electrical plug that has had the ground prong removed.
3. When using compressed air, always check the hoses and connections. Wear safety goggles.
4. Do not use the vehicle lifts in the shop until you have been properly instructed on their use.
5. After lifting a vehicle with a jack, position jack stands under the vehicle before working on it.
6. Always wear the required eye and skin protection when working with chemicals.

Hand Tool Safety

Hand tools need to be clean and in good working order. The most common type of injuries caused by hand tool use is the tool slipping out of your hand or hand tools losing their grip on a cap screw. A poorly maintained ratchet driver may change direction when under torque, causing hand injuries.

Power Tool Safety

The two most common safety concerns with power tools are the electric power cord and plug condition. Frayed power cords and electric plugs with the ground prong missing can cause electrical shocks.

Remember to disconnect the electrical power cord before adjusting or attaching objects or accessories to a piece of equipment.

Compressed Air Equipment Safety

The average technician is exposed to compressed air equipment, commonly referred to as pneumatic tools, in the form of impact tools, air ratchets, air drills, and tire-mounting equipment. Other applications or uses of compressed air include tire inflation and drying and/or cleaning of parts. When working with compressed air, safety glasses or a face shield are used to reduce the possibility of eye injuries.

When using compressed air to dry ball or roller bearings, do not allow the bearing to spin. The spinning of the bearing not only could damage the bearing, but the bearing could disintegrate and cause personal injury because of flying bearing parts.

 Concept Activity 7-3
Compressed Air

1. Shop air pressure is regulated to psi per OSHA regulations.

2. Inspect the air hoses and fittings in the shop area. Note any leaks or other dangerous conditions that might exist. ∎

Lift Safety

When using a vehicle lift or hoist, be sure to adhere to the suggested safety precautions in its operating manual. Be aware of the maximum lifting capacity of the equipment, and do not exceed that value. The hoist should display a tag showing when it was last inspected for proper operation. It is important to engage the hoist locking device before going under the vehicle.

Jack and Jack Stand Safety

The first step in using a hydraulic floor jack is to choose the correct jack for a particular application, based on lifting capacity. When using a hydraulic floor jack to raise a vehicle's front or rear end off the ground, use the proper lifting locations. When the vehicle is raised to the required distance off the ground, place a jack stand under the vehicle to support its weight and then lower the floor jack.

Chain Hoist and Crane Safety

When operating a chain hoist and/or crane hoist, be aware of the lifting capacity. A label attached to the equipment will state the maximum lifting value. The lifting capacity of the crane will change as you change the boom length on the equipment.

Cleaning Equipment Safety

There are several types of automotive parts cleaners available. Choose a type that will not cause damage to the component being cleaned. As an example, placing an aluminum head in a chemical cleaning caustic bath solution will cause severe erosion to the component. Wear the appropriate safety protective clothing for the type of cleaning equipment being used.

Vehicle Operation

1. Move vehicles very slowly in and out of the shop (2 mph maximum).
2. When a vehicle is being brought in or taken out of the shop, do not get close to it or between the vehicle and the wall or a heavy object.
3. Honk the vehicle's horn and check carefully for people or other vehicles before driving out of the shop door.
4. If an engine must be run in the shop, its exhaust must be collected by the shop's ventilation system.
5. Use extra caution if you are working on or around hybrid vehicles. These vehicles, when moving under battery/electric motor power, will be very quiet. Other people working in the shop may not hear the vehicle moving and could accidentally walk into its path. Additionally, hybrid-electric vehicles, if they are not completely shut down, can start the internal combustion engine to recharge the battery pack if the system voltage drops too low.

WORK AREA SAFETY

1. Keep your work area neat and clean. Clean up any spills immediately.
2. Do not leave tools or equipment scattered around on the floor.
3. Return any tools or equipment to where they belong.
4. Do not smoke while working on a vehicle.
5. Always work in a shop with ample ventilation.

Fire Safety

Many materials used in the shop are highly combustible. Always adhere to the following fire safety guidelines:

1. Familiarize yourself with the location and operation of firefighting equipment in your work area.
2. Store welding cylinders in an upright position, attached to a cart to prevent falling, and away from combustible materials and heat.
3. Store all combustible materials and oil- and gasoline-soaked rags in the proper container, such as a self-closing metal safety container.
4. Clean up all spilled grease, oil, and gas immediately.
5. Store volatile liquids in tight metal containers, properly labeled, color-coded, and kept in a metal cabinet labeled "Flammable—Keep Fire Away."
6. Wear clean shop clothes, not clothes that are oil and grease soaked.
7. Keep flames or sparks away from batteries to prevent explosions.
8. If a fire extinguisher is used, report it to your instructor so it may be immediately recharged.
9. Report all unsafe fire conditions to your instructor immediately.
10. Storage of flammable or combustible liquids should not limit use of exits, stairways, or areas normally used for the safe movement of people.

Types of Fires

Fires are commonly classified by the type of burning material:

■ *Class A*—Fires in which the burning material is ordinary combustibles, such as paper, wood, cloth, or trash. To extinguish this type of fire requires drowning it with water or solutions containing a high percentage of water, or smothering it with a multipurpose dry chemical extinguisher.

■ *Class B*—Fires in which the burning material is a liquid, such as gasoline, diesel fuel, oil, grease, or solvents. To extinguish this type of fire requires a smothering action, such as that from foam, carbon dioxide, or a dry chemical extinguisher. *Do not use water on this type of fire.* It will cause the fire to spread.

■ *Class C*—Fires in which the burning material is "live" electrical equipment: motors, switches, generators, transformers, or general wiring. To extinguish this type of fire requires a nonconductive smothering action, such as that from carbon dioxide or dry chemical extinguishers. *Do not use water.*

■ *Class D*—Fires in which the burning material is combustible metals. Special extinguishing agents are required to put out this type of fire.

Fire Extinguishers

Fire extinguishers are available for fighting Class A, B, C, and D fires. Chapter 7 in *Automotive Technology* provides further details on fire extinguishers and safety procedures. Following are some general tips for operating the various types of portable extinguishers, based on the type of extinguishing agent they use:

■ Foam—Do not spray the stream into the burning liquid. Allow the foam to fall lightly on the fire.

- Carbon dioxide—Direct discharge as close to the fire as possible, first at the edge of the flames and gradually forward and upward.

- Soda-acid, gas cartridge—Direct stream at the base of the flame.

- Pump tank—Place foot on footrest and direct the stream at the base of the flame.

- Dry chemical—Direct at the base of the flames. In the case of Class A fires, follow up by directing the dry chemicals at the remaining material that is burning.

 Concept Activity 7-4
Fire Extinguishers

Locate all of the fire extinguishers in the shop. Make a note of what types of fire each kind of extinguisher puts out. ■

MANUFACTURER'S WARNINGS AND GOVERNMENT REGULATIONS

Laws Regulating Hazardous Materials

The Hazard Communication Standard—commonly called the "Right to Know law"—was passed by the federal government and is being administered by Occupational Safety and Health Administration (OSHA). This law requires that any company that uses or produces hazardous chemicals or substances must inform its employees, customers, and vendors of any potential hazards that might exist in the workplace as a result of using the products.

Most important is that you keep yourself informed. You are the only person who can keep yourself and those with whom you work protected from the dangers of hazardous materials. Here is what the law says:

- You have a right to know what hazards you may face on the job.

- You have a right to learn about these materials and how to protect yourself from them.

- You cannot be fired or discriminated against for requesting information and training on how to handle hazardous materials.

- You have the right for your doctor to receive the same hazardous materials information that you receive.

Automotive repair work involves the use of many materials classified as hazardous by both state and federal governments. These materials include such items as solvent and cleaners, paint and body repair products, adhesives, acids, coolants, and refrigerant products.

Hazardous materials are those that could cause harm to a person's well-being. Hazardous materials can also damage and pollute land, air, or water. There are four types of hazardous wastes:

1. Flammable—Materials that will easily catch on fire or explode.

2. Corrosive—Materials that are so caustic that they can dissolve metals and burn skin and eyes.

3. Reactive—Materials that will become unstable (burn, explode, or give off toxic vapors) if mixed with air, water, heat, or other materials.

4. Toxic—Materials that can cause illness or death after being inhaled or contacting the skin.

Exposure

You will be exposed to many hazardous materials while working in a shop. Make sure you always wear the proper protective clothing and eye protection when handling these materials. Exposure to these materials may cause sickness.

Hazardous exposure can take place through inhalation, through your skin or eyes, or from ingesting the materials. Always wash your hands before eating or drinking; it will help prevent the ingestion of chemicals.

Some materials present a potential hazard because they are flammable and/or combustible. Carefully read the label on the container of material. Pay close attention to all warnings and precautions. This is the first step in protecting yourself and those you are working with.

Employer/Employee Obligations

The use of hazardous materials in a shop creates a set of responsibilities for both the owner of the shop and his or her employees. During your training to become a service technician, this relationship involves the school, instructor, and students. An employer or school that uses hazardous materials must follow these rules:

- Provide a safe workplace for employees.

- Educate employees about the hazardous materials they will encounter while on the job **(Figure 7-1)**.

- Recognize, understand, and use warning labels and Materials Safety Data Sheets (MSDSs) (Workplace Hazardous Material Information Systems, or WHMIS, in Canada).

- Provide personal protective clothing and equipment, and train employees to use them properly.

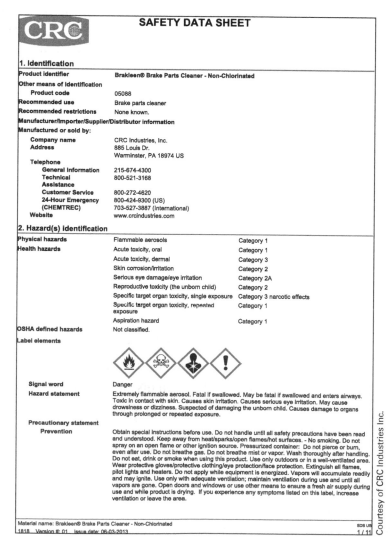

Figure 7-1 An employer's responsibilities to inform employees about hazardous materials they may encounter in the workplace include keeping an accurate, up-to-date hazardous materials inventory roster that includes a material safety data sheet (MSDS) on all hazardous materials.

You, the employee or student, must adhere to these practices:

- Read the warning labels on the materials.
- Follow the instructions and warnings on the MSDS or WHMIS.
- Take the time to learn to use protective equipment and clothing.
- Use common sense when working with hazardous materials.
- Ask the service manager if you have any questions about a hazardous material.

 Concept Activity 7-5
Hazardous Waste

Identify all storage areas for hazardous materials and wastes. Make sure they are properly labeled. ■

 Concept Activity 7-6
MSDS

Locate the MSDS or WHMIS manual in the shop. Look through the MSDS and read through the safety precautions for several commonly used fluids or chemicals. ■

Asbestos

The importance of long-term exposure problems is easily seen in present-day concerns over asbestos. When first introduced on the market as a friction material and heat insulator, asbestos was widely used in brake pads, brake shoes, clutches, and other automotive applications. We now know that asbestos fibers pose a tremendous health risk and that long-term exposure to small amounts of asbestos can cause cumulative health problems. For this reason, asbestos has been phased out of the automotive parts market in favor of safer products.

 REVIEW QUESTIONS

Review Chapter 7 of the textbook to answer these questions:

1. List five safety rules for personal protection in the shop.

   a. _____

   b. _____

   c. _____

   d. _____

   e. _____

2. Describe the proper operation of a vehicle when it is in the shop.

   _____

   _____

   _____

3. List six types of hazardous materials found in the shop.

   a. _____

   b. _____

c. _____

d. _____

e. _____

f. _____

4. Describe two ways to legally dispose of hazardous wastes.

a. _____

b. _____

5. What information is found in an MSDS manual?

_____

_____

_____

_____

6. Technician A says discarded antifreeze can be poured down the shop drains. Technician B says used transmission fluid and used oil can be stored in the same container. Who is correct?

a. Technician A   c. Both A and B

b. Technician B   d. Neither A nor B

7. Technician A says used oil filters must be crushed, then placed in the container holding waste oil and disposed of according to local regulations. Technician B says the ingestion of engine coolant can cause serious illness. Who is correct?

a. Technician A   c. Both A and B

b. Technician B   d. Neither A nor B

8. True or false? The MSDS manual contains information about where different chemicals can be purchased. _____

9. True or false? All of the information about a chemical that is listed on an MSDS or WHMIS is also given on the container's label. _____

10. True or false? If a technician has a car lifted with a floor jack, it is not necessary to use a jack stand when the technician will only be under the vehicle for a brief time. _____

PREVENTIVE MAINTENANCE AND BASIC SERVICES

OVERVIEW

Chapter 8 gives an overview of preventive maintenance and basic services. Though modern cars and trucks do not require the same amount of routine maintenance and service as those of years past, the importance of proper maintenance and service procedures has not decreased. In fact, proper maintenance and service is more important than ever to ensure the proper operation and longevity expected by today's consumers.

REPAIR ORDERS

Repair orders (ROs) are legal documents used to protect the shop and the customer. ROs are used to collect and maintain customer information, provide estimates, and provide a bill and receipt for work performed. Most shops today use computerized RO software to maintain customer databases, inventory tracking, and technician payroll.

Concept Activity 8-1
Repair Orders

Ask your instructor for a practice RO, or use an electronic RO to create an order for a vehicle in the lab. You can have another student role-play the part of the customer. Record all of the necessary information, and verify the vehicle information. ∎

VEHICLE IDENTIFICATION

To properly complete a RO, find service information, or price parts, you must know what exact vehicle you are working on. The best way to accurately identify a vehicle is by using the vehicle identification number, or VIN. The VIN is found on many places on the vehicle, including on the dash behind the driver side A-pillar, on various identification labels, on the firewall, and on stickers attached to body panels **(Figure 8-1)**.

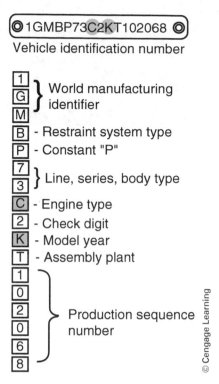

Figure 8-1 This picture shows the alphanumerical breakdown of the digits in a vehicle identification number.

PREVENTIVE MAINTENANCE AND BASIC SERVICES

Preventive maintenance (PM) involves performing certain services to a vehicle on a regularly scheduled basis. Regular inspection and routine maintenance keep cars and trucks running efficiently and safely, and can prevent major breakdowns and expensive repairs.

Perhaps the best-known service is the changing of the engine's oil and filter. Since oil is the lifeblood of the engine, it is critical that the oil be changed routinely. Equally important is using the correct oil for the engine. There are many different brands, weights, and certifications for engine oil. Using the correct oil is paramount for the welfare of the engine. Using incorrect oil can result in serious engine damage. When performing an oil change, you should also do a visual inspection of many of the systems under the hood.

Check the cooling system hoses for signs of leaks or deterioration. Inspect the coolant level, protection factors, and pH level. It is best to inspect the cooling system when the engine is cold. Check the coolant level in the overflow tank (if equipped). Overflow and pressure tanks are normally marked with acceptable hot and cold coolant levels. Remove the radiator cap and check the level in the radiator. Coolant should only by topped off or replaced with the manufacture's recommended coolant. Use of the incorrect coolant can cause severe cooling system and engine damage.

Look at the accessory drive belts for wear and tension. Belts that show excessive cracking, glazing, or damage should be replaced. Rapid wear of the drive belt can occur if it is exposed to fluids leaking from the engine. Always check for engine oil, coolant, and power-steering leaks when replacing a drive belt.

Check the air filter element and housing. Air filter service life is affected by the climate conditions in which the vehicle operates, but a common replacement interval is about 30,000 miles. Inspect and clean the filter housing of any dirt or debris that may have accumulated once the filter is removed.

Visually inspect the battery, battery cables, and hold-down. Light corrosion can be cleaned with baking soda and water or with a battery-cleaning product.

Check the fluid level in the automatic transmission. Most manufacturers recommend that the fluid be checked when hot. This means that the transmission is up to operating temperature, so driving the vehicle may be required to fully warm the fluid.

On some vehicles with manual transmissions, there may be a second smaller master cylinder mounted near the brake master cylinder. This is the clutch master cylinder. This fluid should be checked and topped off in the same way as the brake fluid.

Power steering can usually be checked hot or cold. Many power-steering pumps have a cap that includes a built-in dipstick for checking the fluid. Some power-steering systems use a remotely mounted reservoir, which is often made of a semi-opaque plastic that allows you to see the fluid level inside the reservoir.

Brake fluid is checked at the brake master cylinder. Remove the reservoir cap only if necessary to check the brake fluid level. Clean the cap of the master cylinder before removing to prevent dirt from falling into the reservoir when the cap is removed. Master cylinder caps usually have the recommended brake fluid listed on them.

Also, inspect the windshield washer fluid. Note that some vehicles, particularly minivans, may have two washer fluid reservoirs, one for the front windshield and one for the rear.

Once finished under the hood, inspect the windshield wiper blades. Activate the wiper/washer switch and note how well the blades clean the windshield. Remember to check the rear wiper and washer as well.

Inspect the tires for proper inflation and wear. The recommended tire inflation can be found on a label on one of the doorjambs, in the glove box or center console, or in the owner's manual. Do not inflate the tires to the pressure indicated on the sidewall of the tire. This is the maximum inflation pressure, not necessarily the recommended pressure. If possible, check the air pressure of the spare tire.

Check each tire for excessive wear. If the wear bars or tread wear indicators are easily visible, the tire is worn down to the point of replacement. Visually inspect each tire for signs of bulges, cuts, splits, or tread separation. Any of these indicates the tire should be replaced. A tire rotation may be required if the tires are showing wear, especially the front tires on a FWD car.

Once you are able to check underneath the vehicle, inspect the undercarriage for signs of fluid leaks. Older, high-mileage vehicles often show evidence of fluid leaks. A slight amount of engine oil loss may not present a serious problem, but signs of excessive loss should prompt a complete inspection to determine the fault. If there is coolant or transmission fluid leaking, each system should be thoroughly inspected before any major damage is done.

Visually check the components of the suspension and steering system. Look at sway bar bushings and links, as they are common failure items. If the vehicle has grease or zerk fittings, these should be greased during routine service. Clean each fitting with a shop rag before attaching the grease gun. Clean off any old grease or any excess that comes out of the grease boots.

The brake system cannot usually be inspected easily if the tires are installed, but you can turn each wheel and compare the effort needed. A tire that does not turn easily can alert you to a possible brake problem.

Examine the exhaust system for signs of damage, rust-through, or missing or altered components. An exhaust leak can be dangerous for the passengers inside the vehicle because of the possibility of carbon monoxide exposure.

If the vehicle has a manual transmission, the fluid level is often checked by removing an access plug in the side of the transmission case. The fluid level should be up to the hole of the access plug. Use only the lubricant recommended by the manufacture if the level needs adjusting.

Rear-wheel-drive vehicles and some 4WD vehicles should have the rear/front differential fluid level checked whenever routine service is being performed. To check the fluid, locate and remove the access plug. Some manufacturers use a rubber plug in the differential cover, whereas others use a threaded plug in the side of the differential housing. Similar to that of a manual transmission, the fluid level should be up to the access hole. Always use the correct fluid in a differential.

The goal of preventive maintenance is to find and correct problems before they become larger, more expensive problems.

The following job sheets will prepare you for preventative maintenance services.

☐ JOB SHEET 8–1

Locating the VIN

Name _____ Station _____ Date _____

Objective

Upon completion of this job sheet, you will have demonstrated the ability to locate vehicle identification numbers. This task applies to ASE Education Foundation Tasks MLR 1.A.1 – 8.A.1 (P-1), AST/MAST 1.A.2 – 3.A.2 (P-1), 4.A.1 – 6.A.1 (P-1), 7.A.2 & 8.A.2 (P-1).

VIN no.1 _____

Using service information, use the VIN to determine the following:

Model Year _____ Manufacturer _____

Country of manufacture _____

Engine _____ Serial no. _____

VIN no.2 _____

Using service information, use the VIN to determine the following:

Model Year _____ Manufacturer _____

Country of manufacture _____

Engine _____ Serial no. _____

Problems Encountered

_____

_____

_____

Instructor's Comments

_____

_____

_____

☐ JOB SHEET 8–2

Identifying the Correct Fluids

Name _____ Station _____ Date _____

Objective

Upon completion of this job sheet, you will have demonstrated the ability to locate the correct engine oil, coolant, and transmission fluids for a vehicle. This task applies to ASE Education Foundation Tasks MLR 1.C.5 (P-1), and AST/MAST 1.D.1 (P-1).

Vehicle 1: Year _____ Make _____ Model _____

VIN _____

Engine _____ Transmission _____

Oil Weight _____ API Service Rating _____

ILSAC Rating _____ ACEA Rating _____

Manufacturer specific rating: _____

Recommended fill amount with filter: _____

Coolant base type (ethylene or propylene glycol): _____

Manufacturer specific requirements: _____

Recommended refill amount: _____

Transmission fluid type: _____

Recommended refill amount: _____

Problems Encountered

_____

_____

_____

Instructor's Comments

_____

_____

_____

☐ JOB SHEET 8–3

Measuring Tire Pressure

Name _____ Station _____ Date _____

Objective

Upon completion of this job sheet, you will have demonstrated the ability to locate the recommended tire pressure and measure tire pressure with a tire pressure gauge. This task applies to ASE Education Foundation Tasks MLR 4.D.1 (P-1), and AST/MAST 4.F.1 (P-1).

Vehicle Make _____ Model _____ Year _____

Pressure: LF _____ RF _____ LR _____ RR _____

Spare _____ Specification _____ Spare Spec. _____

Vehicle Make _____ Model _____ Year _____

Pressure: LF _____ RF _____ LR _____ RR _____

Spare _____ Specification _____ Spare Spec. _____

Vehicle Make _____ Model _____ Year _____

Pressure: LF _____ RF _____ LR _____ RR _____

Spare _____ Specification _____ Spare Spec. _____

Problems Encountered

_____

_____

_____

Instructor's Comments

_____

_____

_____

☐ JOB SHEET 8–4

Basic Preventive Maintenance Inspection

Name _____ Station _____ Date _____

Objective

Upon completion of this job sheet, you will have demonstrated the ability to perform basic but necessary checks on various systems of a vehicle. Before beginning, review Chapter 8 of *Automotive Technology*.

Description of Vehicle

Year _____ Make _____ Model _____

VIN _____

Engine _____ Transmission _____

PROCEDURE

1. Inspect the engine oil level.

   Describe the color and texture. _____ _____ _____

   Recommendations: _____ _____

2. Inspect the engine coolant.

   Inspect the coolant hoses.

   Recommendations: _____

3. Inspect the accessory drive belts.

   Recommendations: _____

4. Inspect the air filter and housing.

   Recommendations: _____

5. Inspect the battery, terminals, cables, and hold-down.

   Recommendations: _____

6. Inspect the automatic transmission fluid.

   Describe the color and texture. _____

   Recommendations: _____

7. Inspect the power-steering fluid.

   Describe the color and texture. _____

   Recommendations: _____

8. Inspect the brake fluid.

   Recommendations: _____

9. Inspect the washer fluid level.

10. Inspect and test the wiper/washer.

    Recommendations: _____

11. Inspect the lap/shoulder seat belts for wear and proper operation.

    Recommendations: _____

12. Inspect the exterior lights, turn signals, brake lights, hazard lights, parking, and headlights.

    Recommendations: _____

13. Inspect the condition of the tires.

    Recommendations: _____

14. Inspect for fluid leaks under the vehicle.

    Recommendations: _____

Problems Encountered

_____

_____

_____

Instructor's Comments

_____

_____

_____

 **REVIEW QUESTIONS**

Review Chapter 8 of the textbook to answer these questions:

1. Where should you look to find the proper tire inflation pressure?

   _____

2. Define *preventive maintenance*.

   _____

3. Why is preventive maintenance important to the life of the vehicle?

   _____

4. Technician A says the VIN code contains information about where the vehicle was manufactured. Technician B says the VIN code identifies who the vehicle manufacturer is. Who is correct?

   a. Technician A   c. Both A and B
   b. Technician B   d. Neither A nor B

5. Technician A says lubricating grease fittings is no longer necessary on modern vehicles. Technician B says grease fittings are used on all FWD cars. Who is correct?

   a. Technician A   c. Both A and B
   b. Technician B   d. Neither A nor B

CHAPTER 9

Courtesy of Chrysler LLC

AUTOMOTIVE ENGINE DESIGNS AND DIAGNOSIS

OVERVIEW

Both gasoline and diesel engines have been in use for decades, and their basic operation has not changed much in all those years. What has changed with today's engines is the materials from which they are made, the amount of power they produce, how efficiently they produce power, and a greatly increased operational lifespan.

Suggested Materials: To perform the activities and job sheets contained in this chapter, it is recommend to have the following items:

- 100 cc syringes
- Vacuum hose
- Plastic water bottle, empty
- Rubber bands
- Glass jars
- ½-gallon or 1-gallon milk jug, empty

- Plastic quart bottle of oil, empty
- 1 quart of motor oil
- Differential fluid bottle cap
- Vacuum gauge
- Compression gauge
- Cylinder leak detector

Nearly all reciprocating gasoline engines work the same way, have similar parts, and share similarities in service and repair requirements. To understand how an engine runs, a technician must know what factors are involved in engine operation. The following activities will demonstrate concepts related to pressure, temperature, volume, and compression, all of which are factors in engine operation.

Every day we hear and talk about pressure; air pressure and tire pressure are just two common examples. But what is air pressure? Where does it come from? Does it change? Why is it important to automotive repair? Understanding pressure will help you understand many of the principles utilized in almost every aspect of the automobile.

Concept Activity 9-1
Air Pressure

Air pressure, or atmospheric pressure, is a measurement of the weight of the air from the edge of the atmosphere down to ground level. We do not often think about the air that surrounds us, but the weight of that air has a huge impact on our lives. Air pressure at sea level is about 14.7 pounds per square inch (psi). This means that if we could take a 1-square-inch column of air from the ground up to the edge of space, it would have a weight of approximately 14.7 pounds **(Figure 9-1)**. As you move higher and higher above sea level, air pressure decreases.

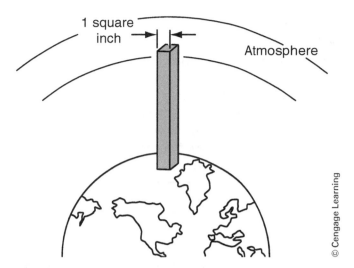

Figure 9-1   A column of air 1 square Inch extending from the earth's surface at sea level to the outer edge of the atmosphere weighs 14.7 pounds.

1. Explain why air pressure decreases with altitude. _____

_____

Because air pressure is a product of our atmosphere, it is subject to changes in the atmosphere. In addition, because air pressure is affected by the weather, it can be easily measured and recorded in your lab on a daily basis by making this simple barometer. Take a glass container and stretch a balloon or similar material over the opening of the jar. If necessary, use a rubber band to seal the balloon around the jar: it must be airtight. Next, attach a straw horizontally to the balloon with a piece of tape. The height of the end of the straw can be marked on a piece of paper to record the changes in air pressure **(Figure 9-2)**. As the local air pressure changes, the straw will rise or drop with the pressure changes. Record the height over several days or weeks to track the changes in air pressure.

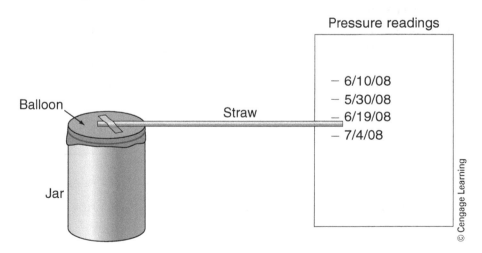

Figure 9-2   Measuring local barometric pressure with self-made barometer.

2. Describe how this simple barometer operates. _____

3. What causes the up or down movement of the straw? _____

4. If the air in the jar were heated or cooled, how would the straw react? _____

5. How does temperature affect the pressure inside the jar? _____

■

**Concept Activity 9-2
Pressure and Temperature**

We saw in the last activity that the air pressure changes, but what can cause the pressure to change? Temperature is one factor that can cause pressure change. When air is heated, it rises, and when air is cooled, it falls. The reason for this is that as air is heated, it weighs less per unit of volume than cooled air. Because the hotter air is less dense, the cooler denser air lifts it. To demonstrate how temperature affects pressure, fill a plastic milk jug approximately one-third of the way full with very hot water. Use caution when working with hot water, to avoid being accidentally burned. Screw the cap on securely, and wait about an hour.

1. What happened to the jug? _____

2. Why did the jug react in the way it did? _____

3. How did the outside air pressure affect the jug? _____

To see how increasing the air temperature will result in an increase air pressure, place a balloon over the opening of a plastic water bottle that is filled about halfway. Carefully heat the water bottle with a heat gun, hair dryer, or other safe and instructor-approved method.

4. Describe what happened to the balloon. _____

5. Why did the balloon react the way it did? _____

6. What does the balloon do as the water and bottle slowly cool? _____

7. Why did this happen? _____

■

 **Concept Activity 9-3
Pressure and Volume**

The pressure of air (or other gas) in a container can be affected by several things, such as the temperature in the container and the volume of the container. In an automotive engine, air and fuel are drawn into the cylinder, which, once sealed, becomes smaller. This decrease in volume causes the pressure and temperature of the gas to increase as the molecules are forced more tightly together.

When the volume of the cylinder is increased, such as when the piston moves from TDC to BDC on the intake stroke, the pressure in the cylinder decreases.

1. Why does the pressure decrease? _____

2. What causes the air to enter the cylinder? _____

Place the small open end of a syringe against one of your fingers, and draw the plunger out until it stops moving.

3. Describe what happens. _____

4. What was the effect within the syringe by pulling out on the plunger? _____

Attach a pressure/vacuum gauge to a syringe with the plunger about halfway pulled out. Pull the plunger back to the end of its travel, and note the reading on the gauge.

5. Gauge reading: _____

6. With the plunger at the end of its travel, push the plunger in, and note the reading on the gauge.

Obtain two syringes of equal size, and pull each plunger out about halfway. Then connect the two together with a piece of vacuum hose.

7. Push on one plunger and note what happens to the other plunger. _____

8. Was the result what you expected? Why or why not? _____

9. What caused the movement of the second plunger? _____

10. Now, pull on the plunger and note what happens. _____

11. Why did the second plunger move? _____

12. How did the change in pressure in the first syringe affect the second syringe?

_____

Obtain from your instructor a vacuum pump and a cone-shaped dispensing lid from a differential fluid bottle. Attach the cap to an empty motor oil quart bottle. Pull vacuum on the bottle via the cap, and watch the oil bottle.

13. What happens to the oil bottle? _____

14. Why did the bottle react the way it did? _____

15. Release the vacuum and note the effect on the bottle. Explain the result. _____

_____

Using a vacuum pump and a vacuum chamber connected to a bottle of oil, apply vacuum and watch the oil.

16. Why did the oil move from the oil bottle? _____

You have now demonstrated a basic physical principle; that pressure moves from a higher pressure to a lower pressure.

Next, place a balloon into an empty water bottle, and leave the balloon opening above the opening of the bottle. Now attempt to inflate a balloon that is inside of the bottle.

17. Explain what happened. _____

18. Why did the balloon not inflate normally? _____

19. What could be done to allow the balloon to properly inflate? _____

The balloon was trying to inflate within the bottle and was exerting pressure against the air already inside the bottle. Because the air in the bottle had nowhere to go, the pressure equalized against the pressure you attempted to place in the balloon, causing a pressure balance. The harder you attempted to blow up the balloon, the more the pressure in the bottle worked against you. Now place a small hole in the bottom of the bottle and repeat the experiment.

20. Describe what happened. _____

21. Why was the balloon able to inflate more than before? _____

22. If you plug the hole in the bottle and release the air from the balloon, what will happen to the bottle?

_____

23. Why? _____

You have seen how pressure can move from high to low and how to increase pressure. Next, we will examine low pressure, also called vacuum. Vacuum is pressure that is less than atmospheric pressure. Pressure can be measured with a pressure gauge, such as a tire pressure gauge, but a tire gauge is calibrated to read atmospheric pressure as zero. This is referred to as psi gauge, or psig. A flat tire will read zero psi on a pressure gauge, but it actually has 14.7 psi (or equivalent local) pressure **(Figure 9-3)**. To read pressures lower than atmospheric, you will need either a gauge that reads psi absolute (psia) or a vacuum gauge. Using a vacuum gauge and a pressure chart, you can easily see what a vacuum-gauge reading reflects in actual air pressure. Connect a vacuum gauge to a vacuum hose, and start the engine.

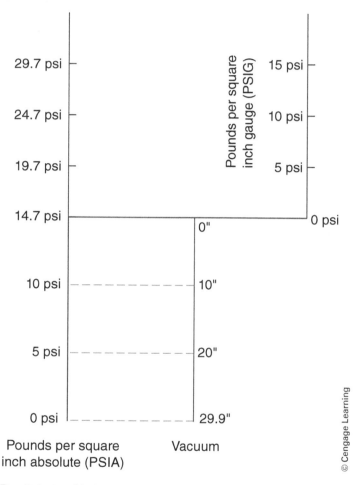

Figure 9-3 Relationship between atmospheric pressure, higher pressure, and vacuum.

24. Record the gauge reading. _____

25. What does this equate to in actual air pressure? _____

26. Why is the engine producing vacuum? _____

27. What does the vacuum gauge read if the engine speed is increased? _____

■

Concept Activity 9-4
Displacement

Engine size is determined by cylinder displacement, which is the total volume of the cylinders and combustion chambers. To understand displacement, let us first look at the volume of a basic cylinder. When the piston is at bottom dead center, the cylinder can hold a certain amount of liquid: let us call the amount V for the moment. To determine the value of V, we need to know a few things such as the radius of the cylinder and its depth. Radius is one-half of the diameter, which is the distance across the cylinder from side to side. Depth is the total height from the top edge to the bottom of the cylinder, which is also the top of the piston **(Figure 9-4)**.

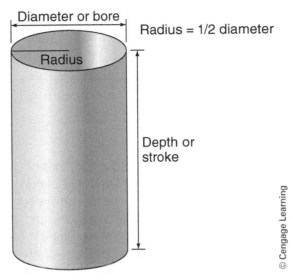

Figure 9-4 Measuring a cylinder.

Suppose we have a cylinder that is 3.5" in diameter and 3.5" deep; this would equal a bore of 3.5" and a stroke of 3.5". *Bore* and *stroke* are the terms used to describe cylinder diameter and depth. To calculate the volume, the bore of 3.5 inches is converted to radius. Because radius is ½ of diameter (or bore), then a bore of 3.5" inches has a radius of 1.75 inches. Using the formula $V = \pi r^2 h$, where V is the total volume; π is the value of pi, or 3.14159; r^2 is the cylinder radius squared; and h is the height (or stroke) of the cylinder, the volume can be determined. Our equation will look like $V = 3.14159 \times 1.75^2 \times 3.5$. The cylinder volume, or displacement, equals about 33.67 cubic inches, or $33.7^{3"}$. If our engine is an eight cylinder, then we multiply $8 \times 33.7^{3"}$ to get a total engine displacement of 269.39 cubic inches.

To calculate displacement using cubic centimeters (cc), simply use the same formula but substitute the metric equivalent measurements for the English ones. In our sample engine, we had a bore and stroke of 3.5 inches each. Because there are 2.54 centimeters (or 25.4 millimeters) per inch, simply convert inches to centimeters by multiplying 3.5 by 2.54. This gives us a cylinder bore and stroke measurement of 8.89 centimeters. Now substitute 8.89 into our equation, $3.14159 \times 8.89 \, cc^2 \times 8.89 \times 8$ for a displacement of 4,412 cc, or 4.4 liters. A nice benefit of using metric units is conversion is a matter of just moving the decimal.

Another method of calculating displacement is by using the formula: $0.785 \times$ Bore $\times$ Stroke $\times$ Number of cylinders. Because 0.785 is one quarter of π, some people find it easier to remember this way because the actual bore measurement can be used.

To measure displacement, obtain from your instructor a cylindrical container that you can measure for the diameter and depth. Using a telescoping gauge and depth gauge, measure the inside diameter and depth of the container, and record your measurements. Calculate the volume of the container, using the formulas above.

1. Inside diameter _____   Depth _____   Volume _____

Using the volume you found, multiply the total by four, six, and eight to get an idea of what the amount equates to for a four-cylinder, six-cylinder, and eight-cylinder engine.

2. Four-cylinder _____ Six-Cylinder _____ Eight-Cylinder _____

   Compare these numbers with the engine sizes found in modern cars and light trucks.

3. How does the displacement compare to the external size of an engine? _____

4. What are some factors that limit engine displacement? _____

5. Can engine wear affect displacement? Why/how? _____

6. If an engine is bored oversize to correct for cylinder wear, what effect will this have on displacement?

   _____

7. How would carbon buildup on a piston affect displacement? _____

■

 ### Concept Activity 9-5
Compression Ratio

Using cylinder displacement, the measure of bore and stroke, cylinder compression ratio can be determined. Compression ratio refers to the volume of the cylinder when the piston is at BDC compared to the volume when the piston is at TDC. The difference in volume is the compression ratio. Suppose we have a clear cylinder, such as a syringe that holds 100 cubic centimeters of liquid and has graduated markings along the length, marking every 10 cubic centimeters, **(Figure 9-5)**. Pull the plunger back, filling the syringe with 100 cubic centimeters of air. This is similar to filling the cylinder of an engine when the piston has moved from TDC to BDC on the intake stroke. Now, cap the opening of the syringe and push the plunger back in, compressing the trapped air inside. Notice that as you compress the air, it becomes more difficult to push the plunger. If you can compress the original 100 cc volume of air until the plunger reaches the 10 cc mark, you have compressed the air by a factor of 10, or by a ratio of 10:1. By taking the original volume of 100 cc and reducing it down to 10 cc, you have compressed the air 10 times. Compression ratio = volume 1/volume 2 or 100/10 = 10 or 10:1.

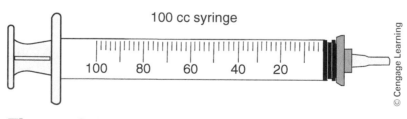

Figure 9-5 A typical 100 cc syringe.

1. Why did the plunger become harder to push as the volume of air decreased? _____

2. What would the compression ratio in the syringe be if you compressed the air from 100 cc to 50 cc?

   _____

3. If you compressed the air in the syringe to the 20 cc mark, what would the compression ratio be?

   _____

For a simple experiment, Boyle's Law can be used to determine the pressure in the syringe. P = pressure; P_i = initial pressure, or 14.7 psi, or atmospheric pressure at sea level; V_i is the initial volume, in this case 100 cc; and V_f is the final volume. Use the formula for final pressure or $P_f = P_i V_i / V_f$ and a compression ratio of 2:1 (100 cc to 50 cc):

4. What would be air pressure in the syringe? _____

5. If the air were compressed 10:1, what would the pressure of the air in the syringe be? _____

6. What design factors affect the compression ratio of an engine? _____

7. Why do some engines have higher or lower compression ratios than others? _____

8. Why is the compression ratio of a gasoline-powered engine different from a diesel engine? _____

_____

The actual pressure in an operating cylinder will be different than what you have experienced with this experiment because of the effects of valve timing, intake design, combustion chamber design, and other factors. These experiments do not take into account combustion chamber size, which also has an effect on displacement and compression ratio.

By completing these activities and experiments, you should have an understanding of some of the basic principles that govern engine operation. The following job sheets apply your understanding of these principles.

☐ JOB SHEET 9–1

Engine Classifications

Name _____ Station _____ Date _____

Objective

Upon completion of this job sheet, you will have demonstrated the ability to properly look up vehicle internal engine operation. This task applies to ASE Education Foundation Tasks MLR 1.A.1 (P-1), and AST/MAST 1.A.2 (P-1).

Using available lab vehicles, determine the following engine design characteristics:

Vehicle 1

Year _____ Make _____ Model _____

Engine Size _____ Cylinder Configuration _____

Valvetrain Configuration _____ Fuel Type _____

Firing Order _____

Vehicle 2

Year _____ Make _____ Model _____

Engine Size _____ Cylinder Configuration _____

Valvetrain Configuration _____ Fuel Type _____

Firing Order _____

☐ JOB SHEET 9–2

Identifying the Correct Cleaning Methods for the Engine Components

Name _____ Station _____ Date _____

Objective

Upon completion of this job sheet, you will have demonstrated the ability to identify the correct engine component cleaning methods. This task applies to ASE Education Foundation Tasks AST/MAST 1.B.2 (P-1), and MAST 1.C.2 (P-1).

Description of Vehicle

Year _____ Make _____ Model _____ Mileage _____

VIN _____ Engine _____

_____

Upon disassembly, the various components and sections of the engine must be cleaned. To prevent damage to engine parts, only the approved cleaning methods should be used as specified by the engine manufacturer.

1. Cylinder block construction material(s): _____

   Recommended cleaning method: _____

   Cleaning equipment used: _____

2. Cylinder head construction material(s): _____

   Recommended cleaning method: _____

   Cleaning equipment used: _____

3. Crankshaft construction material: _____

   Recommended cleaning method: _____

   Cleaning equipment used: _____

4. Piston construction material: _____

   Recommended cleaning method: _____

   Cleaning equipment used: _____

5. Intake manifold construction material: _____

   Recommended cleaning method: _____

   Cleaning equipment used: _____

6. Exhaust manifold construction material: _____

   Recommended cleaning method: _____

   Cleaning equipment used: _____

Problems Encountered

_____

_____

_____

Instructor's Comments

_____

_____

_____

☐ JOB SHEET 9–3

Performing a Cylinder Compression Test

Name _____ Station _____ Date _____

Objective

Upon completion of this job sheet, you will have demonstrated the ability to perform a cylinder compression test. This task applies to ASE Education Foundation Tasks MLR 8.A.4 (P-2), and AST/MAST 8.A.7 (P-1).

Tools and Materials

Spark plug sockets _____ Fender covers _____ Air blowgun _____ Small oilcan _____

Description of Vehicle

Year _____ Make _____ Model _____ Mileage _____

VIN _____ Engine _____

1. Disable the fuel and ignition systems. Explain how this is accomplished. _____

_____

WARNING: *Hybrid vehicles require special procedures for conducting compression tests. Do not attempt to perform this test on a hybrid without following the manufacturer's precautions and procedures.*

2. Use an air blowgun to clean around the spark plugs. Remove all the spark plugs.

3. Block the throttle open. Explain why the throttle should be wide open when performing a compression test.

_____

Note: *Refer to the service information for procedures if vehicle is equipped with electronic throttle control.*

4. Install a battery charger and set to a low charge rate. Explain why a battery charger should be used during a compression test.

_____

5. Install the compression gauge into cylinder number 1. Crank the engine for 5 seconds and note the compression gauge reading. Record the reading in the space below.

6. Repeat for each cylinder and record the readings.

7. If a cylinder's compression reading is more than 10 percent lower than the others, a wet test should be performed. Squirt a small amount of oil into the low cylinder and repeat the compression test.

	Dry Test	Wet Test		Dry Test	Wet Test
Cyl no.			Cyl no.		
Cyl no.			Cyl no.		
Cyl no.			Cyl no.		
Cyl no.			Cyl no.		

Problems Encountered

_____

_____

_____

Instructor's Comments

_____

_____

_____

☐ JOB SHEET 9–4

Performing a Running Compression Test

Name _____ Station _____ Date _____

Objective

Upon completion of this job sheet, you will have demonstrated the ability to perform a cylinder compression test. This task applies to ASE Education Foundation Tasks MLR 8.A.4 (P-2), and AST/MAST 8.A.7 (P-1).

Tools and Materials

Spark plug sockets _____ Fender covers _____ Air blowgun _____ Small oilcan _____

Description of Vehicle

Year _____ Make _____ Model _____ Engine _____

VIN _____ Valvetrain Type _____

Attach a copy of the cylinder firing order.

PROCEDURE

1. Attach test record sheet from previous compression test.

2. Install all plugs but one. Install a standard compression gauge in the remaining hole with the Schrader valve removed.

   Note: *Damage to the gauge can result if the needle is pulled against the stop pin during this test. Depending on the type of gauge being used, leave the Schrader valve in the compression tester and use the bleed valve to bleed air from the gauge during the test.*

3. Start the engine and allow the reading to stabilize. Needle bounce is normal for a running compression test. Cylinder running pressure _____

4. Snap the throttle wide open and return to idle. Note and record the peak reading. This reading should be higher than the idle reading. Snap throttle pressure _____

5. Record your readings for running and snap compression for all cylinders. The running compression reading should be approximately 50–75 psi and snap compression should be about 80 percent of cranking compression.

Running Pressure		Snap Pressure	
1.	5.	1.	5.
2.	6.	2.	6.
3.	7.	3.	7.
4.	8.	4.	8.

Worn cam lobes and weak or broken valve springs can cause low running compression. Higher than normal, over 80 percent of cranking compression pressure, can be caused by a restricted exhaust system.

Problems Encountered

_____

_____

_____

Instructor's Comments

_____

_____

_____

☐ JOB SHEET 9–5

Performing a Cylinder Power Balance Test Using Scan Tool

Name _____ Station _____ Date _____

Objective

Upon completion of this job sheet, you will have demonstrated the ability to perform a cylinder power balance test. This task applies to ASE Education Foundation Tasks MLR 8.A.2 (P-2), and AST/MAST 8.A.6 (P-2).

Description of Vehicle

Year _____ Make _____ Model _____ Mileage _____

VIN _____ Engine _____

PROCEDURE

1. Obtain a scan tool applicable for the vehicle being tested. Scan tool used

   _____

2. Connect the scan tool to the DLC and enter into Powertrain diagnostics. Locate the power balance test.

3. Initial engine rpm:

   _____

4. Follow the procedure indicated by the scan tool and disable each cylinder and note the rpm drop:

   No. 1 _____   No. 5 _____

   No. 2 _____   No. 6 _____

   No. 3 _____   No. 7 _____

   No. 4 _____   No. 8 _____

5. What is the rpm difference between the weakest and strongest cylinders?

   _____

6. Based on the test results, what is the necessary action?

   _____

Problems Encountered

_____

_____

_____

Instructor's Comments

_____

_____

_____

☐ JOB SHEET 9–6

Engine Vacuum Testing

Name _____ Station _____ Date _____

Objective

Upon completion of this job sheet, you will have demonstrated the ability to perform a engine vacuum test. This task applies to ASE Education Foundation Tasks MLR 8.A.2 (P-2), and AST/MAST 8.A.5 (P-1).

Description of Vehicle

Year _____ Make _____ Model _____ Mileage _____

VIN _____ Engine _____

PROCEDURE

1. Connect a vacuum gauge to a suitable vacuum port on the engine. Describe the location of the port.

   _____

2. With the ignition and/or fuel system disabled, crank the engine and record the vacuum reading.

   Cranking vacuum reading:

   _____

   Reconnect the ignition and/or fuel system.

3. Start the engine and allow it to idle. What is the idle vacuum reading?

   _____

   Is the gauge showing a steady needle?   _____ YES   _____ NO

   What would a bouncing or fluctuating needle indicate?

   _____

4. Quickly snap the throttle wide open and then closed. What was the reading at WOT and on deceleration?

   WOT _____ Decel. _____

5. Increase the engine speed to 2,000 rpm and hold. Note the vacuum reading:

   _____

6. Does the vacuum reading decrease after one minute?   _____ YES   _____ NO

7. If the vacuum reading dropped at 2,000 rpm, what would that indicate?

   _____

Problems Encountered

_____

_____

_____

Instructor's Comments

_____

_____

_____

☐ JOB SHEET 9–7

Performing a Cylinder Leakage Test

Name _____ Station _____ Date _____

Objective

Upon completion of this job sheet, you will have demonstrated the ability to perform a cylinder leakage test properly. This task applies to ASE Education Foundation Tasks MLR 8.A.5 (P-2), and AST/MAST 8.A.8 (P-1).

Tools and Materials

Chalk	Radiator coolant (if applicable)
Fender covers	TDC indicator
Jumper lead	Test adapter hose

Description of Vehicle

Year _____ Make _____ Model _____

VIN _____ Engine Type and Size _____

Mileage _____

PROCEDURE

CAUTION: *Very high voltages are present with high-energy ignition systems. Do NOT use this procedure on cars with distributorless ignition systems unless the ignition system is totally disabled.*

1. Inspect the coolant level and fill, if needed. Disable the ignition and fuel injection system. Explain how the ignition and fuel are disabled. _____

_____

2. Use compressed air to clean all foreign matter out of the plug wells. Remove all spark plugs and set them on a workbench or other clean surface in the order in which they were removed.

3. Remove the air cleaner. Block the throttle plate in a wide-open position using a screwdriver or similar tool. Disconnect the PCV hose from the crankcase. Why is the PCV hose disconnected from the crankcase?

_____

_____

4. Install the test adapter hose in the number 1 cylinder spark plughole.

5. Using a socket and ratchet on the crankshaft pulley nut or bolt, slowly rotate the engine in its normal direction until the piston is at TDC of the compression stroke.

6. Connect the tester to the adapter hose. Does the gauge show more than 20% leakage? Air may be leaking from/into the following areas:

   a. _____   c. _____

   b. _____   d. _____

   Record your results on the accompanying "Report Sheet on Cylinder Leakage Test."

7. Disconnect the tester from the adapter hose. Resume rotating engine using socket and ratchet on the crankshaft nut or bolt until the next cylinder is at TDC compression.

8. Remove the adapter from the previously tested cylinder and install it in the plughole of the next cylinder in the firing order.

9. Repeat steps 6, 7, and 8 on each remaining cylinder.

Problems Encountered

_____

_____

_____

Instructor's Comments

_____

_____

_____

Name _____   Station _____ Date _____

REPORT SHEET FOR CYLINDER LEAKAGE TEST

No.	Percentage	Leakage From
1.		
2.		
3.		
4.		
5.		
6.		
7.		
8.		

Conclusions and Recommendations _____

_____

_____

☐ JOB SHEET 9–8

Measuring Engine Oil Pressure

Name _____ Station _____ Date _____

Objective

Upon completion of this job sheet, you will have demonstrated the ability to measure engine oil pressure properly. This task applies to ASE Education Foundation Tasks AST/MAST 1.D.9 (P-1).

Tools and Materials

Adapter fitting   Tachometer or scan tool

Fender covers

Description of Vehicle

Year _____ Make _____ Model _____

VIN _____ Engine Type and Size _____

Mileage _____

PROCEDURE

1. Look up the oil-pressure specifications for this engine in the appropriate service manual or electronic service information resource.

   Oil pressure _____

   Engine speed _____

2. Locate the oil-pressure sending unit and disconnect the wire from the sending unit. Use an appropriate wrench or specialized socket to remove the sender.

   Sending Unit Location _____

3. Tighten the oil-pressure test gauge into the hole in the block where the sender was removed. Use an adapter fitting, if necessary, to make the connection shown in **Figure 9–6**.

4. Check the engine's oil level and fill, if required.

   Oil Level _____

   WARNING: *Be extremely careful when working near a running engine. Always wear safety goggles or glasses with side shields when working around moving machinery and be sure that your clothing is not loose.*

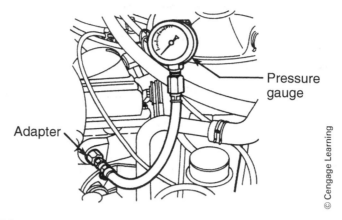

© Cengage Learning

Figure 9-6 Adapter fitting.

5. Start the engine and observe the pressure reading on the gauge. Make sure the engine speed is set to specifications for testing pressure. If necessary, use a tachometer and adjust the engine idle speed. Record the measured oil pressure, and then turn the engine off.

_____

6. Is your measurement below specifications?

_____

7. Consult the appropriate service manual and list your conclusions and recommendations.

_____

_____

8. Remove the test gauge and adapter fitting. Reinstall the oil-pressure sender and connect the wire. Start the engine and check for leaks.

Instructor's Check _____

Problems Encountered

_____

_____

_____

Instructor's Comments

_____

_____

_____

 REVIEW QUESTIONS

Review Chapter 9 of the textbook to answer these questions:

1. What engine components are tested by the cranking compression test?

   _____

   _____

   _____

2. Explain what could cause lower than normal engine vacuum readings.

   _____

   _____

   _____

3. Bubbles are seen in the radiator when performing a cylinder leakage test. Describe two possible causes.

   _____

   _____

   _____

4. List four factors regarding how much air can be drawn into an engine.

   _____

   _____

   _____

5. List four causes of low oil pressure.

   _____

   _____

   _____

   _____

CHAPTER
10

ENGINE DISASSEMBLY AND CLEANING

OVERVIEW

This chapter covers the removal of an engine from a vehicle, basic disassembly, and cleaning of the engine. During disassembly, it is important that all parts be thoroughly inspected and cleaned. The job sheets in this chapter cover the basic procedure for preparing to remove an engine and disassembling a cylinder head.

Most often the engine must be removed from the vehicle to be repaired or rebuilt. However, there are some operations that are done with the engine in place in the vehicle, such as cylinder head service, which is why removal and disassembly of the head is treated separately from the rest of the engine.

REMOVING AN ENGINE

Before beginning to remove an engine from a vehicle, it is important to have a plan of attack. Carefully look over the area surrounding the engine, and identify anything that may get in the way. Also identify everything that should be removed or disconnected from the engine.

Before disconnecting anything, drain all fluids from the engine. Clean up any spills before you continue. Also, disconnect the battery before proceeding. Typically, all wires and hoses connected to the engine should be disconnected. To prevent connecting them to the wrong place during installation of the engine, use tape and a marker to identify where they should be connected. Disconnect all linkages from the engine and remove only those parts that must be removed in order to get the engine out.

Fasten the appropriate lifting tool to the engine, and raise the engine just enough to take the weight off the engine mounts. Then disconnect the mounts. Carefully lift the engine out of the vehicle. Once it is out, mount the engine on an engine stand.

ENGINE DISASSEMBLY AND INSPECTION

Disassembling the engine is not a matter of just taking it apart. Always check the service manual for the correct procedure for disassembling the engine you are working on. You should remove one part at a time. After each part is removed, it should be inspected and set aside. It is wise to keep all removed fasteners with the part that was removed.

 Concept Activity 10-1
Service Procedures

Choose two vehicles, each made by a different manufacturer, and compare the engine disassembly procedures given in the service information. Describe the similarities and the differences, and explain why there are differences. ■

CYLINDER HEAD REMOVAL

For the most part, the procedure for removing a cylinder head will vary according to the type of engine. If the engine is an OHC-type, the timing belt must be removed first. Make sure you follow the procedure for doing this, and install a new belt when reassembling the engine.

It is important that you mark the location of the head bolts. They may be different lengths. Installing a long bolt where a short bolt should be may cause the bolt to bottom out and prevent proper torquing.

On most vehicles it is possible to remove the cylinder heads without removing the entire engine. The procedure for removing a cylinder head from an engine mounted in a vehicle is the combined procedure for removing an engine and removing a cylinder head from a removed engine. Again, before you begin the procedure, refer to your service manual and have a plan of attack.

CLEANING ENGINE PARTS

Many different methods can be used to clean the different engine parts. Often the method used is determined by the equipment available. Normally, the selection of cleaning chemicals and/or equipment is based on the type of dirt or residue you need to remove. Check the textbook for a complete description of each cleaning method. Remember that there are strict laws about hazardous wastes. Used solvents and run-off grease should not be released into a regular sewage drain system.

 **Concept Activity 10-2
Engine Cleaning**

List all of the engine cleaning equipment and solvents that are available in your shop and what they are used to clean. When listing the solvents, include the health-related precautions listed on the solvent container's label or on its MSDS. ■

☐ JOB SHEET 10–1

Preparing Engine for Removal

Name _____ Station _____ Date _____

Objective

Upon completion of this job sheet, you will have demonstrated the ability to prepare an engine for removal. This task applies to ASE Education Foundation Task MAST 1.A.10 (P-3).

Tools and Materials

Drain pans	Penetrating oil
Fender covers	Vacuum plugs
Masking tape	Transmission jack

Description of Vehicle

Year _____ Make _____ Model _____

VIN _____ Engine Type and Size _____

Mileage _____

WARNING: *The technician should always follow the specific engine removal and disassembly procedures given in the service information for the particular vehicle being worked on. If the following procedure differs at any point from the service information procedure, follow the procedure given in the service information.*

PROCEDURE

1. Install fender covers on the fenders. Mask any areas where there is any possibility of scratching the paint.

2. Scribe a line to mark the location of the hood hinges for reference in reassembly. Unbolt the hood hinges and remove the hood.

3. Disconnect the battery cables from the battery.

   CAUTION: *Always disconnect the negative side of the battery first.*

   WARNING: *Always wear eye protection when working around the battery. Any type of spark can cause the battery to explode, resulting in severe injury.*

4. If the battery is under the hood, remove the battery from the vehicle and store it so that it is out of the way and in a place where it is unlikely that something can easily short across the negative and positive battery terminals.

5. Disconnect the ground cable from the cylinder head or engine block. On some engines it will be necessary to remove the separate ground strap from the engine block to the bulkhead or from the engine block to the chassis frame near the engine mounts.

6. Remove the air cleaner, intake air duct, and related tubes to aid visibility and to increase the size of working area.

7. Drain the cooling system by removing the radiator cap, then opening the petcock near the bottom of the radiator **(Figure 10-1)**.

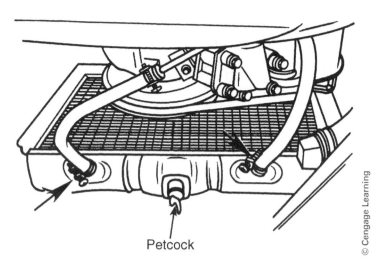

Petcock

Figure 10-1 Petcock near bottom of the radiator.

WARNING: *Do not open the radiator cap or attempt to drain the radiator if the engine is still hot. Allow the engine to cool before opening the cap or severe burns may result.*

8. Disconnect the engine coolant hoses at the top and bottom of the radiator. If the vehicle is equipped with an automatic transmission, disconnect and plug the oil cooler lines at the radiator.

9. Loosen the belt that drives the water pump. Remove the fan shroud, fan assembly, and water pump pulley. Remove the radiator.

10. Disconnect the heater hoses **(Figure 10-2)** from the water pump and engine block or intake manifold.

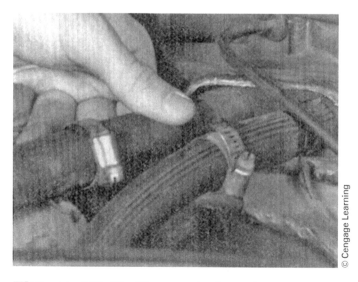

Figure 10-2 Disconnecting heater hoses.

11. Disconnect and plug all fuel lines to prevent fuel loss and keep dirt from entering the system **(Figure 10-3)**. Disconnect the throttle linkage from the throttle body.

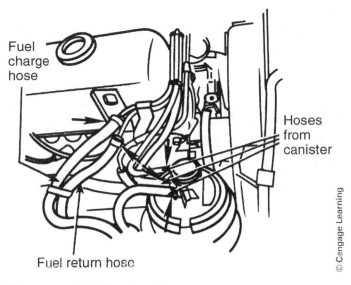

Figure 10-3 Fuel lines.

CAUTION: *Most fuel injection lines are under pressure even when the engine is off. Verify the bleed-down procedure in the service manual before disconnecting a fuel injection line.*

WARNING: *Wear eye protection to prevent gasoline under pressure from causing irritation or blindness. Also, do not allow gasoline to come into contact with any hot engine components.*

12. Remove the carbon canister and any fuel system or emissions-related component that could pose engine removal problems.

13. Before disconnecting any vacuum lines or electrical lines, attach a piece of masking tape to both sides of the parts that are to be disconnected. Mark the same code letter or number on both ends of the connector.

14. Disconnect all vacuum hoses, wiring connections, and harnesses.

15. Remove the generator, distributor, distributor cap, and spark plug wires from the engine. Remove the power-steering drive belt. Remove the bolts that attach the pump bracket to the engine block and set the assembly aside in the engine compartment with the hoses connected. Secure the pump with wire or twine.

16. On models with air conditioners, remove the compressor mounting bracket attaching bolts. Remove the compressor and mounting bracket assembly and set them aside without disconnecting the refrigerant lines.

CAUTION: *If the refrigerant lines are disconnected, the system will require recharging during reassembly.*

WARNING: *If the air-conditioner lines must be disconnected, use only an EPA-approved recovery system to evacuate the refrigerant. Always wear eye protection when servicing the air-conditioning system.*

WARNING: *Improper evacuation procedures can result in frostbite or asphyxiation.*

17. Raise the vehicle following basic safety procedures. Drain the engine's oil.

18. Disconnect the exhaust system from the manifold. To help in removing bolts and studs, soak them with penetrating oil.

19. If so equipped, remove the turbocharger.

20. Remove the flywheel or converter housing cover. If equipped with a manual transmission, remove the bellhousing lower attaching bolts. If equipped with an automatic transmission, remove the converter-to-flywheel bolts, then remove the converter housing attaching bolts.

21. Remove the starter motor.

22. Attach an engine sling to the engine and support the weight of the engine with a hoist. Then disconnect the engine mounts from the brackets on the frame. Remove all bolts attaching the bellhousing to the engine block. If so equipped, disconnect the clutch linkage.

23. Lower the vehicle and position a suitable transmission jack under the transmission **(Figure 10-4)**. Raise the jack enough to support the transmission. It may be necessary to raise the transmission while removing the engine.

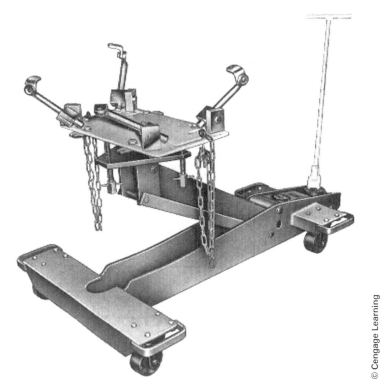

© Cengage Learning

Figure 10-4 A transmission jack.

24. Make a final check to ensure that everything has been disconnected from the engine. The engine is now ready to be lifted from its compartment.

25. Using the accompanying "Report Sheet on Engine Removal," carefully examine the condition of all parts and indicate your findings.

Problems Encountered

Instructor's Comments

_____

_____

_____

Name _____   Station _____ Date _____

REPORT SHEET ON ENGINE REMOVAL			
Component	**OK**	**Repair**	**Replace**
1. Radiator			
2. Radiator hoses			
3. Heater hoses			
4. Hose clamps			
5. Battery cables			
6. Engine mounts			
7. Transmission mounts			
8. Belts (all)			
9. Water pump			
10. Clutch and/or shift linkage			
11. Wire connections			
12. Air filter			
13. Oil filter			
14. Vacuum lines			
15. Oil filler cap			
16. PCV valve hoses			
17. Dipstick			
18. Antifreeze			
19. Other items			

© Cengage Learning 2015

Instructor's Check _____

☐ JOB SHEET 10–2

Removing and Disassembling the Cylinder Head

Name _____ Station _____ Date _____

Objective

Upon completion of this job sheet, you will have demonstrated the ability to properly remove and disassemble a cylinder head. This task applies to ASE Education Foundation Tasks AST/MAST 1.B.1 (P-1).

Tools and Materials

Small magnet	Valve spring compressor
Soft-faced hammer	Valve storage tray

Description of Vehicle

Year _____ Make _____ Model _____

VIN _____ Engine Type and Size _____

Mileage _____

PROCEDURE

Cylinder Head Removal

1. Using the appropriate service information, look up the sequence for loosening and removing the cylinder head bolts. In the space below, sketch the cylinder head and the number sequence for loosening and removing the bolts. If the manual does not illustrate the loosening sequence, reverse the installation sequence.

2. Remove the valve cover or covers. If the engine has an overhead camshaft, remove it so that you have access to the cylinder head bolts **(Figure 10-5)**. Locate and attach the procedure for camshaft removal.

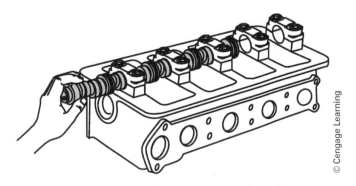

© Cengage Learning

Figure 10-5 Removing an overhead camshaft.

3. Drain the coolant from the engine. Loosen and remove the cylinder head bolts. Make sure you refer to your drawing and bolt-loosening sequence. All of the bolts should be initially loosened one turn. Then remove the bolts and note the location of each on your drawing. This procedure is important because some engines use head bolts with different lengths.

4. Remove the pushrods (if equipped), and place them in a fixture that will keep them in order.

5. Lift the cylinder head off the engine **(Figure 10-6)**. Store the cylinder head gasket for later reference.

Figure 10-6 Removing a cylinder head.

Instructor's Check _____

Cylinder Head Disassembly

6. Place the head on a flat surface with the valve heads down. With a soft-faced hammer, tap on the valve spring retainers to make disassembly easier.

7. Adjust the jaws of a valve spring compressor so that they fit the spring retainers. Put the cylinder head on its side so that you have access to the valve head and stem.

CAUTION: *Never use a valve spring compressor that has bent or distorted jaws. Also, make sure you are wearing safety glasses while removing the valves.*

8. Begin at one end of the cylinder head and remove the valve assemblies by compressing the valve spring and removing the valve locks **(Figure 10-7)**. A small magnet makes removing the valve locks easier.

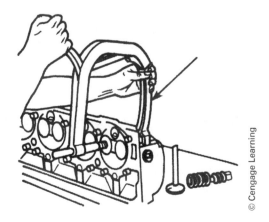

Figure 10-7 Using a valve spring compressor to remove the valve assembly.

9. Slowly release the spring compressor, and remove it from the head. Collect the valve spring and spring retainer, and then pull out the valve.

10. Arrange the valve, spring, and washer in a storage rack so they can be reinstalled in the same locations.

Instructor's Check _____

Problems Encountered

_____

_____

_____

Instructor's Comments

_____

_____

_____

 REVIEW QUESTIONS

Review Chapter 10 of the textbook to answer these questions:

1. What should be done before removing the vacuum and electrical wiring?

   _____

   _____

2. Why is it important to loosen the cylinder head bolts in the recommended order?

   _____

   _____

3. If the service information does not provide a loosening sequence for the cylinder head bolts, what should you do?

   _____

   _____

   _____

   _____

4. List five different methods of cleaning engine components.

   _____

   _____

   _____

   _____

5. Why must a cylinder ridge be removed before attempting to remove the piston?

   _____

   _____

OVERVIEW

This chapter covers the inspection and service of an engine's lower end. The lower end, also called a short block, is the basic mechanical unit of the engine, minus the cylinder heads. While performing services to the short block, you need to make many precise measurements. Make sure you know how to use and read a micrometer before proceeding through the contents in this chapter of the manual.

A basic short block consists of the cylinder block, crankshaft, crankshaft bearings, connecting rods, pistons and rings, and oil gallery and core plugs.

CYLINDER BLOCK

The cylinder, or engine, block houses the cylinders and supports the other parts of the short block. Combustion takes place in the cylinders as the pistons move through their four-stroke cycle. A cylinder block is normally one piece, cast and machined so that all of the parts contained in it fit properly. Blocks are typically made from iron or aluminum.

The cylinder bores are critical areas of service for a technician. The walls of the bore must be the correct size and without flaws, to allow the piston to move smoothly up and down the bore while it maintains a seal against the walls.

CAMSHAFT

The camshaft is driven by the crankshaft by way of belts, gears, or chains. The camshaft rotates at one-half the speed of the crankshaft. The rotation of the camshaft converts rotary motion into reciprocating (up-and-down) motion. It is the reciprocating motion of the camshaft cam or lobe that allows the intake and exhaust valves to open. The amount of valve opening is in direct relationship to the configuration or profile of the camshaft cam or lobe (eccentric).

Several camshaft bearing surfaces support the camshaft in the engine block or cylinder head, depending on the engine design. When dual overhead camshafts are employed, the camshafts are designated as exhaust and intake camshafts. Placement of the camshaft(s) in the engine block or cylinder head determines the amount of valvetrain linkage required.

The valvetrain linkage will be minimal when the camshaft is mounted above the cylinder head. The valvetrain linkage consists of a camshaft lifter or cam follower to open the valve. Conversely, if the camshaft is mounted in the engine block, the valvetrain linkage is composed of the camshaft lifter or follower, valve push rods, and valve rocker arms to open the valve.

The camshaft and valvetrain work together to open and close the intake and exhaust valves of an engine. This interaction must be perfectly timed for the engine to efficiently pass through the four strokes of its pistons.

All current engines have at least one camshaft, and some have more. V-type engines with dual overhead camshafts have four. Regardless of the number of camshafts an engine has, there is one camshaft lobe for each valve. That lobe is responsible for the opening and closing of that valve.

The sides of a camshaft lobe have ramps that control how quickly the valve opens and closes. The highest point of the lobe gives the maximum valve opening. When the valve is fully open, more air/fuel can enter the cylinder or more exhaust gas can exit the cylinder. The base circle of the lobe provides no valve lift, and the valve is closed.

Camshafts rotate on insert bearings similar to those used with the crankshaft. Camshaft bearings for overhead valve (OHV) engines and some overhead camshaft (OHC) engines are full-round insert bearings. The bearings are pressed into bores in the engine block or cylinder head. Many OHC engines use split bearings. These engines have bearings caps that secure the camshaft in the cylinder head or in the camshaft housing mounted to the top of the cylinder.

CAMSHAFT DRIVES

The camshaft is driven by the crankshaft, but at half the speed of the crankshaft. This difference allows the camshaft to control the valves in relation to the position of the piston through one complete combustion cycle. Remember that the crankshaft makes two complete revolutions to complete one four-stroke cycle.

The camshaft in an OHV engine can be driven by meshed gears or a timing chain(s) and sprockets. The camshaft(s) of an OHC engine is driven by either a timing belt(s), timing chain(s) and sprockets, or meshed gears. Regardless of the drive mechanisms, the gears and sprockets must be timed with the crankshaft. Always refer to the appropriate service manual when servicing camshaft drive systems.

CRANKSHAFT

The reciprocating motion of the pistons must be converted to a rotary motion before that movement can be used to drive the wheels of the vehicle. This conversion is accomplished by connecting the pistons to a crankshaft with a connecting rod. The upper end of the connecting rod moves with the piston. The lower end of the connecting rod is attached to the crankshaft and moves in a circle.

The crankshaft is housed by the engine block and is held in place by main bearing caps between the main bearing caps and the crankshaft, and between the cylinder block and the crankshaft are the main bearings.

The crankshaft does not rotate directly on the bearings; rather, it rotates on a film of oil trapped between the bearing surface and the crankshaft journals. The area that holds the oil film is critical to the life of an engine. If the crankshaft journals become excessively worn, the proper oil film will not form. Without the correct oil film, premature wear will occur on the bearings and crankshaft journals. This condition can also result in crankshaft breakage. Therefore, this is another critical area for technicians. The specified gap between the crankshaft journals and the bearing surfaces must exist.

PISTONS AND PISTON RINGS

Combustion takes place between the top of the piston and the cylinder head. A piston is a can-shaped part that is closely fit into a cylinder. The gap between the outside of the piston and the cylinder walls is sealed by piston rings. These rings of steel are formed under tension so that they expand out from the piston to the cylinder walls to form a good seal.

If the piston does not seal well, the efficiency of its four strokes decreases. Consider the intake stroke. The piston moves down to increase the volume of the cylinder and to lower the pressure in the cylinder. If there is leakage around the piston rings, air from the crankcase, which is below the piston, will enter the cylinder. This, in turn, will lower the amount of vacuum formed and reduce the amount of air/fuel mixture that will be drawn into the cylinder. The other strokes are affected the same way.

☐ JOB SHEET 11–1

Measuring Cylinder Bore

Name _____ Station _____ Date _____

Objective

Upon completion of this job sheet, you will have demonstrated the ability to measure cylinder bore. This task applies to ASE Education Foundation Task MAST 1.C.4 (P-2).

Tools and Materials

Outside micrometer   Telescoping gauge

Description of Vehicle

Year _____ Make _____ Model _____

VIN _____ Engine Type and Size _____

Mileage _____

PROCEDURE

1. Locate and record the specifications for cylinder bore size, tolerances for out-of-round, and taper.

   Standard bore size specification _____ _____

   Out-of-round specification _____

   Taper specification _____

2. Inspect the cylinder for ring ridge. Measure the diameter of the cylinder at the ridge and below the ridge. Top Diameter _____ Lower Diameter _____

   What action is necessary based on your measurements? _____

3. Insert the correct size telescoping gauge into the cylinder, with the plungers oriented along the length of the block. Gently rock the gauge back and forth to ensure proper fit, and lock the plungers. Remove the gauge and measure with a micrometer, and record the reading **(Figure 11-1)**.

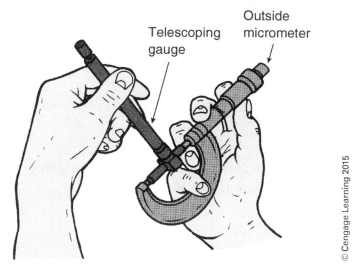

© Cengage Learning 2015

Figure 11-1 The gauge is read with an outside micrometer.

Actual Bore Sizes   1 _____   2 _____   3 _____   4 _____

5 _____   6 _____   7 _____   8 _____

4. Place the telescoping gauge in the cylinder perpendicular to the first measurement **(Figure 11-2)**, and record the reading for each cylinder.

Bore Sizes 90 degree   1 _____   2 _____   3 _____   4 _____

5 _____   6 _____   7 _____   8 _____

© Cengage Learning

Figure 11-2 A telescoping gauge in a cylinder bore.

5. Measure the diameter of the cylinder at the highest and lowest points of ring travel. The difference between the two measurements is the amount of cylinder taper **(Figure 11-3)**. Record the taper measurements for each cylinder.

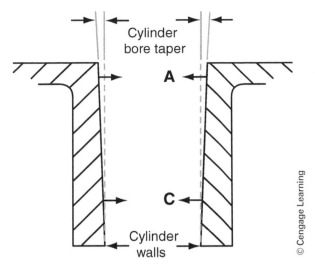

Figure 11-3 To check for taper, measure the cylinder diameter at A and C. The difference between the two readings is the amount of taper.

Bore Taper   1 _____   2 _____   3 _____   4 _____

                               5 _____   6 _____   7 _____   8 _____

6. Examine each cylinder for signs of cracks, for scoring, and for ridge buildup at the top of each cylinder. Note your findings._____

Problems Encountered

_____

_____

_____

Instructor's Comments

_____

_____

_____

☐ JOB SHEET 11–2

Measuring Crankshaft Journals

Name _____ Station _____ Date _____

Objective

Upon completion of this job sheet, you will have demonstrated the ability to measure crankshaft journals for wear, taper, and out-of-roundness. This task applies to ASE Education Foundation Task MAST 1.C.7 (P-1).

Tools and Materials

Outside micrometer

Description of Vehicle

Year _____ Make _____ Model _____

VIN _____ Engine Type and Size _____

Mileage _____

PROCEDURE

1. Look up the specifications for standard crankshaft size and tolerances for normal wear. Record these specifications on the accompanying Report Sheet for Measuring Crankshaft Bearing Journals.

2. Inspect the crankshaft and use proper procedures to clean it before beginning the measurement process. What method was used to clean the crankshaft?

   _____

3. Using the proper size outside micrometer **(Figure 11-4)**, check the number 1 main bearing journal twice at each end of the journal, once horizontal to the crankshaft and once vertical. Record all of the measurements on the report sheet.

   Note: *If these measurements are different, the crankshaft main bearing journal is out-of-round or tapered (**Figure 11-5**) and the crankshaft should be machined.*

Figure 11-4 Measuring the connecting rod journal with a micrometer.

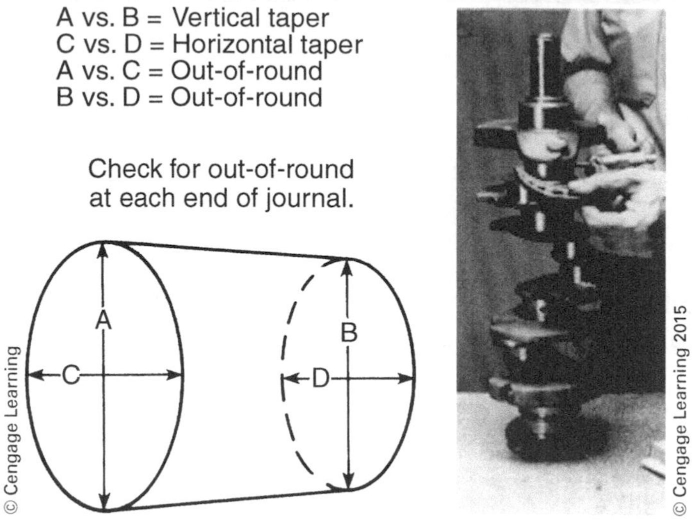

A vs. B = Vertical taper
C vs. D = Horizontal taper
A vs. C = Out-of-round
B vs. D = Out-of-round

Check for out-of-round
at each end of journal.

Figure 11-5 Check for out-of-roundness and taper.

4. Repeat step 3 for each of the remaining main bearing journals.

5. Measure the number 1 connecting rod journal twice at each end of the journal, once horizontal to the crankshaft and once vertical. Record all of the measurements on the report sheet.

   Note: *If these measurements are different, the crankshaft connecting rod journal is out-of-round or tapered, and the crankshaft should be machined.*

6. Repeat step 5 for each of the remaining connecting rod journals.

7. Compare the measurements of the crankshaft journals with the specifications. If the measurements are within specifications, the crankshaft can be reinstalled in the engine. If the measurements are not within factory specifications, the crankshaft must be machined before it is reinstalled in the engine.

Problems Encountered

_____

_____

_____

Instructor's Comments

_____

_____

_____

Name _____   Station _____   Date _____

REPORT SHEET FOR MEASURING CRANKSHAFT BEARING JOURNALS

Main Bearing Journal Number	1	2	3	4	5	6	7
Standard journal size							
Actual journal size							
Taper limits							
Actual taper							
Out-of-round limits							
Actual out-of-roundness							
Connecting Rod Journal Number	1	2	3	4	5	6	
Standard journal size							
Actual journal size							
Taper limits							
Actual taper							
Out-of-round limits							
Actual out-of-roundness							

Conclusions and Recommendations _____

_____

_____

☐ JOB SHEET 11–3

Measuring Crankshaft End Play

Name _____ Station _____ Date _____

Objective

Upon completion of this job sheet, you will have demonstrated the ability to correctly measure crankshaft end play. This task applies to ASE Education Foundation Task MAST 1.C.7 (P-1).

Tools and Materials

Bracket   Light engine oil

Dial Indicator   Two pry bars

Feeler gauge set

Description of Vehicle

Year _____ Make _____ Model _____

VIN _____ Engine Type and Size _____

Mileage _____

PROCEDURE

1. Following the manufacturer's recommended procedures presented in a service manual, install the rear main oil seal in the block and bearing cap.

2. Thoroughly lubricate all bearings and rear main oil seal with engine oil. Install the main bearing halves into the main bearing bores in the block. Make sure the oil holes in the bearing halves line up with the oil holes in the block. What lubricant was used?

   _____

3. Carefully install the crankshaft into the block.

   Instructor's Check _____

4. Install the main bearing caps with bearings, being careful to match the location numbers on the cap with the block.

5. Tighten the main bearing cap bolts one at a time in three steps until the full, specified torque is obtained. Do not torque the thrust bearing cap at this time.

   Torque procedure and specification _____

   Instructor's Check _____

6. Pry crankshaft back and forth to align the thrust surfaces at the thrust bearing. While prying the crankshaft forward, pry the thrust bearing cap rearward. When assured that the surfaces are aligned, torque the cap bolts to specifications in three steps.

7. Using a service manual, locate the minimum and maximum crankshaft end play specifications. Record specifications here.

Minimum specification _____

Maximum specification _____

8. Pry the crankshaft toward the front of the engine. Install a dial indicator and bracket so that its plunger rests against the crankshaft flange and the indicator axis is parallel to the crankshaft axis.

    Note: *This procedure can be accomplished using a feeler gauge if a dial indicator is not available. This check would be made between the thrust bearing and the thrust surfaces on the crankshaft.*

9. Set the dial indicator at zero. Pry the crankshaft rearward. Note the reading on the dial indicator.

10. Record your dial indicator reading. _____

11. If your results are not within manufacturer's specifications, list your conclusions and recommendations.

    _____

    _____

Problems Encountered

_____

_____

_____

Instructor's Comments

_____

_____

_____

☐ JOB SHEET 11–4

Installing Pistons and Connecting Rods

Name _____ Station _____ Date _____

Objective

Upon completion of this job sheet, you will have demonstrated the ability to properly install pistons and connecting rods. This task applies to ASE Education Foundation Task MAST 1.C.14 (P-1).

Tools and Materials

Allen wrench

Anacrobic thread-locking compound

Compressor tool

Feeler gauges

Flat-blade screwdriver

Light engine oil

Plastic mallet

Plastigage

Rubber or aluminum protectors or guides

Description of Vehicle

Year _____ Make _____ Model _____

VIN _____ Engine Type and Size _____

Mileage _____

PROCEDURE

1. Place rubber or aluminum protectors or guides over the threaded section of the rod bolts. Lightly coat the piston, rings, cylinder wall, crankpin, and compressor tool with light engine oil. Do not coat the rod bearings with oil at this time.

2. Be sure that the ring gaps are located on the piston in the positions recommended in the service manual. Attach a diagram of ring positional.

3. Expand the compressor tool around the piston rings. Position the steps on the compressor tool downward. Tighten the compressor tool with an Allen wrench to compress the piston rings.

Instructor's Check _____

4. Rotate the crankshaft until the crankpin is at its lowest level (bottom dead center [BDC]). Place the piston/rod assembly into the cylinder bore until the steps on the compressor tool contact the cylinder block deck. Make sure the piston reference mark is in the correct relation to the front of the engine.

5. Lightly tap on the head of the piston with a mallet handle or block of wood until piston enters the cylinder bore.

6. Push the piston down the bore while making sure the connecting rod fits into place on the crankpin. Remove the rod bolt protective covers.

Instructor's Check _____

7. Using a service manual, locate the oil clearance specification for the connecting rod bearings and torque specifications for the cap bolts. Record these specifications on the accompanying Report Sheet for Installing Connecting Rods and Pistons.

8. Determine the rod bearing oil clearance using a plastigage in the same manner as when checking the oil clearance for main bearings. Record the results on the report sheet. If out of specification, consult with your instructor.

9. Remove all of the cap bolts, and push the piston rod assembly a slight distance from the crankshaft. Lubricate all bearings with light engine oil.

10. Seat the connecting rod yoke onto the crankshaft. Apply a drop of anaerobic thread-locking compound to each thread of the rod bolts.

11. Position the matching connecting rod cap and finger-tighten the rod nuts. Make sure the connecting rod blade and cap markings are on the same side. Gently tap each cap with a plastic mallet as it is being installed to properly position and seat it.

12. Torque the rod nuts to specifications.

   CAUTION: *Use new nuts because they are self-locking.*

   Instructor's Check _____

13. Repeat the piston/rod assembly procedure for each assembly.

14. Using a service manual, locate the connecting rod side clearance specification. Record it on the report sheet.

15. With a feeler gauge, measure the side clearance by spreading the connecting rods with a large, flat-blade screwdriver.

Problems Encountered

_____

_____

_____

Instructor's Comments

_____

_____

_____

Name _____ Station _____ Date _____

REPORT SHEET FOR INSTALLING CONNECTING RODS AND PISTONS

Cylinder No.	1	2	3	4	5	6	7	8
Specified oil clearance								
Actual oil clearance								
Specified side clearance								
Actual side clearance								
Cap bolt torque								

Conclusions and Recommendations _____

_____

_____

☐ JOB SHEET 11–5

Measuring Camshaft Lobes/Cams and Bearing Journals

Name _____ Station _____ Date _____

Objective

Upon completion of this job sheet, you will have demonstrated the ability to evaluate a camshaft by performing a series of precise measurements. This task applies to ASE Education Foundation Tasks MAST 1.B.13 (P-3), 1.B.14 (P-3), and 1.C.6 (P-3).

Tools and Materials
Outside micrometer

Description of Vehicle

Year _____ Make _____ Model _____

VIN _____ Engine Type and Size _____

Mileage _____

PROCEDURE

1. Using a service manual, look up the specifications for standard camshaft size and tolerances for normal wear. Record these specifications on the accompanying Report Sheet for Measuring Camshaft Lobes/Cams and Bearing Journals.

2. Inspect the camshaft for signs of case hardening failure; if present, discard the camshaft. Note camshaft vision condition.

_____

_____

3. Use proper procedures to clean it before beginning the measurement process. What process was used to clean the camshaft?

_____

_____

4. Using the proper size outside micrometer, inspect the front camshaft bearing journal twice at each end of the journal, once horizontal to the camshaft and once vertical. Record all of the measurements on the report sheet.

   Note: *If these measurements are different, the camshaft bearing journal is out-of-round or tapered, and the camshaft should be replaced.*

5. Repeat step 3 for each of the remaining camshaft bearing journals.

6. Measure the first camshaft lobe three times at each end of the surface, once horizontal to the camshaft and twice vertically at each edge. Record all of the measurements on the report sheet.

Note: *The measurement difference between the horizontal and the vertical is the camshaft lift for that lobe.*

7. Determine whether this lobe is an exhaust or intake, and record the measurements.

8. Repeat step 6 for each of the remaining exhaust and intake lobes.

9. Compare the measurements of the camshaft bearing journals with the specifications. Next compare the measurements of the camshaft lobes with the specifications to determine the wear of the lobe. If the measurements are within specifications, the camshaft can be reinstalled in the engine. If the measurements are not within factory specifications, the camshaft must be replaced.

Problems Encountered

_____

_____

_____

Instructor's Comments

_____

_____

_____

Name _____ Station _____ Date _____

REPORT SHEET FOR MEASURING CAMSHAFT LOBES/CAMS AND BEARING JOURNALS							
Bearing Journal Number	1	2	3	4	5	6	7
Standard bearing journal size*							
Maximum tolerances							
Actual journal size							
Taper limits							
Out-of-round limits*							
Actual out-of-roundness							
Camshaft Lobe Number	1 in	1 ex	2 in	2 ex	3 in	3 ex	
Base circle measurement							
Base circle actual size*							
Lobe nose size 1							
Lobe nose size 2							
Actual lobe nose size*							
Taper limits*							
Actual taper							
Camshaft Lobe Number	4 in	4 ex	5 in	5 ex	6 in	6 ex	
Base circle measurement							
Base circle actual size*							
Lobe nose size 1							
Lobe nose size 2							
Actual lobe nose size*							
Taper limits*							
Actual taper							

Conclusions and Recommendations _____

_____

_____

*Specification size from shop manual.

☐ JOB SHEET 11–6

Inspecting and Replacing Camshaft Drives

Name _____ Station _____ Date _____

Objective

Upon completion of this job sheet, you will have demonstrated the ability to inspect and replace camshaft drives. This task applies to ASE Education Foundation Tasks AST/MAST 1.B.5 (P-1).

Description of Vehicle

Year _____ Make _____ Model _____ _____

VIN _____ Engine Type and Size _____

Mileage _____

PROCEDURE

1. Disconnect the negative battery cable.

2. Remove the covers protecting the camshaft drive gears.

3. Inspect for proper alignment of the timing marks on the timing gears.

   Instructor's Check _____

4. Remove the timing gears according to the service manual.

5. Clean the timing gears.

6. Inspect the gears for wear or cracks, and replace necessary parts. Note the results of the inspection here.

   _____

   _____

7. After replacing any worn or damaged gears and drives, set the timing marks according to the service manual. Attach the timing diagram.

   Instructor's Check _____

8. Rotate the engine through two complete rotations of the crankshaft. Check to see if the timing marks on the gears are still in proper alignment.

   Instructor's Check _____

9. Reinstall all covers removed for access.

10. Reconnect the negative battery cable, and start the engine.

11. Set the ignition timing to the manufacturer's specifications.

    Instructor's Check _____

Problems Encountered

_____

_____

_____

Instructor's Comments

_____

_____

_____

REVIEW QUESTIONS

Review Chapter 11 of the textbook to answer these questions:

1. What tools can be used to measure crankshaft end play?

   _____

   _____

   _____

2. Describe the pattern that should be present after the final honing of a cylinder wall.

   _____

   _____

   _____

3. Describe how to check for crankshaft bearing journal wear.

   _____

   _____

   _____

4. List two reasons why a block may need to be line bored.

   _____

   _____

   _____

5. Explain why a camshaft lobe is tapered.

   _____

   _____

   _____

OVERVIEW

The cylinder head and valves are the focus of this chapter. Included in the chapter is a complete inspection of the cylinder head and the valves and their related parts. Also included are those services that may be performed while rebuilding an engine.

The cylinder head mounts on top of the engine block and serves as the uppermost seal for the combustion chamber. To aid in that sealing, a head gasket is sandwiched between the cylinder head and engine block. The head also contains the valves that must open to allow the air/fuel mixture to come in and allow the exhaust gases to go out. When the valves are closed, they too need to provide a positive seal. The cylinder head may also be the mounting spot or serve as the housing for a camshaft.

Much of the work by a technician on a cylinder head is done to ensure a good seal. The cylinder head and valves are another area where the ability to use precise measurement devices is important.

INTAKE AND EXHAUST VALVES

An engine's valves close into a valve seat. The interference fit between the face of the valve and the valve seat provide for the seal. Because this seal is made by two metal surfaces, it is critical that the surfaces be smooth and perfectly round.

The valves move in a valve guide that is typically pressed into the cylinder head. The clearance between the stem of the valve and the inside diameter of the guide provides some lubrication for the valves' movement. Excessive valve-to-guide clearance can allow oil to enter into the combustion chamber from the valvetrain.

Some oil always moves down the valve guide from the upper portion of the cylinder head. This oil is used to lubricate the valve stems. The amount of oil, however, is minimal. To prevent excessive oil from passing through the guides, oil seals are installed on the valve guides or valve stems.

The opening of the valve is controlled directly or indirectly by a camshaft. The closing of the valve is the primary job of a valve spring. The valve spring forces the valve closed when the camshaft ceases to put pressure on the valve.

The valve spring is held in place by a spring retainer and valve locks or keepers. This assembly also works to hold the valve in place in the cylinder head.

CYLINDER HEADS

A good seal is formed between the deck of the block and the cylinder head when both surfaces are flat and parallel to each other. This not only provides a seal for the combustion chamber, but it also allows passages in the head to positively connect with passages in the engine block. These passages are for engine coolant and lubricating oil flow.

☐ JOB SHEET 12–1

Inspecting Cylinder Head for Wear

Name _____ Station _____ Date _____

Objective

Upon completion of this job sheet, you will have demonstrated the ability to check a cylinder head for warpage, inspect it for cracks, and measure stem-to-guide clearance. This task applies to ASE Education Foundation Tasks AST/MAST 1.B.2 (P-1) and MAST 1.B.9 (P-1) and 1.B.10 (P-3).

Tools and Materials

Dial indicator

Dye penetrant or magnetic inspection equipment

Feeler gauge

Micrometer

Straightedge

Telescoping gauge/small-hole gauge

Description of Vehicle

Year _____ Make _____ Model _____

VIN _____ Engine Type and Size _____

Mileage _____

PROCEDURE

1. Look up the specifications for stem-to-guide clearance for your engine, and record them on the accompanying Report Sheet for Cylinder Head Inspection.

2. Look up the specification for maximum warpage limit for your engine, and record it on the report sheet.

3. Perform a visual inspection of the cylinder head. Note your findings on the report sheet.

4. Position a straightedge over the cylinder head's sealing surface. Try to slide a feeler gauge strip the size of the maximum limit under any part of the straightedge **(Figure 12-1)**. If the feeler gauge goes under the straightedge, the cylinder head is warped.

Instructor's Check _____

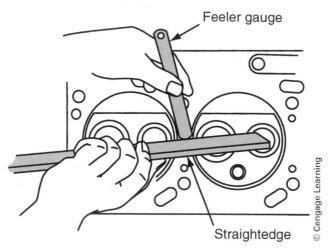

Feeler gauge

Straightedge

© Cengage Learning

Figure 12-1   Measuring cylinder head warpage.

5. Use a dye penetrant or magnetic inspection equipment to look for cracks in the cylinder head around the valve seats. Note problems found. _____

_____

_____

6. To measure valve stem guides, first make sure the guides are clean. Install a small hole gauge or telescoping gauge into one of the guides and expand it. Remove the gauge and measure it with an outside micrometer. Measure the guide in three places: top, middle, and bottom. Note your largest reading on the report sheet.

7. From your storage tray, select a valve that fits in this guide. Use an outside micrometer to measure the part of the stem that rides in the guide. Take measurements in three places: top, middle, and bottom. Note your smallest reading on the report sheet.

8. The clearance for this guide is found by comparing the largest guide measurement with the smallest stem measurement. Subtract the stem measurement from the guide measurement. Record the clearance on the report sheet.

9. Repeat this procedure for each valve guide, and record your results on the report sheet.

10. If you wish to use a dial indicator to check the clearance, mount the dial indicator on cylinder head **(Figure 12-2)**.

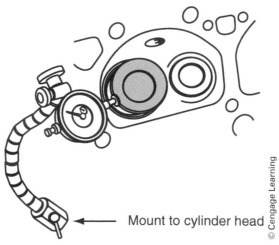

Mount to cylinder head

© Cengage Learning

Figure 12-2   Measuring valve guide clearance with a dial indicator.

11. Place a valve that fits in the guide into the guide, and adjust the dial indicator to contact the valve head. Zero the dial indicator. Rock the valve against the dial indicator. The indicator will show the clearance.

12. Repeat this operation for each valve, and record the results on the report sheet. Compare your clearance readings to the specifications to determine whether the valve guides require servicing.

Problems Encountered

_____

_____

_____

Instructor's Comments

_____

_____

_____

Name _____ Station _____ Date _____

REPORT SHEET FOR CYLINDER HEAD INSPECTION

1. Visual Inspection

	Serviceable	Not serviceable
Threaded holes		
Core-hole plugs		
Machined surfaces		
Rocker studs or pedestal		
Water jackets		

Conclusions and Recommendations _____
_____
_____

2. Head Warpage
Specified limit _____
Actual _____

3. Valve Stem and Guide Clearance Specification
Intake valve _____
Exhaust valve _____

Cylinder		Guide	Stem	Clearance
1.	Intake			
	Exhaust			
2.	Intake			
	Exhaust			
3.	Intake			
	Exhaust			
4.	Intake			
	Exhaust			
5.	Intake			
	Exhaust			
6.	Intake			
	Exhaust			
7.	Intake			
	Exhaust			
8.	Intake			
	Exhaust			

Conclusions and Recommendations _____
_____
_____

☐ JOB SHEET 12–2

Inspecting and Testing Valve Springs for Squareness, and Free Height Comparison

Name _____ Station _____ Date _____

Objective

Upon completion of this job sheet, you will be able to test valve springs for squareness, and free height. This task applies to ASE Education Foundation Task MAST 1.B.7 (P-2) and 1.B.11 (P-3).

Tools and Materials
Straightedge

Square

Feeler gauge

Description of Vehicle

Year _____ Make _____ Model _____

VIN _____ Engine Type and Size _____

Mileage _____

PROCEDURE

1. Clean all of the valve springs. Which cleaning method was used?

   _____

2. Visually inspect all of the valve springs for cracks or signs of wear. Record condition of springs.

   _____

   _____

   _____

3. Perform a squareness test: Set a spring upright against a square as shown in **Figure 12-3**. Turn the spring until a gap appears between the spring and the square. Measure the gap with a feeler gauge. If the gap is greater than 0.060 inch (1.52 mm), the spring should be replaced.

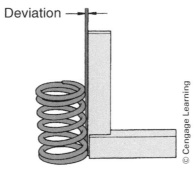

Figure 12-3 Measuring valve spring squareness deviation.

4. Perform a freestanding height test: Line up all of the valve springs on a flat surface. Place a straight-edge across the tops of the springs. Any height discrepancy will be obvious. There should be no more than 1/16 inch (1.59 mm) of variance between the heights of the valves.

5. Measure the height of the shortest and the tallest valve spring, and compare those measurements to the specifications. Based on the results of this measurement, you should be able to determine which other valves should be measured. Replace any spring that is not within specifications.

Problems Encountered

_____

_____

_____

Instructor's Comments

_____

_____

_____

☐ JOB SHEET 12–3

Replacing Valve Stem Seals and Inspecting Valve Spring Retainers, Locks, and Lock Grooves in Vehicle

Name _____ Station _____ Date _____

Objective

Upon completion of this job sheet, you will have demonstrated the ability to replace valve stem seals and inspect valve spring retainers, locks, and lock grooves on an assembled engine in a vehicle. This task applies to ASE Education Foundation Task MAST 1.B.8 (P-3).

Tools and Materials

Adaptor to pressurize cylinder with compressed air

Valve spring compressor

Small magnet

Valve stem seals

Description of Vehicle

Year _____ Make _____ Model _____ Mileage _____

VIN _____ Type of Engine _____

PROCEDURE

1. Remove the spark plugs, and store them in the proper order for reinstallation.

2. Disconnect any electrical, vacuum, or cooling components necessary to remove the valve cover.

3. Remove the valve cover.

   Instructor's Check _____

4. Locate the rocker arm or follower removal procedure. Remove the rocker arms or cam followers. Keep them in the order in which they were removed.

5. Describe how to position the engine so cylinder number one is at top dead center (TDC) compression stroke.

   _____

   Viewed from the crankshaft pulley, which direction does the engine rotate? _____

   Install the compressed air adaptor into the spark plughole.

6. Attach an air hose to the adaptor and pressurize the cylinder.

   Instructor's Check _____

7. Using the valve spring compressor, compress the valve spring and remove the retainer, valve keeper locks, and spring. **Figure 12-4** shows a typical valve spring, retainers, and keeper assembly.

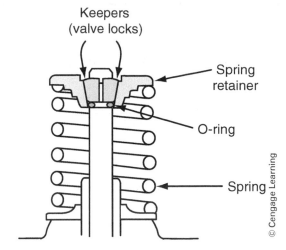

Figure 12-4 Typical valve spring and retainers.

8. Remove the valve guide seal. Inspect the seal for wear, damage, hardening, or cracking. Record seal type and condition. **Figure 12-5** shows common styles of valve seals.

_____

_____

Figure 12-5 Three types of valve stem seals.

9. Visually inspect the valve retainer and keepers. Record condition of retainer, keepers, and valve key grooves and your recommendations.

_____

_____

_____

10. Install a new valve guide seal. Reinstall the spring and retainer. Carefully reinstall the valve keepers and release the valve spring. Tap the valve retainer with a plastic hammer to ensure proper retainer seating.

Instructor's Check _____

11. The amount of engine rotation will vary depending on the number of cylinders. Record the crankshaft degrees of rotation needed to bring the next cylinder to TDC compression. _____

    Rotate the crankshaft according to the service information to bring the next cylinder to TDC compression. Repeat procedures for each cylinder until all valve guide seals are replaced.

12. Reinstall the valve cover. Reconnect all electrical, vacuum, and cooling components disconnected previously.

Instructor's Check _____

13. Reinstall and torque the spark plugs to specifications.

Torque Spec _____

Problems Encountered

_____

_____

_____

Instructor's Comments

_____

_____

_____

☐ JOB SHEET 12–4

Inspecting Valve Lifters, Pushrods, and Rocker Arms

Name _____ Station _____ Date _____

Objective

Upon completion of this job sheet, you will have demonstrated the ability to properly inspect valve lifters, pushrods, and rocker arms. This task applies to ASE Education Foundation Tasks AST/MAST 1.B.3 (P-2), and MAST 1.B.12 (P-2).

Tools and Materials

Inside and outside micrometer

Small, flat-blade screwdriver or pick

Description of Vehicle

Year _____ Make _____ Model _____

VIN _____ Engine Type and Size _____

Mileage _____

PROCEDURE

1. Visually inspect each valve lifter where it contacts a camshaft lobe. Note any signs of wear.

   _____

   _____

2. With your finger or a pushrod, firmly press down on the plunger of the lifter. Make sure it moves with some resistance. If there is no movement or if the plunger moves easily, replace the lifter.

3. Roll each pushrod across a surface plate or drill press table. Inspect oil passage of each pushrod and the tips for galling and wear.

4. Visually inspect the contact surface of each rocker arm. Note any signs of wear.

   _____

   _____

5. If your engine has a rocker shaft, look for evidence of wear in the area where the rocker arm fits. Use inside and outside micrometers to determine the clearance between the rocker shaft and rocker arm. Compare the measurement to specifications. If the clearance is excessive, the rocker arms, shaft, or bushing may have to be replaced.

   Rocker arm shaft diameters _____ Spec _____

   Rocker arm internal diameter _____ Spec _____

Problems Encountered

_____

_____

_____

Instructor's Comments

_____

_____

_____

↰ REVIEW QUESTIONS

Review Chapter 12 of the textbook to answer these questions:

1. Define rocker arm ratio.

   _____

   _____

2. Which areas of the cylinder head are most prone to cracking?

   _____

   _____

3. Explain the benefits of variable valve timing.

   _____

   _____

4. Explain the operation of cam phasors.

   _____

   _____

5. Explain why an exhaust valve requires more stem to guide clearance than an intake valve.

   _____

   _____

Courtesy of Federal-Mogul
Corporation

OVERVIEW

This chapter covers the final assembly of an engine. It includes the proper installation of the major parts on the engine and their associated gaskets. It also includes the proper use of chemicals during the assembly. Procedures for starting and breaking-in a newly rebuilt engine are also covered.

Careful machining and rebuilding of the engine's parts are necessary for a successful rebuild of an engine. These can be a waste of time if the engine is not assembled properly. Proper reassembly includes the correct tightness of all fasteners and the correct installation of gaskets. After parts are installed on the engine, the fasteners should provide a secure mounting and, in many cases, a proper seal.

FASTENERS

Various fasteners are used to hold the engine together and to mount accessories and parts to the engine. These fasteners are specifically designed to do the job they are intended to do. Never substitute a different type of fastener for the type that was originally used.

Most bolts used to assemble an engine have a torque specification and some have a tightening sequence. Both of these specifications must be met during the assembly of the engine. Failure to do so will result in part failure and/or leaks.

GASKETS

Gaskets are used predominantly to prevent gas or liquid leakage between two parts that are bolted together. Gaskets also serve as spacers, heat insulators, and vibration dampers. There are basically two types of gaskets: soft and hard. The application of each of these depends on what is being sealed and the environment the gasket must work in.

OIL SEALS

Oil seals around a rotating shaft keep fluid from escaping. The seal may be a one- or two-piece design, and the seal lip requires prelubrication on installation. The proper direction of the seal lip on an installation of oil seals determines whether the seal functions as expected. The lip of the seal always faces toward the fluid it is retaining.

CHEMICALS

Often chemicals are used to help a gasket seal. These chemicals are applied to the gasket. Although adhesives do not contribute to the sealing ability of a gasket, they ensure proper alignment or positioning by preventing the gasket from moving out of place while it is being tightened in place. Often a particular gasket should not be installed with a chemical sealant. Always refer to the service or the instructions given with the gasket before using a sealant with a gasket.

Some parts are sealed together without a gasket; these unions rely on a chemical sealer for sealing. Sealers are also used to join the ends of two gaskets. There are many types of silicone sealants used on the engine, the most common of which is room temperature vulcanizing (RTV).

RTV silicone sealants fall into two main categories: aerobic and anaerobic. The chemical characteristics and applications of these two sealants are quite different. Aerobic sealants cure in the presence of air, and their most common usage is that of sealing and filling minute gaps in part surfaces. Anaerobic sealants cure in the absence of air. The application for this sealant category is to seal between parted surfaces of rigid casting. When the anaerobic sealants are considered for thread locking, many levels of holding ability are available. It is important to be aware of the various holding abilities in order to be able to choose the product that best suits the application.

Antiseize Compounds

Antiseize compounds are used in areas of a vehicle where oxidation of threaded bolts and capscrews are a concern. Adhere to the vehicle manufacturers' recommendations concerning the use of this chemical.

ENGINE ASSEMBLY

The assembly of the engine is a major responsibility for the technician and requires care and following the manufacturers recommendations as to assembly procedures, including fastener torque values and sequences.

WARNING: *Failure to double check a torque value may have a catastrophic result for the engine.*

INSTALLING THE ENGINE

The installation of an engine in the vehicle means connecting the engine to a number of subsystems such as fuel, cooling, electrical, air conditioning, and exhaust. It is important to follow the manufacturers' recommendations for the starting and break-in procedures in the operation of the engine.

☐ JOB SHEET 13–1

Installing Camshaft Drive Belt and Verifying Timing

Name _____ Station _____ Date _____

Objective

Upon completion of this job sheet, you will have demonstrated the ability to verify engine crankshaft-to-camshaft time. This task applies to ASE Education Foundation tasks MLR 1.A.5 (P-1), AST 1.B.6 (P-1), and MAST 1.B.5 (P-1) and 8.A.11 (P-1).

Description of Vehicle

Year _____ Make _____ Model _____

VIN _____ Engine Type and Size _____

Mileage _____

    Engine crank-to-cam timing is crucial for proper engine performance operation. Some engines may allow a slight degree of timing jump and still run, but will have drivability concerns. Many engines are interference engines and serious internal damage will result if the crank-to-cam timing is allowed to change significantly.

PROCEDURE

1. Determine whether the engine is an interference engine.   YES _____ NO _____

2. Determine engine rotation direction.

   IMPORTANT NOTE: *Engine must only be rotated in the direction as stated by the manufacturer. Failure to follow this procedure can result in catastrophic engine damage.*

   Direction of rotation: _____

3. Align timing marks on camshaft(s), crankshaft, and any other sprockets driven by the timing belt. Attach a picture of the correct timing mark alignments.

4. With the tensioner pulley held retracted, install the timing belt.

5. Rotate the engine crankshaft in the specified direction. Align crankshaft timing mark as shown in repair information. Note location of camshaft timing mark. Are marks properly aligned?

   YES _____ NO _____

6. If NO was checked in number five, it is necessary to realign the timing marks to their proper position.

7. Realign timing marks. Rotate engine using the crankshaft a minimum of two complete revolutions. Do all timing marks align properly?

   YES _____ NO _____

Problems Encountered

_____

_____

_____

Instructor's Comments

_____

_____

_____

☐ JOB SHEET 13–2

Applying RTV Silicone Sealant

Name _____ Station _____ Date _____

Objective

Upon completion of this job sheet, you will have demonstrated the ability to properly apply an RTV silicone sealant. This task applies to ASE Education Foundation Tasks MLR 1.A.4 (P-1) and AST/MAST 1.A.5 (P-1).

Tools and Materials

Dry towel

RTV silicone sealant

Description of Vehicle

Year _____ Make _____ Model _____

VIN _____ Engine Type and Size _____

Mileage _____

PROCEDURE

1. Using the service manual, determine the correct type of RTV sealer to use.

   Recommended Sealant Type _____

   CAUTION: *The use of the wrong type of sealer may result in failure of the oxygen sensor and/or catalytic converter.*

2. Make sure that the mating surfaces are free of dirt, grease, and oil. Apply a continuous 1/8 inch (3.18 mm) bead on one surface only. Be sure to circle all bolt holes.

   WARNING: *Make sure there is adequate ventilation when applying RTV sealers.*

   Instructor's Check _____

3. Adjust the shape of the RTV before skin forms (in about 10 minutes). Remove all excess RTV silicone with a dry towel or paper towel.

4. Press the parts together. Do not slide them together.

5. Tighten all retaining bolts to the manufacturer's specified torque.

   Torque Specs _____

Problems Encountered

_____

_____

_____

Instructor's Comments

_____

_____

_____

☐ JOB SHEET 13–3

Adjusting Valves on an OHC Engine

Name _____ Station _____ Date _____

Objective

Upon completion of this job sheet, you will have demonstrated the ability to adjust the valves on an OHC engine. This task applies to ASE Education Foundation Tasks MLR 1.B.1 (P-1) and AST/MAST 1.B.4 (P-1).

Tools and Materials

Feeler gauge

Remote starter button

Allen wrench set

Description of Vehicle

Year _____Make _____ Model _____

VIN _____ Engine Type and Size _____

Mileage _____

Describe how valve lash is adjusted on this engine.

_____

PROCEDURE

1. Choose the appropriate service information to determine whether the valves are supposed to be adjusted with a cold engine or when the engine is at normal operating temperature.

   Cold or Hot _____

2. Remove necessary hoses and wires to allow the removal of the cam cover (valve cover).

3. List the order in which the values are adjusted. _____

4. Adjust the valves according to the procedure described in the service manual for the engine you are working on.

   Instructor's Check _____

5. Reinstall the cam cover, and connect any wires or hoses that were disconnected.

6. Start the engine to test the operation and inspect for oil leaks.

Problems Encountered

_____

_____

_____

Instructor's Comments

_____

_____

_____

 # REVIEW QUESTIONS

Review Chapter 13 of the textbook to answer these questions:

1. What is the difference between a hard and a soft gasket?

   _____

   _____

2. Explain anaerobic sealant.

   _____

   _____

3. Explain the service differences between TTY bolts and ordinary bolts.

   _____

   _____

4. List four methods of valve lash adjustment.

   _____

   _____

   _____

5. What can result if proper engine prelubrication is not performed?

   _____

   _____

LUBRICATION AND COOLING SYSTEMS

OVERVIEW

The lubrication and cooling systems have perhaps the most unappreciated roles in modern engine operation. Without proper lubrication or cooling system operation, the engine would quickly self-destruct. Although lubricants and coolants have changed greatly over the years, the basic principles and functions of each remain the same.

The lubrication system not only supplies oil to the various moving parts of the engine, it also removes heat, filters out contaminates, and cleans the internal components. Even though the functions of the lubrication system have not changed significantly, the main component, motor oil, has changed greatly in recent years.

If you have purchased motor oil, you probably noticed that oil is available in different categories, such as 0W20, 5W20, 5W30, and many others. These numbers classify the oil by its weight. Oil weight has to do with viscosity, which is a measure of a liquid's resistance to flow. The smaller the first number, the thinner the oil is when cold and the easier it will flow. The larger the second number, the thicker the oil is when hot and the more resistant it is to flow. For many years, oils were sold only in straight weights, like SAE 30. Modern engines require multiviscosity oils like those listed above. Multiviscosity oils allow the oil to be thin enough when cold to provide adequate engine lubrication in cold weather. Oil that is too thin when hot will not provide sufficient lubrication either, so multiviscosity oil changes as its temperature increases. Straight-weight oils cannot adapt for wide temperature ranges and do not provide protection as well as modern lubricants.

Concept Activity 14-1
Viscosity

To provide an example of the viscosities of various liquids, obtain a Styrofoam or similar cup and, using a needle or other small object, poke a small hole in the bottom of the cup. Next, place a measured quantity of water in the cup, and note the frequency of the drips from the hole. Once the water is tested, repeat the test with samples of motor oil, automatic transmission fluid, and differential lubricant. Record your results below.

1. Water drops per minute _____

2. Motor oil drops per minute _____

3. ATF drops per minute _____

4. Differential lubricant drops per minute _____

5. If the oil was tested at a temperature below 32°F (0°C), what effect do you think that would have on the frequency of the drops? _____

6. If the oil was tested at a temperature above 180°F (82.2°C), what effect do you think that would have on the frequency of the drops? _____

   If possible, cool a container of engine oil, and retest.

7. Oil temperature _____ Drips per minute _____

    Now, if possible, obtain engine oil that is hot, and retest.

8. Oil temperature _____ Drips per minute _____

9. Describe the results of testing the oil both cold and hot. _____

_____

_____

_____

10. Did the results of the cold/hot oil test coincide with your predictions in questions 5 and 6?

_____

11. Explain why the correct oil viscosity is important for the internal engine components.

_____

_____

_____

12. What other problems, other than component lubrication, could result from using the incorrect weight of oil?

_____

_____

_____

_____

■

    Viscosity is an important principle to understand when servicing modern automobiles. Using the incorrect oil weight and type can cause issues such as increased engine oil consumption and, in some situations, major engine damage.

    When performing the experiments on viscosity, you noticed that some liquids flow faster than others. You have probably noticed that after it rains, the oil on the road rises to the top of the water collecting on the road. This is because of the density, or specific gravity, of the oil and water. Water is denser than oil, diesel fuel, or gasoline. Because these petroleum products are less dense, they will float on water.

Concept Activity 14-2
Density

Place into a glass jar or similar container an equal amount of water and oil. To perform a more environmentally friendly version of this experiment, use vegetable oil instead of motor oil. Once both liquids are in the jar, let them settle for a few minutes, and observe what happens to the two liquids.

1. Describe how the water and oil appear. _____

_____

_____

2. Why do the oil and water separate? _____

_____

_____

To test the density of liquids, you will need three jars or beakers for water, vegetable oil, and corn syrup. You will also need paper clips, raisins, and small balls of paper. Carefully place a paper clip into each jar, attempting to float the paper clip on each liquid (check all that apply).

3. Paper clip floats on: Water _____ Vegetable oil _____ Corn syrup _____

Now try to get the raisin to float on each liquid.

4. Raisin floated on: Water _____ Vegetable oil _____ Corn syrup _____

Last, ball up some small pieces of paper and see if they will float.

5. Paper balls floated on: Water _____ Vegetable oil _____ Corn syrup _____

6. Were the results of the tests what you expected? _____

_____

_____

The density of water changes as the temperature of water changes. Water is at its most dense at about 39°F (4°C) and decreases in density up to its boiling point of 212°F (100°C). Repeat the experiment above by trying to float objects on water, but this time use hot and cold water.

7. Water temperature: _____ Did the paper clip float? _____

8. Water temperature: _____ Did the paper clip float? _____

Why is it important to understand specific gravity and how it is affected by temperature? Both cooling systems and batteries are tested by using hydrometers, which determine the concentrations of coolant and battery acid based on measurements of specific gravity.

Specific gravity tests are performed on engine coolant to determine the concentration of coolant to water. Coolant should be mixed 50/50 with water for proper cooling system operation. One method of determining this mixture is by using a coolant hydrometer **(Figure 14-1)**. Coolant hydrometers, both the floating disc/ball-and-dial type, rely on the coolant's density to indicate coolant effectiveness. But as you saw in the experiments above, temperature has an effect on water's specific gravity. To accurately use a hydrometer to test coolant, the reading must be temperature compensated. Another problem with coolant testing is that newer propylene glycol–based coolants cannot be tested with a hydrometer.

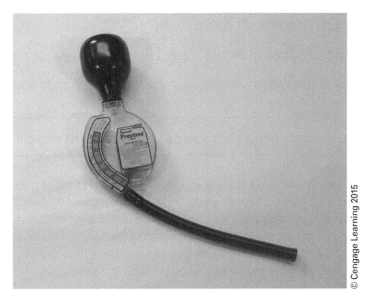

© Cengage Learning 2015

Figure 14-1 A hydrometer is used to check the coolant for protection against freezing.

For many years batteries have been tested with hydrometers. The concentration of sulfuric acid and water, when corrected for temperature, can provide an indication of battery charge and condition. Many modern batteries are sealed, however, and hydrometer readings are not possible.

A refractometer should be used to more accurately determine engine coolant condition **(Figure 14-2)**. A refractometer can measure ethylene glycol– and propylene glycol–based coolants, as well as battery acid. Have you noticed that if you look at a stick, a pole, or even yourself in a body of water, the object (or you) appears to have bent at an angle below the surface of the water? The reason for that effect is that light passes through liquids more slowly than it does through air. A refractometer measures the amount that light bends as it passes through a liquid, providing a reading of the percentage of concentration. However, just as reading with a hydrometer must be temperature compensated, so must readings with a refractometer. Fortunately, because a refractometer needs only a drop or so of liquid for testing, by placing the sample of the liquid on the tester, it will soon assume the temperature of the tester.

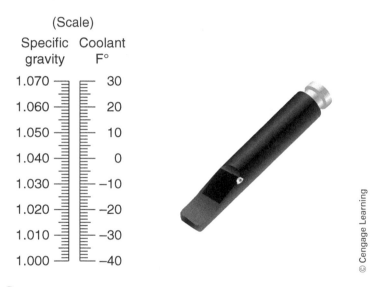

Figure 14-2 A refractometer shows specific gravity as well as the protection factor.

9. Measure the temperature of the water being tested; then place a drop of water on a refractometer, and record its specific gravity.

Temperature _____ Specific Gravity _____

10. Perform the same test with a sample of coolant.

Temperature _____ Specific Gravity _____

11. Retest the sample of coolant, using a hydrometer, and record the indicated specific gravity. _____

12. Note any difference in readings. _____

As you may see, temperature affects the specific gravity of the water and coolant. ■

Another property of heat that affects the cooling system is thermal expansion. When the engine is running, heat is produced and has to be removed by the oil and cooling systems. As the engine block, cylinder head, coolant, and other components become hotter, they expand. Some components, depending on what material they are made of, expand more than others. This can cause issues for gaskets and seals. As components expand, they move relative to each other. When the engine is shut off and begins to cool, the components contract and move against each other again.

Concept Activity 14-3
Engine Construction Materials

Inspect a vehicle in your lab, and note the materials that make up the major engine components.

1. Engine block material _____

2. Cylinder head material _____

3. Intake manifold material _____

4. Which of the materials listed above will expand and contract the most over the engine's operational temperature range?

5. List the gaskets and seals that are affected by these components expanding and contracting.

_____

As the engine warms, the coolant expands. When the engine is shut off and cools, the coolant contracts and is drawn back into the engine. This is why cooling systems are equipped with coolant reservoir bottles. On systems with pressure tanks, space is provided to allow for coolant expansion. Both systems have both hot and cold full-fill markings.

The combination of heat and expansion creates pressure in the cooling system. As the pressure increases, the boiling point of the coolant increases.

6. At what temperature does water boil? _____

Locate a cooling system cap from several lab vehicles.

7. What is the system pressure listed on the caps? _____

For every 3 psi increase in pressure, the boiling point will increase about 10°F.

8. For each cooling system cap, determine the boiling point based on the pressure listed on the cap.

_____

With a pressure cap installed on the radiator or pressure tank, system pressure increases up to a certain amount, often between 13 and 16 psi, though some newer systems are approaching 30 psi. If the preset pressure limit is reached, the cap will allow the coolant to vent.

Before continuing, obtain permission from your instructor operate a lab vehicle. With the engine cold, remove the radiator cap, and ensure the coolant level is full. Reinstall the cap. Next, check the coolant level in the overflow reservoir. Place a mark on the reservoir with a grease pencil or similar marker. Start the engine, and allow it to reach operation temperature. Monitor coolant temperature via the temperature gauge on the dashboard or a computer scanner. Once the engine is fully warmed, shut the engine off, and note the coolant level in the reservoir. Place a mark at the level of the coolant. Determine the volume of coolant displaced by expansion into the reservoir. This can be done in a couple of ways: by direct measurement of the amount of coolant or by approximating the size of the reservoir and figuring the displacement.

9. How much coolant is displaced by expansion? _____

Find the cooling system capacity in the service information or manual.

10. What percentage of the total cooling system capacity is displaced when the engine is hot?

_____

11. What factors do you think affect how much coolant is displaced?

12. Explain why the coolant reservoir or the pressure tank is not filled to the top with coolant.

_____

As the engine and cooling system heat up, coolant is displaced into the reservoir or recovery tank. As the engine and cooling system cool, the displaced coolant is returned to the cooling system.

13. What cooling system component is responsible for allowing the coolant to flow out and back from the reservoir? _____

14. How would a leak in the cooling system affect the operation of the coolant recovery system?

_____

_____

■

Concept Activity 14-4
Heat Transfer

The ability of the cooling system, and the lubrication system as well, to transfer heat is critical. Without sufficient heat transfer from either system, the engine would quickly be destroyed. Engine oil and coolant are both used to transfer heat from various components.

1. Describe how heat is transferred and removed by engine oil.

_____

_____

_____

2. Describe how heat is transferred and removed by coolant.

_____

_____

_____

3. How does the condition of the fins on the radiator affect its ability to transfer heat?

_____

_____

_____

4. Describe other factors that can affect the ability of the lubrication and cooling systems to remove heat from the engine.

_____

_____

_____

_____

■

LUBRICATION

The engine relies on the circulation of engine oil to lubricate and cool moving parts and to minimize wear. Always use the quality and grade of oil recommended by the manufacturer.

Oil is delivered under pressure to all moving parts of the engine. Engine damage will occur if dirt gets trapped in small clearances designed for an oil film or if dirt blocks an oil passage. Periodic oil and filter changes ensure that clean oil circulates throughout the engine.

A screen (strainer) on the oil pickup tube **(Figure 14-3)** initially filters the engine's oil. The oil pump draws oil out of the oil pan through the pickup tube. The oil pan serves as a reservoir for the oil. Oil drawn from the pan is circulated through the engine. After circulating, the oil drips back into the oil pan.

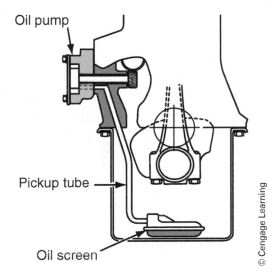

Oil pump

Pickup tube

Oil screen

© Cengage Learning

Figure 14-3   Oil pump pickup.

The oil pump is driven by the crankshaft or camshaft. The pump is responsible for the circulation of pressurized oil throughout the engine. The pressurized oil travels through oil passages or galleries in the engine, crankshaft, connecting rods, and other parts. The key parts of an engine's lubrication system are shown in **Figure 14-4.**

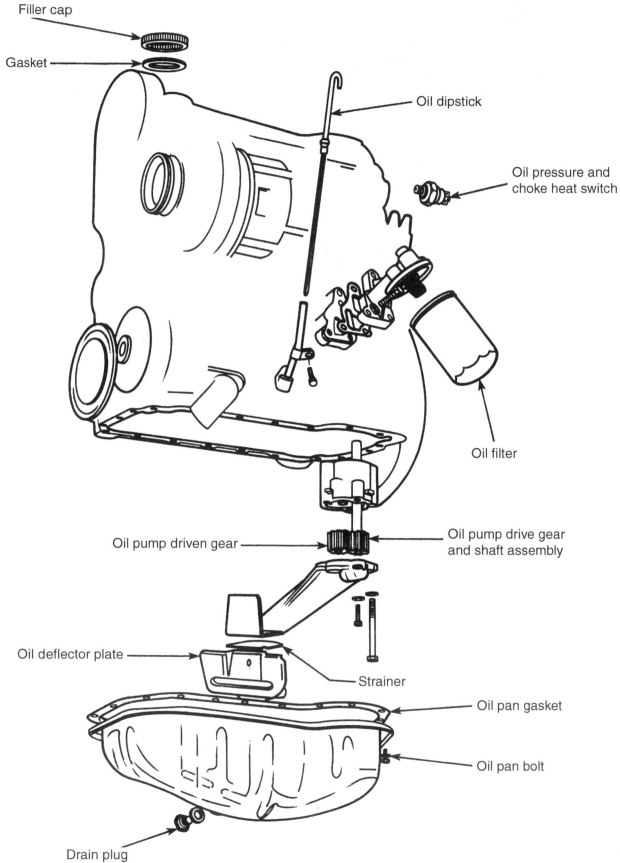

Filler cap

Gasket

Oil dipstick

Oil pressure and choke heat switch

Oil filter

Oil pump driven gear

Oil pump drive gear and shaft assembly

Oil deflector plate

Strainer

Oil pan gasket

Oil pan bolt

Drain plug

© Cengage Learning

Figure 14-4 Major components of the lubrication system.

Problems with the engine's lubrication system will cause engine problems. As an aid to diagnosis, **Table 14-1** is included in this chapter.

Oil Pump Inspection and Service

The oil pump inspection and service is an important task because the oil pump is the lifeline of the engine. It is important to carefully follow the manufacturers' service manual procedures to determine whether the pump is serviceable. If there is any doubt as to the ability of the pump to perform, the replacement of the oil pump is necessary.

Oil Pump Installation

Installation of the oil pump should follow the recommended procedures set forth in the service manual. Oil pump priming and correct placement of the oil pump pickup are two examples of the items that are included in the oil pump installation procedures.

TABLE 14-1 Lubrication System Diagnosis.

Symptom	Possible Cause
External oil leaks	Cylinder head cover RTV sealant broken or improperly seated
	Oil filter cap leaking or missing
	Oil filter gasket broken or improperly seated
	Oil pan gasket broken or improperly seated, or opening in RTV sealant
	Oil pan gasket broken or improperly seated
	Oil pan rear oil seal broken or improperly seated
	Timing case cover oil seal broken or improperly seated
	Excess pressure because of restricted PCV valve
	Oil pan drain plug loose or has stripped threads
	Rear oil gallery plug loose
	Rear camshaft plug loose or improperly seated
	Distributor base gasket damaged
Excessive oil consumption	Oil level too high
	Oil with wrong viscosity being used
	PCV valve stuck closed
	Valve stem oil deflectors (or seals) damaged or missing
	Valve stems or valve guides worn
	Poorly fitted or missing valve cover baffles
	Piston rings broken or missing
	Scoffed piston
	Incorrect piston ring gap
	Piston rings sticking or excessively loose in consumption grooves
	Compression rings installed upside-down
	Cylinder walls worn, scored, or glazed
	Piston ring gaps not properly staggered
	Excessive main or connecting rod bearing clearance
No oil pressure	Low oil level
	Oil-pressure gauge, warning light, or sending unit inaccurate
	Oil pump malfunction
	Oil-pressure relief valve sticking
	Oil passages on pressure side of pump obstructed
	Oil pickup screen or tube obstructed
	Loose oil inlet tube

(Continued)

TABLE 14-1 *(Continued)*

Symptom	Possible Cause
Low oil pressure	Low oil level
	Inaccurate gauge, warning light, or sending unit
	Oil excessively thin because of dilution, poor quality, or improper grade
	Excessive oil temperature
	Oil-pressure relief spring weak or sticking
	Oil inlet tube and screen assembly restriction or air leak
	Excessive oil pump clearance
	Excessive main, rod, or camshaft bearing clearance
High oil pressure	Improper oil viscosity
	Oil-pressure gauge or sending unit inaccurate
	Oil-pressure relief valve sticking closed

COOLING SYSTEMS

Extreme heat is created in an engine during the combustion process. If this heat is not removed from the engine, the engine will self-destruct **(Figure 14-5)**.

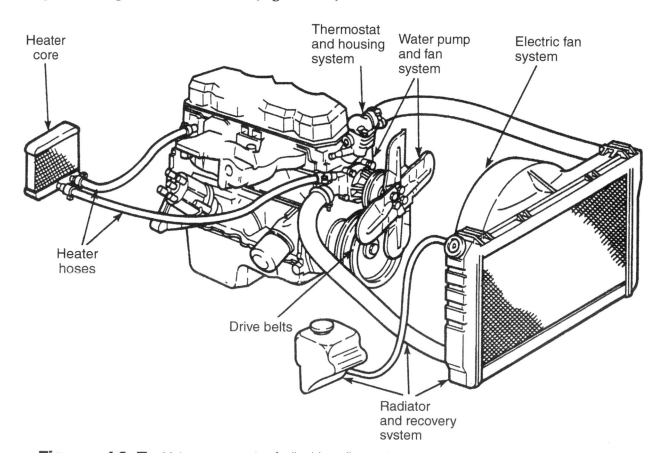

Figure 14-5 Major components of a liquid-cooling system.

Engine coolant is pumped through the engine and, after absorbing the heat of combustion, flows into the radiator, where the heat is transferred to the outside air. The cooled liquid is then returned to the engine to repeat the cycle. The flow of the coolant and the amount of heat that is given off to the outside air are controlled to keep the engine within a specified temperature range.

A thermostat is used to control the temperature by controlling the amount of coolant that moves into the radiator. When the engine is cold, the thermostat is closed, and no coolant flows to the radiator. This allows the engine to warm up to the correct operating temperature. As the coolant warms, the thermostat opens and allows coolant to flow through the radiator.

Engine coolant is actually a mixture of antifreeze/coolant and water. This mixture lowers the freezing point of the coolant and raises its boiling point. The pressure of the system also increases the boiling point of the coolant. The cooling system is a closed system. The pressure in the system is the result of the operation of the water pump and the increase in temperature. As temperature increases, so does pressure.

All cooling systems are designed to operate under a specific pressure. This pressure is maintained by the radiator pressure cap. The cap is designed to keep the cooling system sealed until a particular pressure is reached. At that time, the cap allows some of the pressure to be vented from the system. This action prevents excessive cooling system pressures.

Cooling system problems can not only affect the durability of an engine, they can also cause many driveability problems. **Table 14-2** is included in this chapter to serve as an aid in diagnosing engine cooling problems. Cooling system problems can be averted by following the manufacturers' recommendations for cooling system servicing. Preventive maintenance provides continued vehicle operation and greatly reduces the possibility of vehicle breakdown.

The following job sheets will prepare you for lubricating and cooling system diagnosis and service.

TABLE 14-2 Cooling System Diagnosis.

Symptom	Possible Cause
High-temperature gauge indication: overheating	Coolant level low
	Fan belt loose
	Radiator hose collapsed
	Radiator airflow blocked
	Faulty radiator cap
	Air trapped in cooling system
	Heavy-traffic driving
	Faulty thermostat
	Water pump shaft broken or impeller loose
	Radiator tubes clogged
	Cooling system clogged
	Casting flash in cooling passages
	Brakes dragging
	Excessive engine friction
	Antifreeze concentration over 68%
	Cooling fan not working
	Faulty gauge or sending unit
	Loss of coolant flow caused by leakage or foaming
	Viscous fan drive failed
Low-temperature gauge indication: undercooling	Thermostat stuck open
	Faulty gauge or sending unit
Coolant loss: boilover (also refer to overheating causes)	Overfilled cooling system
	Quick shutdown after hard (hot) run
	Air in system resulting in occasional "burping" of coolant
	Insufficient antifreeze, allowing coolant boiling point to be too low
	Antifreeze deteriorated because of age or contamination

(Continued)

TABLE 14-2 *(Continued)*

Symptom	Possible Cause
	Leaks caused by loose hose clamps, nuts, bolts, or drain plugs
	Faulty head gasket
	Cracked head, manifold, or block
	Faulty radiator cap
Coolant entry into crankcase or cylinder	Faulty head gasket
	Crack in head, manifold, or block
Coolant recovery system inoperative	Coolant level low
	Leak in system
	Pressure cap not tight or seal missing or leaking
	Pressure cap defective
	Overflow tube clogged or leaking
Noise	Fan contacting shroud
	Loose water pump impeller
	Glazed fan belt
	Loose fan belt
	Rough surface on drive pulley
	Water pump bearing worn
	Belt alignment

☐ JOB SHEET 14–1

Servicing and Installing Oil Pump and Oil Pan

Name _____ Station _____ Date _____

Objective

Upon completion of this job sheet, you will have demonstrated the ability to properly service, inspect, and install the oil pump and oil pan. This task applies to ASE Education Foundation Task MAST 1.D.13 (P-2).

Tools and Materials

Feeler gauge   Straightedge

Measuring scale   Torque wrench

Outside micrometer

Description of Vehicle

Year _____ Make _____ Model _____

VIN _____ Engine Type and Size _____

Mileage _____

Describe the type of oil pump.

PROCEDURE

1. Using a service manual, locate the required specifications for the items listed on the accompanying Report Sheet for Oil Pump Service and enter them on the report.

   WARNING: *Always wear safety goggles or glasses with side shields when doing this task. The oil-pressure relief valve is controlled by spring pressure, and the retainer may fly out of the housing, causing severe injury.*

2. Disassemble the oil pump and relief valve assembly in clean solvent, and allow them to dry.

3. Place a straightedge across the pump cover, and try to push a 0.002 inch (0.050 mm) feeler gauge between the cover and straightedge. If the gauge fits, the cover is worn. Record your results on the report sheet.

4. Measure the thickness and diameter of the rotors or gears **(Figure 14-6)**. Compare the measurements with specifications. Measurements smaller than the specifications mean the pump must be replaced. Record your results on the report sheet.

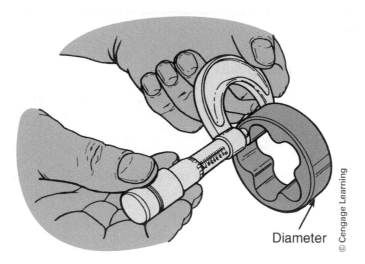

© Cengage Learning

Diameter

Figure 14-6 Measuring the outer rotor with an outside micrometer.

5. Assemble the gears or rotors into the pump body. Measure the clearance between the pump body and the outer rotor or gear with a feeler gauge **(Figure 14-7).** Record your results on the report sheet. If the clearance is greater than the specifications, replace the oil pump.

© Cengage Learning 2015

Figure 14-7 Determining the clearance between the outer rotor and the pump body.

6. Measure the clearance between the inner rotor and outer rotor lobes with a feeler gauge **(Figure 14-8)**. Record your finding on the report sheet. If the clearance is greater than the specifications, replace the oil pump.

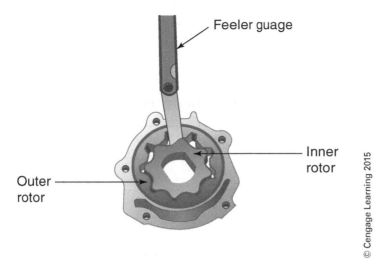

Feeler guage

Inner rotor

Outer rotor

© Cengage Learning 2015

Figure 14-8   Measuring clearance between the inner and outer rotor lobes.

7. Position a straightedge across the gears or rotors and measure the clearance with a feeler gauge. If the measurement is larger than the specifications, replace the pump. Record your results on the report sheet.

8. Measure the relief valve spring height, and record your findings on the report sheet.

9. Lubricate all pump parts and reassemble pump. Tighten all bolts to specifications.

10. Install the pump drive or extension and install the pump on the engine. Torque the mounting bolts to specifications.

Problems Encountered

_____

_____

_____

Instructor's Comments

_____

_____

_____

Name _____ Station _____ Date _____

REPORT SHEET FOR OIL PUMP SERVICE		
Measuring	**Specifications**	**Actual**
Cover warpage		
Rotor diameter		
Rotor thickness		
Rotor clearance (backlash)		
Gear (rotor)-to-cover clearance		
Relief spring free length		
Torque Specifications		
Oil pump cover bolts		
Oil pump mounting bolts		
Oil pan bolts		
Oil pan drain plug		

Conclusions and Recommendations _____

_____

_____

☐ JOB SHEET 14–2

Inspecting, Replacing, and Adjusting Drive Belts and Pulleys

Name _____ Station _____ Date _____

Objective

Upon completion of this job sheet, you will have demonstrated the ability to inspect and replace drive belts. This task applies to ASE Education Foundation Tasks MLR 1.C.2 (P-1) and AST/MAST 1.D.3 (P-1).

Tools and Materials

Belt tension gauge   Straightedge or belt alignment tool

Description of Vehicle

Year _____ Make _____ Model _____

VIN _____ Type of Engine _____

Mileage _____

PROCEDURE

1. Determine belt type(s). _____

2. Draw and label the belt routing(s) and tensioning points for each belt.

3. Locate belt-tensioning points (automatic spring loaded or accessory pivot). Remove belts(s), and note wear.

   _____

   _____

   Instructor check with belt removed from vehicle _____

4. Determine bolt torque and belt tension specification, and record.

   _____

   _____

5. Check for proper pulley alignment, using either a straightedge or pulley alignment tool. Record pulley alignment findings.

   _____

6. Inspect pulleys for damage, sharp edges of grooves, or bending. Record findings.

   _____

7. Reinstall belt(s), and perform final checks. Start engine, and record belt operation.

_____

_____

_____

Instructor Check _____

Problems Encountered

_____

_____

_____

Instructor's Comments

_____

_____

_____

☐ JOB SHEET 14–3

Identifying Causes of Engine Overheating

Name _____ Station _____ Date _____

Objective

Upon completion of the job sheet, you will have demonstrated the ability to identify causes of engine overheating. This task applies to ASE Education Foundation Tasks AST/MAST 1.D.2 (P-1).

Tools and Materials

Thermometer   Scan tool

Description of Vehicle

Year _____ Make _____ Model _____

VIN _____ Engine Type and Size _____

Mileage _____

PROCEDURE

1. Perform a visual inspection of the cooling system, and note the following:

   a. Coolant type and condition _____

   b. Coolant level in radiator and/or pressure tank _____

   c. Coolant hose condition _____

   d. Drive belt condition _____

   e. Cooling fan type _____

   f. Signs of coolant loss _____

   g. Radiator air flow restrictions _____

2. If the engine uses a belt-driven fan and fan clutch, with the engine off, turn the fan. Does the fan spin

   easily or is it tight? _____ If the engine temperature is cold, should the fan

   clutch be loose or tight? _____ Why? _____

   _____

3. Install a scan tool and note the coolant temperature reading with the key on engine off. Coolant tem-

   perature _____

4. With the radiator or tank cap off, start the engine, and note the coolant level. Does the coolant level

   drop, or does it circulate? _____

5. If the coolant is circulating when the engine starts, what could be the problem?

   _____

6. Reinstall the radiator/tank cap.

7. As the engine warms up, note the coolant temperature gauge and the temperature reading on the scan tool. Are the two readings similar? _____

8. Most thermostats are fully open around 190°F (88°C). Carefully check the temperature of the upper and lower radiator hoses when the temperature gauge or scan tool indicates approximately 190°F.

9. What would the symptoms be if the thermostat were stuck closed? _____

   _____

10. If the thermostat opens correctly, continue to monitor coolant temperature.

11. Refer to the service information to find electric cooling fan on and off temperatures. Cooling fan on at _____ and off at _____

12. Does the electric cooling fan turn on and off at the appropriate temperatures? _____

13. If the engine uses a belt-driven fan, turn the engine off, and try to rotate the fan. How does the fan's resistance compare to when the engine was cold? _____

14. Based on your observations, describe the condition of the cooling system. _____

    _____

Problems Encountered

_____

_____

_____

Instructor's Comments

_____

_____

_____

☐ JOB SHEET 14–4

Cooling System Pressure Test

Name _____ Station _____ Date _____

Objective

Upon completion of this job sheet, you will have demonstrated the ability to perform a cooling system pressure test. This task applies to ASE Education Foundation Tasks MLR 1.C.1 (P-1) and AST/MAST 1.D.1 (P-1).

Tools and Materials

Cooling system pressure tester

Description of Vehicle

Year _____ Make _____ Model _____

VIN _____ Engine Type and Size _____

Mileage _____

PROCEDURE

1. Connect the pressure tester to the radiator filler neck or pressure tank.

2. Apply pressure until the same pressure value as stamped on the radiator cap is present. Cooling system pressure rating _____

3. Observe and record the pressure reading.

   _____

4. Does the pressure value reading remain constant or does it drop?

   _____

5. If the pressure remains constant, the cooling system has passed the pressure test. If the pressure drops, look for signs of coolant leak from external sources. List the external locations for coolant leaks.

   _____

   _____

   _____

   _____

6. If the pressure drops and no external leaks are present, suspect an internal engine coolant leak. List the areas for internal coolant leaks.

   _____

   _____

   _____

WARNING: *Extreme caution must be taken when removing the pressure tester from the filler neck of the radiator if the engine is at operating temperature. Personal injury may occur from the hot coolant of the radiator.*

Note: *The radiator cap testing requires an adapter included in the pressure testing kit. Proceed with the testing of the radiator cap.*

8. Test the radiator cap to ensure it holds the specific pressure indicated. Is the vacuum valve portion of the pressure working properly?

_____

_____

Problems Encountered

_____

_____

_____

Instructor's Comments

_____

_____

_____

REVIEW QUESTIONS

Review Chapter 14 of the textbook to answer these questions:

1. List three possible causes of low engine oil pressure.

_____

_____

2. List two effects on driveability of having the wrong thermostat installed.

_____

_____

3. Explain how the radiator cap increases cooling system boiling point.

_____

_____

4. Explain how cooling system electrolysis occurs.

_____

_____

5. Explain why special cooling system service precautions are necessary when working on hybrid vehicles.

_____

_____

 ASE PREP TEST

1. Technician A says that piston slap is a knocking noise that goes away as the engine warms up. Technician B says that piston slap is more noticeable when the engine is idling at operating temperature. Who is correct?

    a. Technician A

    b. Technician B

    c. Both A and B

    d. Neither A nor B

2. Technician A removes all of the spark plugs before performing a cylinder leakage test. Technician B disconnects the PCV hose from the crankcase before performing a cylinder leakage test. Who is correct?

    a. Technician A

    b. Technician B

    c. Both A and B

    d. Neither A nor B

3. Technician A says that excessive wear on the crankshaft and its bearings can cause low oil pressure. Technician B says that restrictions in the oil passages can cause low oil pressure. Who is correct?

    a. Technician A

    b. Technician B

    c. Both A and B

    d. Neither A nor B

4. Technician A says the valve spring locks or keepers are removed when the valve springs are compressed. Technician B uses a small magnet to remove valve keepers. Who is correct?

    a. Technician A

    b. Technician B

    c. Both A and B

    d. Neither A nor B

5. Technician A says the cylinder ring ridge is removed before the piston is removed. Technician B says the cylinder ring ridge tool is driven counterclockwise. Who is correct?

    a. Technician A

    b. Technician B

    c. Both A and B

    d. Neither A nor B

6. Technician A replaces the camshaft and lifters when the lifters are worn. Technician says only the worn lifters need to be replaced. Who is correct?

    a. Technician A

    b. Technician B

    c. Both A and B

    d. Neither A nor B

7. Technician A says the rings of the piston are located in the "piston skirt." Technician B says the pistons must be installed on the connecting rods in a certain direction. Who is correct?

    a. Technician A

    b. Technician B

    c. Both A and B

    d. Neither A nor B

8. Technician A says that the maximum allowable cylinder bore out-of-roundness is 0.006 inch (0.15 mm). Technician B replaces the thrust bearing when crankshaft end play exceeds the specified limit. Who is correct?

    a. Technician A

    b. Technician B

    c. Both A and B

    d. Neither A nor B

9. Technician A uses a small hole gauge and an outside micrometer to check valve stem-to-guide clearance. Technician B uses a feeler gauge for the same measurement. Who is correct?

   a. Technician A
   b. Technician B
   c. Both A and B
   d. Neither A nor B

10. Technician A says that a minimum amount of valve stem-to-guide clearance is needed for lubrication and expansion of the valve stem. Technician B says excessive stem-to-guide clearance cans cause increased oil consumption. Who is correct?

    a. Technician A
    b. Technician B
    c. Both A and B
    d. Neither A nor B

11. Technician A says a torque angle gauge is used on torque-to-yield bolts. Technician B says torque-to-yield bolts may require multiple tightening steps. Who is correct?

    a. Technician A
    b. Technician B
    c. Both A and B
    d. Neither A nor B

12. Technician A says that crankshaft bearings are not normally replaced during engine rebuilding unless they show signs of wear. Technician B says that main bearing wear can be checked with a dial indicator. Who is correct?

    a. Technician A
    b. Technician B
    c. Both A and B
    d. Neither D nor B

13. Technician A says that the camshaft must be removed from the engine to compare cam lobe height with an outside micrometer. Technician B says that premature lobe and lifter wear is usually the result of inadequate lubrication. Who is correct?

    a. Technician A
    b. Technician B
    c. Both A and B
    d. Neither A nor B

14. Technician A says that rocker arms with adjustable screws are typically used for lash adjustment with overhead-valve engines. Technician B says that some engines use shims to set the valve adjustment. Who is correct?

    a. Technician A
    b. Technician B
    c. Both A and B
    d. Neither A nor B

15. Technician A makes a gasket to replace the original one for the end housing of a disassembled oil pump. Technician B measures the thickness of the pump's inner rotor with an outside micrometer. Who is correct?

    a. Technician A
    b. Technician B
    c. Both A and B
    d. Neither A nor B

16. Technician A says that a cooling system pressure tester is used to check for both internal and external leaks. Technician B says that a drop in pressure on the cooling system tester always indicates an external leak. Who is correct?

    a. Technician A
    b. Technician B
    c. Both A and B
    d. Neither A nor B

17. Technician A says molded synthetic rubber lip–type seals are used on many new engines. Technician B says synthetic rubber seals may be available for older vehicle applications.

    a. Technician A
    b. Technician B
    c. Both A and B
    d. Neither A nor B

18. Technician A says many head bolts are tightened in two to three stages. Technician B says torque-to-yield bolts cannot be reused and must be replaced.

    a. Technician A
    b. Technician B
    c. Both A and B
    d. Neither A nor B

19. Technician A says that silicone formed-in-place sealants cannot take the place of cork or rubber gaskets. Technician B says that an RTV silicone sealant is impervious to most automotive fluids. Who is correct?

    a. Technician A
    b. Technician B
    c. Both A and B
    d. Neither A nor B

20. Technician A removes all of the spark plugs to perform a cranking compression test. Technician B says a running compression test can help locate valvetrain problems. Who is correct?

    a. Technician A
    b. Technician B
    c. Both A and B
    d. Neither A nor B

CHAPTER

15

Courtesy of Robert Bosch GmbH

BASICS OF ELECTRICAL SYSTEMS

OVERVIEW

The electrical system forms the backbone on which nearly all other systems operate on the modern automobile. Without the electrical system, there would not be computerized engine or transmission controls, anti-lock brakes, or any of the modern conveniences and extras found in today's cars and trucks. Without a thorough understanding of electrical and electronic systems, a technician will have a very difficult time servicing a modern vehicle.

Even though the presence of electricity, particularly static electricity, has been known for over 2,000 years, electricity did not become a topic of serious study until the 1700s. The term *electricity* is actually a somewhat general name for several related concepts, including electrical charge, electrical potential, current flow, and electromagnetism. Each is discussed in the following chapters.

When talking about electricity, what is being discussed is, simply put, electron flow. All matter is composed of atoms. Atoms are comprised of positively charged protons, neutral neutrons, and negatively charged electrons orbiting the protons and neutrons at the nucleus. Some atoms easily combine with other atoms to form new substances. When two hydrogen atoms combine with one oxygen atom, water is formed **(Figure 15-1)**. Some materials, such as copper and other metals, under certain conditions, can lose and regain an electron from their outer ring, called the valence ring. This movement of electrons is the basis of electricity.

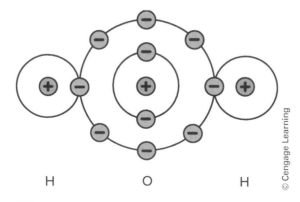

H   O   H

© Cengage Learning

Figure 15-1 A molecule of water.

Concept Activity 15-1
Static Electricity

Almost everyone is familiar with two very common electrical occurrences: lightning and static electricity. Lightning is actually an electrical discharge due to static electricity, just like when you get shocked touching a doorknob on a dry winter day. The difference is the energy in a lightning bolt is many, many times greater.

1. Describe what you think are the conditions necessary for getting a static electricity shock.

   _____

2. Why do you think a lightning discharge is so much more powerful than a static electricity shock?

   _____

The term *static electricity* refers to a buildup of an electrical charge where there is typically poor electrical conductivity. When there is an electrical charge, there is the potential for electron flow. When an electron leaves an atom, it leaves behind an opening. The atom from which the electron left will now have a positive charge due to the loss of the negatively charged electron. The atom the electron moved to now has a negative charge. Whenever an imbalance occurs, there is the potential for electrical current flow. When an electrical force is present, and electrons move from atom to atom, electrical current flow exists. If a material does not easily accept the movement of electrons, it is an insulator or a poor conductor. Metals such as copper, silver, gold, and aluminum can readily give up and transfer electrons, and are considered good electrical conductors **(Figure 15-2)**.

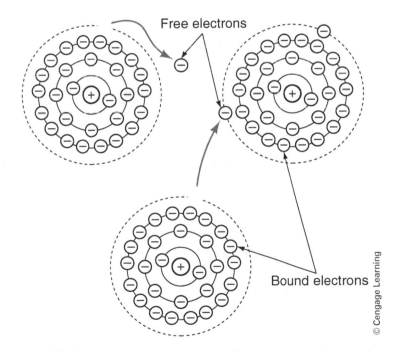

Figure 15-2   Conductors are atoms with loosely bound outer electrons.

If during this static electricity transfer, dissimilar charges are present, the charges will attract. If the charges are similar, the charges will repel—just as two magnets will repel if the north poles are placed close together and attract if the north and south poles are near each other. If you have ever pulled a wool sweater on over your head and had your hair stand on end, you have experienced this phenomenon. When you pulled the sweater over your head, electrons transferred from the wool to your hair. Similarly, rubbing a balloon against a sweater can cause enough of a static charge to allow the balloon to "stick" to a wall.

3. Inflate two balloons and attach a length of string to each knotted end. Rub one balloon against a piece of clothing. Then hold the string so that the balloons hang down near each other. Describe what takes place.

_____

_____

_____

4. Rub the balloons against your clothing, and attempt to stick them to various surfaces. What surfaces stuck the best? _____ Which were the worst?_____

5. Describe why you think some surfaces hold the balloons better or worse than others.

_____

_____

_____

6. What environmental factors affect static electricity?

_____

_____

_____

_____

■

To discuss electricity, it is necessary to include magnetism because the two are interrelated and inseparable. Electromagnetism refers to the interaction of electricity and magnetism. When a current flows through a conductor, a magnetic field develops around the conductor. Likewise, when a conductor is moved through a magnetic field, current is induced into the conductor. This means that electricity can be generated by moving a conductor through a magnetic field. It is this principle that is the basis of electrical power generation both under the hood of a car with a generator and at a power generation facility, like a hydroelectric dam or a nuclear power plant. Electromagnetism also defines how relays, electric motors, and ignition coils operate.

Concept Activity 15-2
Magnetic Fields

Obtain a magnet and some iron shavings from your instructor. Place a piece of construction paper on top of the magnet. Then sprinkle some of the shavings on the paper over the magnet.

1. What shapes do the shavings show on the paper?

_____

_____

2. Describe why the shavings appear in the patterns you see?

_____

_____

Ask your instructor for a magnetic compass. To use the compass as an electrical current detector, place it along the positive or negative battery cable of a vehicle. Place the compass so the needle is parallel with the cable.

3. Turn on the headlights and describe the reaction of the compass.

_____

_____

4. Why did the compass react the way that it did?

_____

_____

5. How would the compass react to a smaller electrical load than the headlights?

_____

_____

_____

■

As you have demonstrated, electrical current flow generates magnetic fields. In Chapter 18, you will see how magnetic fields are used in motors. Chapter 19 will show how magnetic fields are used to generate electricity to charge the battery and power the vehicle's electrical system.

 Concept Activity 15-3
Ohm's Law

Using Ohm's Law, depicted in Figure 15-3, solve the problems in Figures 15-4 through 15-7.

© Cengage Learning

Figure 15-3 Ohm's Law.

Problem 1: Series Circuit

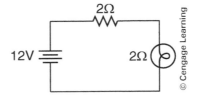

© Cengage Learning

Figure 15-4 Series circuit.

Calculate the following values.

1. Total circuit resistance: _____

2. Total circuit amperage: _____

3. Voltage drop of lamp: _____

4. Voltage drop of resistance: _____

5. Current through lamp: _____

6. Current through resistance: _____

7. What effect does the unwanted resistance have on the circuit?

_____

_____

_____

8. What steps would you take to diagnose this concern?

_____

_____

_____

9. What are three possible causes for this concern?

_____

_____

_____

Problem 2: Parallel Circuit

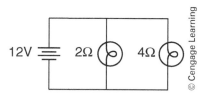

Figure 15-5   Parallel circuit.

Referring to **Figure 15-5**, calculate the following values.

1. Total circuit resistance: _____

2. Total circuit amperage: _____

3. Voltage drop of Lamp 1: _____ Lamp 2: _____

4. Current through Lamp 1: _____ Lamp 2: _____

5. Why is the current flow through each lamp different?

_____

_____

_____

6. How is the total circuit current flow affected by the resistances?

_____

_____

_____

Problem 3: Series-Parallel Circuits

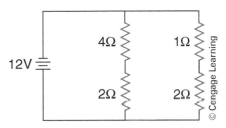

Figure 15-6 Series-parallel circuit.

Referring to **Figure 15-6**, calculate the following values.

1. Total circuit resistance: _____

2. Total circuit amperage: _____

3. Voltage drop of Lamp 1: _____ Lamp 2: _____ Lamp 3: _____ Lamp 4: _____

4. Current through Lamp 1: _____ Lamp 2: _____ Lamp 3: _____ Lamp 4: _____

5. Current through Branch 1: _____ Branch 2: _____

6. Why are the voltage drops for each resistance different?

Problem 4: Series-Parallel Circuits

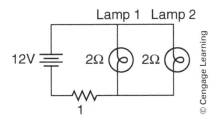

Figure 15-7 Series-parallel circuit.

Referring to **Figure 15-7**, calculate the following values.

1. Total circuit resistance: _____

2. Total circuit amperage: _____

3. Voltage drop of Lamp 1: _____ Lamp 2: _____ Resistance: _____

4. Current through Lamp 1: _____ Lamp 2: _____ Resistance: _____

5. Why is the current flow through each lamp different?

6. How is the total circuit current flow affected by the resistance?

7. If this were a taillight circuit, how would the unwanted resistance affect the operation of the circuit?

8. What steps would you take to diagnose this concern?

■

 Concept Activity 15-4
Circuit Operation

Obtain from your instructor several automotive exterior lights with sockets. Bulb types 194, 1156, and 1157 (or similar) work well. Construct circuits using the diagrams below and record your findings.

1. **(Figure 15-8)** Series circuit.

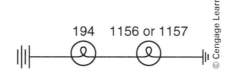

Figure 15-8   Series circuit.

Battery voltage: _____ Circuit resistance: _____ Voltage drops: _____

Ohm's Law predicted amperage: _____ Measured amperage: _____

2. **(Figure 15-9)** Series circuit.

Figure 15-9   Parallel circuit.

Battery voltage: _____ Circuit resistance: _____ Voltage drops: _____

Ohm's Law predicted amperage: _____ Measured amperage: _____

3. **(Figure 15-10)** Parallel circuit.

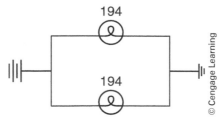

Figure 15-10   Parallel circuit.

Battery voltage: _____ Circuit resistance: _____ Voltage drops: _____

Ohm's Law predicted amperage: _____ Measured amperage: _____

4. **(Figure 15-11)** Parallel circuit.

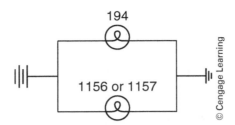

Figure 15-11   Parallel circuit.

Battery voltage: _____ Circuit resistance: _____ Voltage drops: _____

Ohm's Law predicted amperage: _____ Measured amperage: _____

5. **(Figure 15-12)** Series-parallel circuit.

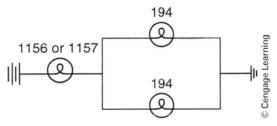

Figure 15-12   Series-parallel circuit.

Battery voltage: _____ Circuit resistance: _____ Voltage drops: _____

Ohm's Law predicted amperage: _____ Measured amperage: _____

6. **(Figure 15-13)** Series-parallel circuit.

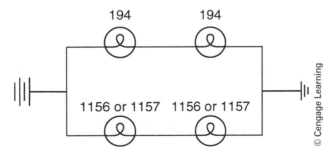

Figure 15-13   Series-parallel circuit.

Battery voltage: _____ Circuit resistance: _____ Voltage drops: _____

Ohm's Law predicted amperage: _____ Measured amperage: _____

7. **(Figure 15-14)** Series-parallel circuit.

194   1156 or 1157

1156 or 1157   1156 or 1157

1156 or 1157

Figure 15-14   Series-parallel circuit.

Battery voltage: _____ Circuit resistance: _____ Voltage drops: _____

Ohm's Law predicted amperage: _____ Measured amperage: _____   ■

Concept Activity 15-5
Schematic Components

Identify the following wiring diagram components:

_____

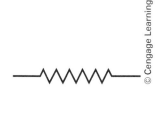

_____

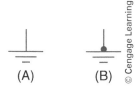

(A)   (B)

_____

_____

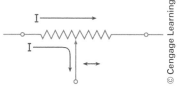

_____

From power source

To load

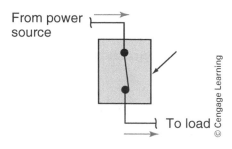

_____

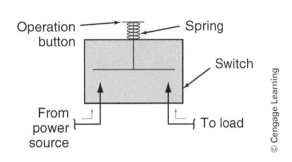

© Cengage Learning

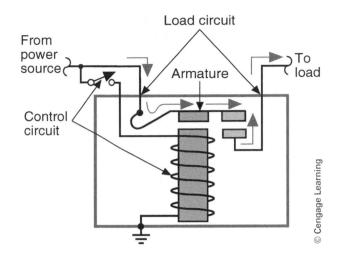

© Cengage Learning

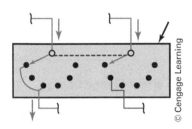

© Cengage Learning

© Cengage Learning

Concept Activity 15-6
Measurement Conversion

Convert the following values:

1. 3.7 ohms = _____ ohms

2. 641 mv = _____ volts

3. 0.783 K ohms = _____ ohms

4. 0.445 volts = _____ mv

5. 20 ma = _____ amps

6. 337 ma = _____ amps

7. 1.2 M ohms = _____ ohms

8. 8375 ohms = _____ K ohms

9. Using a DMM provided by your instructor, list the various setting ranges for the following. If the meter has auto-ranging, list the scales available by pressing the scale button.

    a. Voltage ranges: _____, _____, _____, _____

    b. Resistance ranges: _____, _____, _____, _____

    c. Amperage ranges: _____, _____, _____, _____

Concept Activity 15-7
Fuse Identification

1. Match the fuse amperage rating to its correct color:

4A _____   Red

5A _____   Pink

10A _____   Yellow

15A _____   Light blue

20A _____   Light green

25A _____   Tan

30A _____   Natural

2. Why is it important to replace a blown fuse with one of the correct voltage and amperage ratings?

_____

_____

_____

3. Describe what could happen if a blown fuse were replaced with one of too high an amperage rating?

_____

_____

_____

_____

■

Concept Activity 15-8
Relay Identification

Referring to **Figure 15-15**, identify the components of the relay:

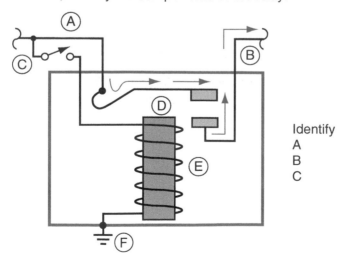

Identify

A   D

B   E

C   F

© Cengage Learning

Figure 15-15   Relay identification.

1. Define the purpose of a relay.

Obtain a relay from your instructor.

2. Measure the resistance of the relay coil: _____ ohms

3. Is the coil resistance acceptable?

4. What could cause the relay coil resistance to be lower than specifications?

5. What could be the result of the relay coil resistance being too low?

Concept Activity 15-9
Fuse Block Identification

Refer to **Figures 15-16** and **15-17** to complete the following table:

TABLE 15-1 Fuse Block Identification.

Fuse/Relay Location #	Fuse Size	Circuit Description	Fuse/Relay Type	Fuse Color
38				
		Radio		
	40			
6				
		Fuel pump	Micro relay	
		Spare		
25				
				Orange
			Diode	
12				

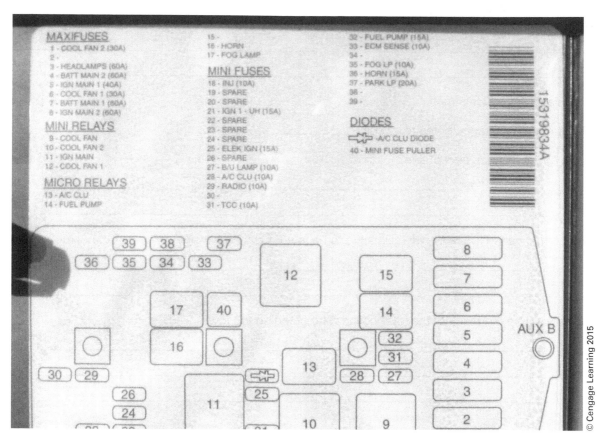

Figure 15-16 Identification figure for fuses in Table 15-1.

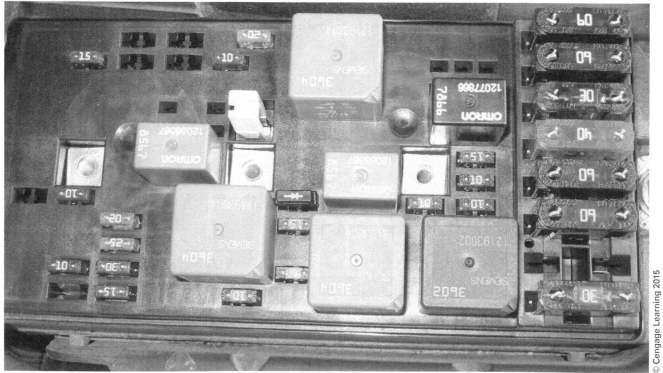

Figure 15-17 Identification figure for fuses in Table 15-1.

REVIEW QUESTIONS

Review Chapter 15 of the textbook to answer these questions:

1. Explain how to test voltage drop.

2. In a circuit powered by AC current, the RMS voltage reading will be _____ times the peak-to-peak voltage reading.

3. All of the following are examples of circuit protection devices *except*:

   a. Fuse link   c. Circuit breaker

   b. Relay   d. Voltage limiter

4. Which of the following is not a factor in a wire's resistance?

   a. Length of the wire   c. Diameter of the wire

   b. Wire temperature   d. Insulation material

5. A taillight circuit is being voltage drop tested. Technician A says excessive voltage drop on the light's power circuit could cause the fuse to blow. Technician B says excessive voltage drop on the light's ground circuit could cause the light to be dim. Who is correct?

   a. Technician A   c. Both A and B

   b. Technician B   d. Neither A nor B

© Cengage Learning 2015

GENERAL ELECTRICAL SYSTEM DIAGNOSTICS AND SERVICE

OVERVIEW

Diagnosing electrical and electronic automotive systems requires a broad base of skills, knowledge, and understanding. These skills include reading and comprehending service manual electrical schematics, as well as the ability to use a variety of test equipment. A logical and systematic approach is essential to troubleshooting system faults.

ELECTRICAL PROBLEMS

Automotive electrical concerns involve one of three conditions: open circuits, short circuits, and circuits with high resistance.

An open circuit is one where the path from power to ground is broken. If you have ever experienced a set of partially working Christmas tree lights because of a burned-out bulb, you have dealt with an open circuit. Open circuits can be a result of a poor connection, a wire rubbing against another object and breaking the wire, or contacts within a switch wearing out and no longer making contact. Open circuits can often be easily diagnosed by following the circuit from a schematic and checking for power at logical points along the circuit. When voltage is no longer present, the open circuit has been located.

A shorted circuit means that an unwanted connection to either power or ground has occurred. Short circuits can cause unwanted circuit operation. If the contacts within a multifunction switch (turn signal, wiper/washer, and high-beam lights) short together, then the wipers could activate when the driver turns on the turn signal.

Short circuits can also overload a circuit and cause the fuse to open. If a power supply wire shorts to ground from rubbing against an edge or corner, the current for that part of the circuit can bypass the intended load and go straight to ground. Since the short to ground will have very low resistance, the current flow will greatly increase. The increased current flow should open the fuse or circuit breaker, preventing an electrical fire. In rare instances, a short does not open a fuse or circuit breaker. When this occurs, the short circuit is extremely dangerous, sometimes causing electrical fires and severe vehicle damage.

High resistance in a circuit can cause a lightbulb to be dim, a motor to turn slowly, or a component not to operate at all. Excessive resistance in a circuit results in an unwanted voltage drop. The load in the circuit gets a reduced amount of voltage because the unwanted voltage drop consumes some of the power. This extra resistance can take several forms, but corroded terminals and loose ground connections are common problems.

ELECTRICAL SCHEMATICS

Electrical diagrams or schematics indicate the power, ground, load, and control portions of a circuit. Reading a schematic is similar to reading a road map. By following the paths of the wires and connections, the power flow through a circuit can be deduced.

Concept Activity 16-1
Schematic Symbols

Referring to **Figure 16-1**, match the following schematic symbols with the correct name:

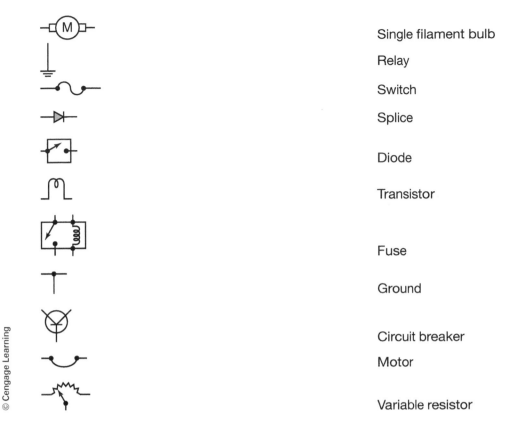

	Single filament bulb
	Relay
	Switch
	Splice
	Diode
	Transistor
	Fuse
	Ground
	Circuit breaker
	Motor
	Variable resistor

© Cengage Learning

Figure 16-1 Identify the schematic symbols.

ELECTRICAL TESTING EQUIPMENT

Special tools are used to properly diagnose electrical system concerns.

The simplest electrical tester is an unpowered test light **(Figure 16-2)**. The test light has a clip on the end of a length of wire, a pointed probe end, and a lightbulb wired between the probe and the wire. When the wire is clipped to ground and the probe is touching power, the light should illuminate, showing the presence of voltage. When the clip is placed on a power source and the probe is touching ground, the bulb should again illuminate, showing the presence of ground. For quickly checking for power at a fuse or other location, a test light is often used. It can also be used to check for a path to ground. Test lights should never be used when testing any electronic or computer-controlled circuits unless specifically directed to by the service information.

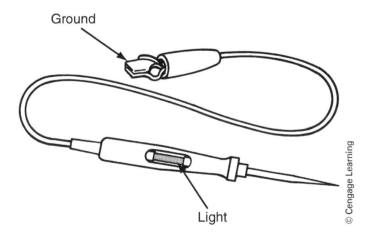

Ground

Light

© Cengage Learning

Figure 16-2 A test light can be used to check for voltages in various parts of an electrical circuit.

Concept Activity 16-2
Test Light Diagnosis

Use the diagram shown in **Figure 16-3** to answer the following:

1. If a test light were attached to the battery positive terminal and point X, would it light?

2. If a test light were attached to the battery negative terminal and point Y, would it light?

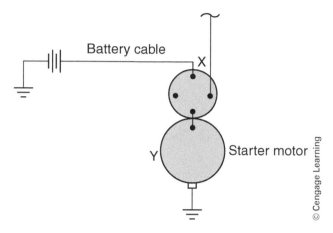

Figure 16-3 Using a test light.

To service modern automobiles, a technician needs more than a test light. A good quality digital multimeter (DMM) is essential for electrical repairs. Also known as digital volt-ohm meters (DVOM), or just multimeters, these testers allow for safe and accurate measurement of voltage, amperage, and resistance. Many multimeters will also measure and display frequency, duty cycle, temperature, and test diodes.

Multimeters are used to measure battery voltage, voltage drops, circuit resistance, component resistance, and current flow, usually up to 10 amps. To measure voltage, the meter is placed in parallel with the circuit. To measure amperage, the meter must be placed in series into the circuit. To measure resistance the power in the circuit must be removed and the meter placed at the ends of the circuit.

Concept Activity 16-3
DMM Measurement Setup

Label the electrical measurement in each of the three drawings shown in **Figure 16-4**.

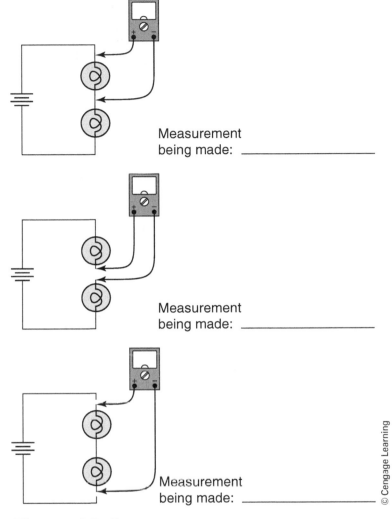

Measurement
being made: _____

Measurement
being made: _____

Measurement
being made: _____

© Cengage Learning

Figure 16-4 DMM setup.

Concept Activity 16-4
DMM Symbols

Match the multimeter symbols with the correct units.

– V	Amperage
~ V	Resistance
Ω	DC volts
A	AC volts

Concept Activity 16-5
Electrical Diagnosis

Use **Figures 16-5** and **16-6** to answer the questions about circuit faults.

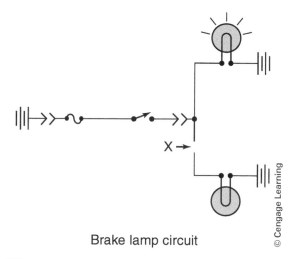

Brake lamp circuit

© Cengage Learning

Figure 16-5 Brake light circuit diagnosis.

1. Based on the open at point X, what would be the symptom or customer complaint?

_____

2. How would you test this circuit on the vehicle? _____

_____

3. What steps would you have to take to narrow down the cause of the problem? _____

_____

_____

_____

4. How would the resistance at point X affect the circuit? _____

_____

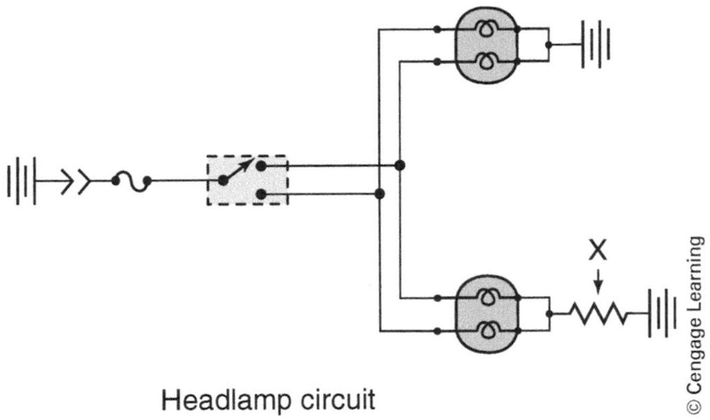

Headlamp circuit

© Cengage Learning

Figure 16-6 Headlamp circuit diagnosis.

5. Would the resistance at point X affect both the right and left headlights? _____

_____

6. Explain your answer to question 5. _____

_____

7. What steps would you take to diagnose this type of problem? _____

_____

The following job sheets will prepare you for electrical system diagnosis and service.

☐ JOB SHEET 16–1

Checking Continuity in a Circuit with a Test Light

Name _____ Station _____ Date _____

Objective

Upon completion of this job sheet, you will have demonstrated the ability to test continuity in a circuit with a test light. This task applies to ASE Education Foundation Tasks MLR 6.A.6 (P-2) and AST/MAST 6.A.5 (P-1).

> **WARNING:** *Never use a regular test light on an electronic circuit; use only a high-impedance test light on these circuits.*

Tools and Materials

Test light

Wiring diagram

Description of Vehicle

Year _____ Make _____ Model _____

VIN _____ Engine Type and Size _____

Mileage _____

PROCEDURE

1. Determine if the circuit to be tested is a power supply circuit or a ground circuit.

   Power_____ Ground _____

2. If the circuit to be tested is a power supply circuit, connect the clip end of the test light to a ground on the vehicle.

   Ground point _____

3. Turn the key to the on position.

4. With the light end, use the probe to check the wire in the circuit at each connector in the specific circuit. If the test light comes on, there is continuity between the power supply and the point where you are testing.

   Circuit tested _____

5. If the circuit to be tested is a ground circuit, connect the clip end of the test light to a power supply, such as the battery.

6. With the light end, use the probe to check the wire in the circuit at each connector in the specific circuit. If the test light comes on, there is continuity between the ground and the point you are testing.

   Circuit tested _____

Problems Encountered

_____

_____

_____

Instructor's Comments

_____

_____

_____

☐ JOB SHEET 16–2

Understanding DMM Controls

Name _____ Station _____ Date _____

Objective

Upon completion of this job sheet, you will have demonstrated the ability to prepare a DMM for taking various electrical measurements. This task applies to ASE Education Foundation Tasks MLR/AST/MAST 6.A.3 (P-1).

> **Tools and Materials**
> DMM
> User's Guide for the DMM

PROCEDURE

1. Describe the DMM you are using. _____

   Manufacturer _____

   Model _____

   Input impedance _____

   Ohms _____

   Docs the meter have an auto range feature? _____

2. Put a check after each of the features the DMM has.

   DC volts _____

   AC volts _____

      Root mean square (RMS) _____

      Average responding _____

   Ohms _____

   Amps _____

      Maximum amps for the meter is _____.

   MIN/MAX _____

   rpm _____

   Duty cycle _____

   Pulse width _____

   Diode test _____

   Frequency _____

   Temperature _____

Problems Encountered

_____

_____

_____

Instructor's Comments

_____

_____

_____

☐ JOB SHEET 16–3

Testing for Voltage with a DMM

Name _____ Station _____ Date _____

Objective

Upon completion of this job sheet, you will be able to measure available voltage and voltage drop with a DMM. This task applies to ASE Education Foundation Tasks MLR/AST/MAST 6.A.3 (P-1).

> **WARNING:** *Use only a high-impedance DMM on late-model vehicles.*

Tools and Materials

A high-impedance DMM

Wiring diagram for vehicle

Basic hand tools

Description of Vehicle

Year _____ Make _____ Model _____

VIN _____ Engine Type and Size _____

Mileage _____

PROCEDURE

1. Set the DMM to the appropriate scale to read 12 volts DC.

2. Connect the meter across the battery (positive to positive and negative to negative). What is the reading on the meter? _____ volts

3. With the meter still connected across the battery, turn on the headlights of the vehicle. What is the reading on the meter? _____ volts

4. Keep the headlights on. Connect the positive lead of the meter to the point on the vehicle where the battery's ground cable attaches to the frame. Keep the negative lead where it is.

   What is the reading on the meter? _____ volts

   What is being measured?

   _____

5. Disconnect the meter from the battery, and turn off the headlights.

6. Refer to the correct wiring diagram, and determine what wire at the right headlight delivers current to the lamp when the headlights are on and low beams are selected.

   Color of the wire _____

7. From the wiring diagram identify where the headlight is grounded.

   Place of ground _____

8. Connect the negative lead of the meter to the point where the headlight is grounded.

9. Connect the positive lead of the meter to the power input of the headlight.

10. Turn on the headlights. What is your reading on the meter? _____ volts

   What is being measured? _____

11. What is the difference between the reading here and the battery's voltage?

   _____ volts

12. Explain why there is a difference.

   _____

   _____

   _____

   _____

Problems Encountered

_____

_____

_____

Instructor's Comments

_____

_____

_____

☐ JOB SHEET 16–4

Testing Voltage Drop across Connectors

Name _____ Station _____ Date _____

Objective

Upon completion of this job sheet, you will have demonstrated the ability to test voltage drop across a connector. This task applies to ASE Education Foundation Tasks MLR/AST/MAST 6.A.3 (P-1).

Tools and Materials

A high-impedance DMM

Wiring diagrams

Description of Vehicle

Year _____ Make _____ Model _____

VIN _____ Engine Type and Size _____

Mileage _____

PROCEDURE

1. Circuit being tested: _____

2. Turn on the ignition to supply power to the circuit, if needed.

3. Set the DMM to the proper voltage scale.

    Scale selector _____

4. Connect the negative lead (the black lead) of the meter to a ground on the vehicle.

    Ground connection point _____

5. Use the positive lead (the red lead) to check the voltage at one side of the connector. _____

    Wire color or terminal number _____

6. What is the reading on the meter? _____ volts

7. Use the positive lead (the red lead) to check the voltage at the other side of the connector.

8. What is the reading on the meter? _____ volts

9. If there is a difference, subtract the lower voltage from the higher voltage.

10. Compare this difference with the specifications in the service manual to determine whether the difference is within specifications.

11. What could cause a reading that is greater than the specification?

    _____

12. What effect would an increased voltage drop have on the operation of the circuit? _____

Problems Encountered

_____

_____

_____

Instructor's Comments

_____

_____

_____

☐ JOB SHEET 16–5

Inspecting and Testing Fuses, Fusible Links, and Circuit Test Breakers

Name _____ Station _____ Date _____

Objective

Upon completion of this job sheet, you will have demonstrated the ability to inspect and test fuses, fusible links, and circuit breakers. This task applies to ASE Education Foundation Tasks MLR/AST/MAST 6.A.9 (P-1).

Tools and Materials

A high-impedance DMM

Test light

Description of Vehicle

Year _____ Make _____ Model _____

VIN _____ Engine Type and Size _____

Mileage _____

PROCEDURE

1. Locate a fuse block diagram.

2. Using a test light to test a fuse, ground the clip of the test light. Touch the probe to each end of the fuse. If the light comes on at both ends of the fuse, the fuse is good. If the light only works when touched to one end of the fuse, it is bad. Fuse test results the fuse _____

    Ground location _____

3. Using a test light to test a fusible link, ground the clip of the test light. Touch the probe to each end of the fusible link. If the light comes on at both ends of the fusible link, the fuse is good. If the light only comes on at one end of the fusible link, it is bad. Fuse link test results _____

    Ground location _____

4. Using a DMM to test a fuse or fusible link can be done in two different ways. Using the voltage scale, connect the black lead (the negative lead) to a ground. Touch the red lead (positive) lead to each end of the fuse or fusible link. If the voltage reading is the same at both ends, the fuse or fusible link is good. If you only get a voltage reading at one end, the fuse or fusible link is bad.

    Results _____

5. If you are using the ohmmeter to test a fuse or circuit breaker, the procedure is different. Remove the fuse or circuit breaker. Touch the leads from the meter to each end of the fuse or circuit breaker. If there is continuity, the fuse or circuit breaker is good. If there is no continuity, the fuse or circuit breaker is bad.

    Ohm meter results _____

Problems Encountered

_____

_____

_____

Instructor's Comments

_____

_____

_____

☐ JOB SHEET 16–6

Part Identification on a Wiring Diagram

Name _____ Station _____ Date _____

Objective

Upon completion of this job sheet, you will have demonstrated the ability to locate different parts on a wiring diagram. This task applies to ASE Education Foundation Tasks MLR 6.A.3 (P-1) and AST/MAST 6.A.7 (P-1).

Tools and Materials

A wiring diagram in a service manual (assigned by the instructor)

Service information used ___

Description of Vehicle

Year _____ Make _____ Model _____

VIN _____ Engine Type and Size _____

Mileage _____

PROCEDURE

1. Look over the wiring diagram, and then answer the following questions:

   a. Are circuit grounds clearly marked in the wiring diagrams? List the grounds shown.

   _____

   b. What is represented by lines that cross other lines?

   _____

   _____

   c. Are the switches shown in their normally open or normally closed position?

   _____

   d. Do all wires have a color code listed by them?

   _____

   e. Is the internal circuitry of all components shown in their schematical drawing?

   _____

2. **List the section or figure numbers** (the location) where the following electrical components are shown in the wiring diagram. Then **draw the schematical symbol** used by this wiring diagram to represent the part.

Component	Location	Drawing
Windshield wiper motor	_____	_____
Dome (courtesy) light	_____	_____
A/C compressor clutch	_____	_____
Turn signal flasher unit	_____	_____
Defroster grid	_____	_____
Fuel gauge sending unit	_____	_____

Problems Encountered

_____

_____

_____

Instructor's Comments

_____

_____

_____

REVIEW QUESTIONS

Review Chapter 16 of the textbook to answer these questions:

1. Describe the difference between voltage and current.

_____

_____

_____

2. Why should only high-impendance meters be used on modern vehicles?

_____

_____

_____

3. Explain how high resistance can affect circuit operation.

_____

_____

_____

4. Explain how a test light can be used to test for power and ground.

_____

_____

_____

5. Describe the procedure to perform a voltage drop test.

_____

_____

_____

_____

BATTERIES: THEORY, DIAGNOSIS, AND SERVICE

OVERVIEW

The battery supplies current to operate the starting motor during engine cranking, functions as a voltage stabilizer for the electrical system, and supplies current for the lights, radio, and other accessories when the engine is off.

A battery stores electrical energy for use by a device. Without batteries, our modern world would not have cell phones, laptop computers, flashlights, or countless other devices that depend on storage batteries for power.

 Concept Activity 17-1
Batteries

1. Look around your lab, and note some of the different battery-powered items and the types of batteries used in these devices.

_____

_____

_____

■

_____

All batteries supply power through a chemical reaction between two dissimilar plates that react with an electrolyte solution. In most automobile batteries, lead plates are submerged in a solution of sulfuric acid and water, known as battery acid. Battery acid is extremely corrosive and toxic; always follow these battery safety precautions:

■ Do not handle batteries with your bare hands; always use a battery carrier.

■ Always wear protective glasses or shields when working with batteries.

■ Never place tools or parts on top of a battery. If any conducting object contacts the battery terminals, the battery may explode.

■ When batteries are charging or discharging, they emit hydrogen gas, which can be highly explosive. Keep all sources of sparks and flames away from batteries.

■ Do not charge a battery with too high of a rate of charge. Overheated batteries are an explosion hazard.

■ Do not attempt to charge a frozen battery. If the battery acid has frozen, allow the battery to completely warm up before attempting to charge.

■ When disconnecting a battery, always disconnect the negative cable first. When reinstalling a battery, install the positive cable first.

■ Inspect the battery case for cracks and swelling. A battery that is cracked must be replaced. A battery that is swelled may have been overcharged or have plugged vents, and should be replaced.

If battery acid gets on your clothes, it will eat through the material. Battery acid can be neutralized with baking soda and water or with chemicals made for battery acid.

Concept Activity 17-2
Build a Battery

Batteries can be made from simple items, such as vegetables, wire, and nails. To construct an organic battery, connect three potatoes, lemons, or other fruits or vegetables with pieces of copper wire. Use a galvanized nail or similar item as one battery terminal and a penny or other piece of copper for the other battery terminal. Place the end of each terminal into the organic battery. Next, connect several batteries together, positive to negative, so that you are left with the two end batteries each having an unconnected lead **(Figure 17-1)**. Connect the voltmeter leads to the unattached leads at each end of the battery.

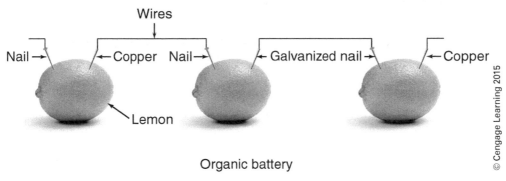

Organic battery

© Cengage Learning 2015

Figure 17-1   Building an organic battery.

1. How much voltage is produced by this battery?

_____

2. How could more voltage be produced?

_____

_____

_____

3. What would be the result if the batteries were connected in parallel instead of in series?

_____

_____

_____

4. Why are the dissimilar metals needed to make the battery function?

_____

_____

_____

5. What allows for the reaction between the metals and the vegetables or fruits?

_____

_____

_____

6. What acts as the electrolyte in this battery?

_____

_____

∎

Batteries can be constructed of many different types of materials and electrolytes. The composition of the plate materials and the type of chemicals used determine what a particular battery will be used for. Automobiles use one of four different types of lead acid batteries: the wet cell lead acid battery, the gel cell, the absorbed glass mat (AGM), and the valve-regulated lead acid battery (VRLA).

 Concept Activity 17-3
Battery Types

1. Examine the batteries used in lab vehicles. List the different types of batteries you find and what vehicles they are installed in.

_____

2. What are some reasons why a vehicle manufacturer may install one type of battery in one vehicle and a different type of battery in another vehicle?

_____

_____

∎

Most automotive batteries are the lead acid type. This means the primary plate component is lead, and the electrolyte is an acid, specifically a combination of sulfuric acid and water. Some low-maintenance batteries have removable caps that allow a technician to check the acid level and the batteries' condition, using a battery hydrometer.

A hydrometer tests the specific gravity, or density, of the battery acid. Pure water has a rating of 1.0. Motor oil, which can easily be seen floating on top of water, typically has a specific gravity of around 0.9. A number lower than 1.0 indicates that oil is less dense than water. Because it is less dense, it will float on the water. Sulfuric acid has a specific gravity of approximately 1.83, nearly twice the density of water.

In a fully charged lead acid battery, the electrolyte consists of approximately 64% water and 36% acid. When the battery is fully charged, the specific gravity will be 1.270. As a battery discharges, the sulfuric acid combines with the lead in the cell plates, decreasing the specific gravity of the electrolyte. When a battery is charging, the sulfuric acid leaves the cell plates, returning to the electrolyte and once again increasing the specific gravity.

See Chapter 14, Concept Activity 14-2, for specific gravity activities.

Batteries are rated by several methods, including the Battery Council International (BCI) group number, amp-hour rating, watt-hour rating, reserve capacity, cranking amps, and cold-cranking amps. A battery's physical size and post arrangement determine its BCI group rating. Both the amp-hour and watt-hour ratings are a measure of a battery's ability to put out current at a certain voltage over an amount of time. Reserve capacity is determined by a battery's ability to maintain a constant current output over an amount of time, without a functioning charging system. Job Sheet 17-1 addresses how to locate battery rating information.

For those who live in areas where cold weather is an issue, the cold-cranking amps rating is very important. The cranking amps (CA) and cold-cranking amps (CCA) ratings both apply to a battery's ability to produce current to crank an engine. However, the CCA rating is measured at 0°F, compared to the CA rating at 32°F. The difference between CA and CCA can be the difference between an engine that starts and an engine that fails to crank over. A typical automotive lead acid battery can lose up to 30% of its cranking capacity in cold temperatures.

Concept Activity 17-4
Batteries and Temperature

Obtain two flashlights that use the same size, type, and number of batteries, such as two AA or AAA powered lights. Note the time and turn on both lights. Place one light in a freezer and leave the other at room temperature. Check the lights every half-hour, and note their light output. Continue periodic checking until one light is out.

1. Length of time until first light became dim:

_____

2. Length of time until first light went out:

_____

3. Which light became dim or went out first?

_____

4. If the light in the freezer went out first, how do you think the temperature affected the light?

_____

_____

_____

5. If you could repeat this activity with insulation on the flashlight placed in the freezer, what effect do you think the insulation would have on the results?

_____

_____

_____

_____

Automotive batteries, though much larger than the organic battery discussed earlier, operate under the same principles of chemical reactions to produce current. Batteries are affected not only by cold temperatures but also by high temperatures. Many automotive batteries are no longer located in the engine compartment. This helps protect the battery from the high temperatures present during hot weather.

The following job sheets will prepare you for battery diagnosis and service.

☐ JOB SHEET 17–1

Determining Battery Type and Ratings

Name _____ Station _____ Date _____

Objective

At the completion of this job sheet, you will be able to determine battery type and ratings. This task applies to ASE Education Foundation Tasks MLR/AST/MAST 6.B.2 (P-1).

Description of Vehicle

Year _____ Make _____ Model _____

VIN _____ Engine Type and Size _____

Mileage _____

PROCEDURE

Using lab vehicles, determine the battery BCI group number, CA, CCA, reserve capacity, and amp-hour or watt-hour ratings, as able.

1. Vehicle 1: Year _____ Make _____ Model _____

   Battery Manufacturer _____ BCI group _____

   CA _____ CCA _____ Amp-hour/Watt-hour _____

   Reserve Capacity _____

   Vehicle 2: Year _____ Make _____ Model _____

   Battery Manufacturer _____ BCI group _____

   CA _____ CCA _____ Amp-hour/Watt-hour _____

   Reserve Capacity _____

   Vehicle 3: Year _____ Make _____ Model _____

   Battery Manufacturer _____ BCI group _____

   CA _____ CCA _____ Amp-hour/Watt-hour _____

   Reserve Capacity _____

2. Explain why different battery types are used in these vehicles.

   _____

   _____

   _____

3. Describe what factors you think are involved in determining battery requirements for a vehicle.

_____

_____

_____

Problems Encountered

_____

_____

_____

Instructor's Comments

_____

_____

_____

☐ JOB SHEET 17–2

Removing, Cleaning, and Replacing a Battery

Name _____ Station _____ Date _____

Objective

Upon completion of this job sheet, you will have demonstrated the ability to properly and safely remove, clean, and replace a battery. This task applies to ASE Education Foundation Tasks MLR/AST/MAST 6.B.4 (P-1).

Tools and Materials

Baking soda	Battery strap or carrier
Battery clamp puller	Conventional wire brush and rags
Battery cleaning wire brush	Fender covers

Description of Vehicle

Year _____ Make _____ Model _____

VIN _____ Engine Type and Size _____

Mileage _____

PROCEDURE

1. Place a fender cover around the work area.

2. Install a memory saving device. List systems and/or components for which a memory saving device would be necessary.

   _____

   _____

   _____

3. Remove the negative (–) battery terminal. **(Figure 17-2A)**. Always grip the cable while loosening the nut to prevent unnecessary pressure on the terminal post that could break it or loosen its mounting in the battery. If the connector does not lift easily off the terminal when loosened, use a clamp puller to free it **(Figure 17-2B)**. Prying with a screwdriver or bar strains the terminal post.

4. Loosen the positive (+) battery cable, and remove it from the battery with a battery terminal puller. If both battery cables are the same color, it is wise to mark the positive cable so that you will connect it to the correct terminal later.

5. Loosen and remove the battery hold-down straps, cover, and heat shield **(Figure 17-2C)**. Note one condition of the hold-down components, cover, and heat shield.

   _____

6. Lift the battery from the battery tray, using a battery strap or carrier **(Figure 17-2D)**.

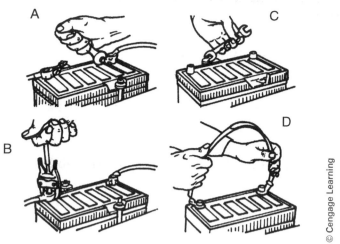

© Cengage Learning

Figure 17-2   Removing a battery. A. Disconnecting the negative terminal. B. Using a terminal clamp puller. C. Removing the holddown and cover. D. Lifting the battery out with a carrying strap.

7. Inspect the battery tray, and clean it using a solution of baking soda and water. Note battery tray condition.

_____

8. Clean the battery cable terminals with baking soda solution and a battery terminal brush. Use the external portion to clean the post **(Figure 17-3A)** and internal portion for the terminal ends **(Figure 17-3B)**. Note terminal and cable condition.

_____

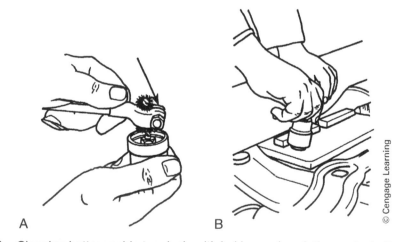

© Cengage Learning

Figure 17-3   Cleaning battery cable terminals with baking soda solution and a battery terminal brush. A. Cleaning the inside of the battery terminal. B. Cleaning the battery post.

9. Using a battery strap or carrier, install the replacement battery (new or recharged) in the battery tray.

10. Install the battery cover or hold-down straps, and tighten their attaching nuts and bolts. Be certain the battery cannot move or bounce, but do not over-tighten.

11. If so equipped, reinstall heat shield.

12. Beginning with the positive cable, both terminal connectors. Do not over-tighten because this could damage the post or connectors.

Instructor's Check _____

13. Test the installation by starting the engine.

Problems Encountered

_____

_____

_____

Instructor's Comments

_____

_____

_____

☐ JOB SHEET 17–3

Inspecting a Battery for Condition, State of Charge, and Capacity

Name _____ Station _____ Date _____

Objective

Upon completion of this job sheet, you will have demonstrated the ability to inspect and test a battery. This task applies to ASE Education Foundation Tasks MLR/AST/MAST 6.B.1 (P-1) and 6.B.2 (P-1).

Tools and Materials

A vehicle with a 12-volt battery

A DMM

Starting/charging system tester (VAT-40 or similar)

Description of Vehicle

Year _____ Make _____ Model _____

VIN _____ Engine Type and Size _____

Mileage _____

PROCEDURE

1. Describe the general appearance of the battery.

   _____

   _____

   _____

   _____

2. Describe the general appearance of the cables and terminals.

   _____

   _____

   _____

   _____

3. Check the tightness of the cables at both ends. Describe their condition.

   _____

   _____

4. Connect the positive lead of the meter (set on DC volts) to the positive terminal of the battery.

5. Put the negative lead on the battery case, and move it all around the top and sides of the case. What readings do you get on the voltmeter?

_____

_____

6. What is indicated by the readings?

_____

_____

7. Measure the voltage of the battery. Your reading was _____ volts.

8. What do you know about the condition and the state of charge of the battery based on the visual inspection and the above tests?

_____

_____

_____

_____

9. Connect the starting/charging system tester to the battery.

10. Locate the CCA rating of the battery. What is the rating? _____

11. Based on the CCA, how much load should be put on the battery during the capacity test? _____ amps

12. Conduct a battery load test. Battery voltage decreased to _____ volts after _____ seconds.

13. Describe the results of the battery load (capacity) test. Include in the results your service recommendations and the reasons for them.

_____

_____

_____

_____

_____

Problems Encountered

_____

_____

_____

Instructor's Comments

_____

_____

_____

☐ JOB SHEET 17–4

Charging a Maintenance-Free Battery

Name _____ Station _____ Date _____

Objective

Upon completion of this job sheet, you will have demonstrated the ability to charge a maintenance-free battery. This task applies to ASE Education Foundation Tasks MLR/AST/MAST 6.B.5 (P-1).

Tools and Materials

Baking soda

Battery charger and cables and adapters
   for side terminal batteries

Battery clamp puller

Battery cleaning wire brush

Battery strap or carrier

Conventional wire brush and rags

Fender covers

Voltmeter

Description of Vehicle

Year _____ Make _____ Model _____

VIN _____ Engine Type and Size _____

Mileage _____

PROCEDURE

CAUTION: *Special care must be given to charging maintenance-free batteries and AGM batteries. Always follow the batteries manufacturer's service procedure.*

1. Place fender covers around the work area. Remove battery from vehicle as described in Job Sheet 17-1. It is also possible to charge a battery in the vehicle. If performing an in-vehicle charge, consult the service manual for any precautions associated with computer controls.

2. Check that the charger is turned off. Connect the positive (+) cable from the charger to the positive (+) battery terminal. Connect the (–) cable from the charger to the negative (–) battery terminal. Be sure you have a good connection to prevent sparking.

3. If at no-load the battery reads below 12.2 volts, charge the battery according to the table below.

Battery voltage reading _____

Battery Charging Table

Reserve capacity rating	20 amp-hour rating	5 amps	10 amps	20 amps	30 amps	40 amps
80 minutes or less	50 amp-hours or less	10 hours	5 hours	2.5 hours	2 hours	
80–125 minutes	50–75 amp-hours	15 hours	4.5 hours	3.25 hours	2.5 hours	2 hours
125–170 minutes	75–100 amp-hours	20 hours	10 hours	5 hours	3 hours	2.5 hours
170–250 minutes	100–150 amp-hours	30 hours	15 hours	7.5 hours	5 hours	3 hours

If the voltage of the battery at room temperature is 12.2 volts, charge the battery for half the time shown under "5 amps." If the voltage is 12.4 volts, charge the battery for one-fourth the "5 amps" time.

Recommended charge time _____

Turn the clock control on the charger to the desired charging time. Do not exceed the manufacturer's battery-charging limits, which generally appear on the battery.

Charge the battery to a voltage of at least 12.6 volts or until the green ball appears. Shaking or tipping the battery may be necessary to make the green ball appear. Never overcharge a battery.

WARNING: *Never smoke around a charging battery. The hydrogen gas produced is highly explosive.*

4. When the battery is fully charged, turn off the power switch, and disconnect the two battery charger cables from the battery.

Battery voltage reading _____

Problems Encountered

_____

_____

_____

Instructor's Comments

_____

_____

_____

REVIEW QUESTIONS

Review Chapter 17 of the textbook to answer these questions:

1. Define a parasitic load and how to test for it.

   _____

   _____

   _____

   _____

   _____

2. Batteries emit _____ gas when charging.

3. When removing and installing a battery, remove the _____ cable and then the _____ cable.

   When installing the battery, install the _____ cable first, then the _____ cable.

4. List five things that should be checked during a battery inspection.

   _____

   _____

   _____

   _____

   _____

5. Battery efficiency can decrease/increase by _____ percent in cold weather.

STARTING AND MOTOR SYSTEMS

OVERVIEW

The starting system typically spends less time in operation than any other system on the vehicle, but it is certainly one of the most important systems for vehicle operation. Starting motors, first introduced on Cadillac vehicles in 1912, have not changed in operation. Though modern starting motors are smaller and more powerful, the operating principles remain the same.

Modern vehicles use a wide variety of DC motors. A typical DC motor consists of a rotating armature, electromagnetic field coils or permanent magnets, and a set of brushes. Power windows, power door locks, moon roofs, and radiator and interior climate control fans are all examples of the DC brush motor. These motors rely on the attraction and repulsion of north and south magnetic fields **(Figure 18-1)**.

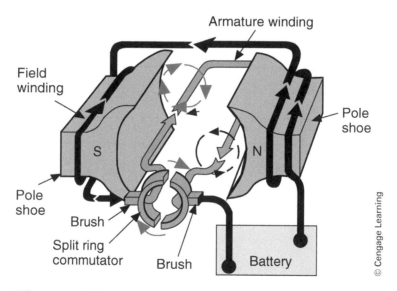

Figure 18-1 A simple DC motor.

Concept Activity 18-1
Magnetic Fields

Obtain several magnets from your instructor. Round magnets with a hole in the center are preferred; transmission oil pan magnets work well. Place a pencil through the magnets. Arrange the magnets so they are repulsing each other.

1. How many magnets can you place over the pencil without the magnets touching each other?

_____

2. Approximately how far apart do the magnetic fields keep each magnet separated?

_____

Next, take two or three of the magnets from the pencil and place them on a wooden (nonmagnetic) bench top. Arrange the magnets so that they are near each other, but not in contact. Use the magnet's repulsion to move a second magnet.

3. How close do the magnets have to be for the magnetic repulsion to cause movement?

_____

Join two magnets together and attempt to move a third magnet.

4. How did using two magnets affect the distance necessary to move the third?

_____

Now, move two magnets close together so that they attract each other. Be careful not to pinch your fingers.

5. How close can the magnets get before the magnetic attraction brings them together?

_____

6. How does this distance compare to the distance for magnetic repulsion?

_____

Place a magnet upright on its edge. Next, try to position two additional magnets near the first magnet and attempt to make the upright magnet rotate or move.

7. Describe what happens when trying to get the magnet to rotate.

_____

_____

_____

8. What do you think would be necessary to make one magnet rotate in response to two other magnets placed in close proximity?

_____

_____

_____

9. Describe your conclusions about magnetic attraction and repulsion.

_____

_____

_____

_____

■

As you have discovered in the previous activity, it requires more than just a magnetic field to make a spinning motor. DC brush motors require that the magnetic fields constantly interact in attraction/repulsion. A switch, called a commutator, accomplishes this constant change in polarity. The commutator is a section of the armature. Carbon brushes carry current to the armature via the commutator. A magnetic field is generated when current is supplied to the windings of the armature; this field is either attracted or repulsed by the field coils or permanent magnets, causing the armature to spin. Once the armature rotates, different commutator segments will align with the brushes, allowing another winding to become attracted or repulsed, continuing the rotation. Without the switching of armature windings, the armature would stop spinning once the north and south magnetic fields aligned.

Concept Activity 18-2
Toy Motor Building

Obtain from your instructor two transmission pan–type magnets, a two- to three-foot length of insulated single-strand wire like that used for armature windings on a fan motor, paper clips or cotter pins, a Styrofoam cup or small box, and a 9-volt battery. Follow these steps to build a simple DC electric motor:

■ Strip the ends of the wire to remove about one-half of the insulation along the length of the last 2 inches of the wire. Each end should have bare copper exposed around half of the wire's circumference and insulation around the other half. This will allow an on–off connection through the wire.

■ Form the wire (coil) into a tightly wound loop approximately one inch in diameter, leaving the two ends protruding out from the middle **(Figure 18-2)**.

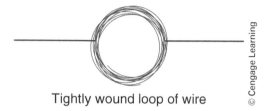

Tightly wound loop of wire

Figure 18-2   A tightly wound loop of wire.

■ Mount the paper clips or cotter pins into the cup or box so that they can form a holding fixture for the wire.

■ Place the magnets so that they hold each other to the top section of the cup or box between the paper clips or cotter pins.

■ Place the wire into the paper clips or cotter pins. Adjust the height so that the coil can rotate very close to the magnet without touching.

■ Attach the 9-volt battery so that the positive is connected to one paper clip or cotter pin, and the negative is connected to the other paper clip or cotter pin **(Figure 18-3)**.

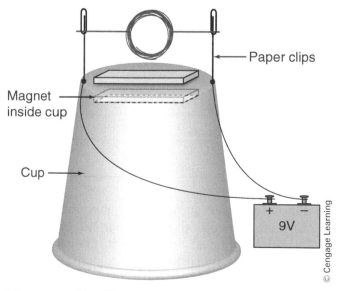

Paper clips

Magnet
inside cup

Cup

9V

Figure 18-3   A simple DC motor.

■ Some adjusting may be required, but once a good connection is made, the coil should easily spin on its own.

1. Note the coil's direction of rotation.

2. Reverse the battery polarity and note the motor rotation.

3. Why did the motor respond as it did to the battery voltage change in polarity?

_____

_____

_____

_____

4. List the electrical motors used in modern automobiles that operate so that they reverse their direction of rotation.

5. List the electrical motors used in modern automobiles that rotate in only one direction.

_____

6. What would be the result if a motor, such as for the cooling fan that only rotates one way, were hooked to the battery in reverse polarity?

_____

7. Match the motor components with their electrical function.

Paper clip or cotter pin	Armature
Wire	Brush
Magnets	Commutator
Wire insulation	Field coils

8. Explain why the wire's insulation is not completely removed.

_____

_____

_____

9. What effect would using a smaller battery, such as a D cell, have on the motor, and why?

_____

_____

_____

10. What effect would using more magnets have on the motor, and why?

_____

_____

_____

11. Why do you think the electric motors used in automobiles have more commutator segments than the simple motor you built?

_____

_____

_____

12. What are the wear points on a DC motor?

_____

_____

_____

13. What are some examples of problems that can affect the operation of a DC motor?

_____

_____

_____

_____ ■

As you have learned, basic DC motor operation is fairly simple to understand and demonstrate. Hybrid vehicles use AC motor/generators to both drive the wheels and recharge the high-voltage battery pack. The AC motor/generators used in hybrids require expensive and complex electronic controls to operate.

The following job sheets will prepare you for electrical motor diagnosis and service.

☐ JOB SHEET 18–1

Testing Starter Current Draw

Name _____ Station _____ Date _____

Objective

Upon completion of this job sheet, you will have demonstrated the ability to test the starter current draw. This task applies to ASE Education Foundation Task MLR/AST/MAST 6.C.1 (P-1).

Tools and Materials

VAT-40 or equivalent current tester   Remote starter switch

Description of Vehicle

Year _____ Make _____ Model _____

VIN _____ Engine Type and Size _____

Mileage _____

PROCEDURE

1. Locate and record the starter current draw specifications.

   _____

   _____

   _____

2. Connect the tester leads to the battery terminals and the inductive clamp around a battery cable. Note that the arrow on the clamp indicates the direction of current flow.

3. Disable the fuel and ignition system and record how this was accomplished.

   _____

   _____

   _____

4. Connect the remote starter switch to the starter control circuit connection.

5. Crank the engine for 5 seconds, and record the initial peak current draw and the draw after 5 seconds.

   Peak current _____   Steady current _____

6. Compare your readings to the specifications. Are the readings acceptable? _____

7. Why is the starter draw initially higher than after turning for several seconds?

   _____

   _____

   _____

   _____

8. List two possible causes for higher than normal current draw.

_____

_____

_____

_____

9. What could be a cause of lower than normal current draw?

_____

_____

_____

_____

Problems Encountered

_____

_____

_____

Instructor's Comments

_____

_____

_____

☐ JOB SHEET 18–2

Starter Circuit Voltage Drop Testing

Name _____ Station _____ Date _____

Objective

Upon completion of this job sheet, you will have demonstrated the ability to perform a circuit voltage drop test. This task applies to ASE Education Foundation Tasks MLR/AST/MAST 6.C.2 (P-1).

Tools and Materials

DMM   Remote starter switch

Description of Vehicle

Year _____ Make _____ Model _____

VIN _____ Engine Type and Size _____

Mileage _____

PROCEDURE

1. Print and attach or copy the starter circuit testing procedures and specifications.

2. Connect the DMM to the battery, and record the battery open circuit voltage. _____

3. Disable the fuel and ignition system. Record how this was performed.

   _____

   _____

   _____

4. Locate the battery connection from the starter solenoid to the starter. Place the negative DMM lead at that point. Place the positive DMM lead on the battery positive terminal.

5. Use the remote starter switch to crank the engine for 5 seconds, and note the voltage reading on the DMM. _____

6. Insulated power circuit voltage drop specification. _____ If no specification is available, up to 0.5-volt voltage drop is considered good. Is the amount measured within the manufacturer's specification or under 0.5 volt? _____

7. What could cause excessive voltage drop on this portion of the starter circuit? _____

8. How can excessive voltage drop on the insulated power circuit affect the starter's operation?

   _____

9. Connect the negative DMM lead to the starter case. Connect the positive DMM lead to the battery negative terminal.

10. Use the remote starter switch to crank the engine for 5 seconds, and record the voltage reading on the DMM.

   _____

11. The insulated ground circuit may have up to 0.5-volt voltage drop. Is the amount measured within the manufacturer's specification or under 0.5 volt? _____

12. What could cause excessive voltage drop on this portion of the starter circuit?

   _____

   _____

   _____

13. How can excessive voltage drop on the insulated ground circuit affect the starter's operation?

   _____

   _____

   _____

Problems Encountered

_____

_____

_____

Instructor's Comments

_____

_____

_____

☐ JOB SHEET 18–3

Testing the Starter Control Circuit

Name _____ Station _____ Date _____

Objective

Upon completion of this job sheet, you will have demonstrated the ability to test voltage drops and switches, connectors, and wires of the starter control circuit. This task applies to ASE Education Foundation Tasks MLR/AST/MAST 6.C.3 (P-1) and 6.C.5 (P-2).

Tools and Materials

DMM   Remote starter switch

Description of Vehicle

Year _____ Make _____ Model _____

VIN _____ Engine Type and Size _____

Mileage _____

PROCEDURE

1. Locate starter circuit testing procedures and specifications, and attach to this job sheet.

2. Connect the DMM to the battery, and record the battery open circuit voltage. _____

3. Disable the fuel and ignition system. Record how this was performed.

   _____

   _____

   _____

4. Locate the control circuit connection at the starter solenoid. Place the DMM negative lead on this terminal. Place the DMM positive lead on the battery positive terminal.

5. Use the remote starter switch to crank the engine for 5 seconds, and note the voltage reading on the DMM._____

6. Compare the voltage drop to specifications. If no specification is available, generally the voltage drop should be less than 0.5 volt.

7. If the voltage drop is higher than specifications, refer to a wiring diagram for the starter circuit.

8. Test between the following circuit points, and note the voltage drops.

   a. Output of safety switch to starter solenoid _____

   b. Input of safety switch to output of safety switch _____

   c. Output of ignition switch to input of safety switch _____

    d. Input of ignition switch to output of safety switch _____

    e. Battery positive to input of ignition switch _____

9. A voltage drop of more than 0.1 volt across any of these points usually indicates a problem.

Problems Encountered

_____

_____

_____

Instructor's Comments

_____

_____

_____

REVIEW QUESTIONS

Review Chapter 18 of the textbook to answer these questions:

1. List and define four components of the starting system.

_____

_____

_____

_____

2. Compare and contrast AC and DC motors and their operation.

_____

_____

_____

_____

3. List five components of a typical starter motor.

_____

_____

_____

_____

4. Describe three starting circuit voltage drop tests.

_____

_____

_____

_____

5. Describe two causes of higher than normal starter current draw.

_____

_____

CHARGING SYSTEMS

OVERVIEW

Once the battery and starting system get the engine started, the charging system must recharge the battery and supply power for all of the necessary electrical systems and accessories.

Modern AC generators (alternators) are electromechanical devices that convert mechanical energy into electrical energy. The engine's crankshaft drives the generator via a drive belt. Inside the generator, rotating magnetic fields cut across stationary conductors, inducing AC current into the conductors. The AC current is converted into DC current by diodes. Diodes are electrical one-way check valves, allowing current to flow in only one direction. Limiting the strength of the magnetic field inside the generator controls its output. Either a voltage regulator or the engine's computer accomplishes the precise control of the current flow generating the field.

Induction is the term for producing voltage with a magnetic field. When a conductor moves through a magnetic field perpendicular to the magnetic lines of flux, voltage is induced into the conductor.

Concept Activity 19-1
Inducing Voltage

Obtain a bar or cylindrical magnet and a length of wire from your instructor. Attach the leads of a DMM to the ends of the wire. Set the DMM to read AC voltage. Next, move the magnet so that the lines of force from the north to south poles cut across the wire. Move the magnet quickly back and forth as close to the wire as you can.

1. How much voltage did the DMM read?

   _____

2. Move the magnet across the wire more slowly than before. How much voltage was generated?

   _____

3. Move the magnet so that the north and south poles move parallel with the wire. How much voltage was generated?

   _____

   Next, obtain two permanent magnet–type wheel-speed sensors or crank/cam position sensors. Attach the leads of the DMM to one of the sensors. Next, move the sensing tip of the second sensor quickly back and forth very close to the sensing tip of the first sensor.

4. How much voltage did the DMM read?

   _____

5. Was the voltage generated by the sensors greater than the voltage generated by the single magnet and piece of wire?

   _____

6. If the voltage was greater, why do you think that was so?

_____

_____

_____

7. What conclusions can you make about induced voltage?

_____

_____

_____

_____

■

The generator uses a rotor to generate a magnetic field. The rotor is comprised of the driveshaft, two pole pieces, a winding of wire within the pole pieces, and two slip rings. The front of the drive shaft has the drive pulley and a fan for air circulation. The pole pieces are easily magnetized and increase the strength of the magnetic field produced by the rotor winding **(Figure 19-1)**. The winding is a coil of heavy gauge wire sandwiched between the pole pieces. The slip rings supply a connection for the coil to the brushes **(Figure 19-2)**.

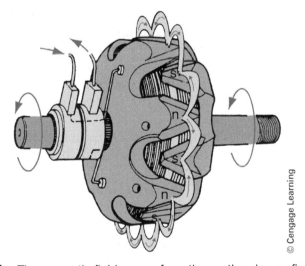

© Cengage Learning

Figure 19-1 The magnetic field moves from the north poles, or fingers, to the south poles.

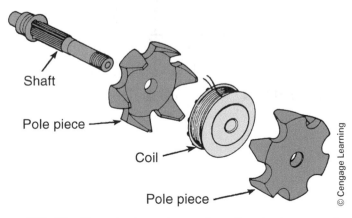

Shaft

Pole piece

Coil

Pole piece

© Cengage Learning

Figure 19-2 The rotor is made up of a coil, pole pieces, and a shaft.

When the generator is spinning and current is supplied to the rotor, magnetic lines of force spin and cut across a set of stationary conductor windings called the stator. The stator is made of three sets of wire rigidly mounted in housing that allows the rotor to spin within the stator **(Figure 19-3)**. The clearance between the rotor and stator is very small. This is because the strength of a magnetic field decreases with the square of the distance. This means that if the distance from a magnetic field is doubled, the strength is reduced by four times. The rotor and stator must be close together to achieve the greatest efficiency possible. Bearings at the front and rear of the drive shaft keep the rotor centered in the stator and allow it to spin two to three times of crankshaft speed.

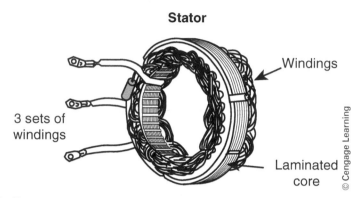

Figure 19-3 Generators use a stator, or stationary set of windings, to generate voltage.

A voltage regulator controls the current to the rotor. Some voltage regulators control the current supplied on the positive side of the circuit, whereas others control the ground side of the resistance. Regardless of style, the voltage regulator is responsible for maintaining sufficient generator output to power the electrical system and keep the voltage level from rising too high. If the generator voltage output is too high, damage to electronic components may result. The powertrain control module, or PCM, controls charging system output on many systems. The PCM can turn the generator on and off very quickly, thereby controlling the output.

Because the battery and electrical system operate on DC current and the generator produces AC current, the AC must be converted into DC; this is the function of diodes. A diode acts as an electrical one-way valve. By placing three sets of two diodes after the stator windings, portions of the AC current are blocked from passing, thus converting the AC into DC **(Figure 19-4)**.

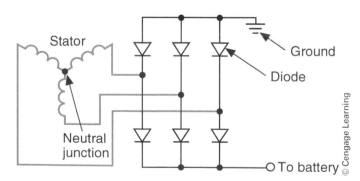

Figure 19-4 A pair of diodes is used for each stator winding.

Concept Activity 19-2
Generator Components

Match the following generator components with their function.

Diodes	Generates magnetic field
Coil	Supplies power to the coil
Fan	Converts AC into DC
Voltage regulator	Has induced AC current
Stator	Remove heat
Slip rings	Connection for brushes

The following job sheets will prepare you for charging system diagnosis and service.

☐ JOB SHEET 19–1

Visually Inspecting the Charging System

Name _____ Station _____ Date _____

Objective

Upon completion of this job sheet, you will have demonstrated the ability to visually inspect the components of a charging system. This task applies to ASE Education Foundation Tasks MLR 6.D.1 (P-1) and AST/MAST 6.D.3 (P-1).

Tools and Materials

Belt tension gauge

Description of Vehicle

Year _____ Make _____ Model _____

VIN _____ Engine Type and Size _____

Mileage _____

PROCEDURE

1. Inspect the condition of the drive belt, and note your observations.

    _____

    _____

2. Look up the drive belt tension specification in the service manual, and note it here.

    _____

    _____

3. Using a belt tension gauge, measure the tension of the generator's drive belt, and describe your findings.

    _____

    _____

4. Visually inspect the battery and battery cables, and describe their condition.

    _____

    _____

5. Visually inspect the wires connected to the generator and regulator, and describe their condition.

    _____

    _____

6. Start the engine and listen for noisy belts, bad generator bearings, and bad diodes (a whirring sound). Describe what you hear.

    _____

    _____

Problems Encountered

_____

_____

_____

Instructor's Comments

_____

_____

_____

☐ JOB SHEET 19–2

Inspecting, Replacing, and Adjusting Drive Belts and Pulleys

Name _____ Station _____ Date _____

Objective

Upon completion of this job sheet, you will have demonstrated the ability to inspect, replace, and adjust drive belts and pulleys. This task applies to ASE Education Foundation Tasks MLR 6.D.2 (P-2) and AST/MAST 6.D.3 (P-1).

Tools and Materials

Two vehicles, one with a serpentine belt and the other with V-belts

Belt tension gauge

Description of Vehicle

Year _____ Make _____ Model _____

VIN _____ Engine Type and Size _____

Mileage _____

PROCEDURE

1. On the vehicle with a serpentine belt, carefully inspect the belt, and describe its general condition.

   _____

   _____

   _____

2. With the proper belt tension gauge, check the tension of the belt. Belt tension should be _____. You found _____. Based on these figures, what are your recommendations?

   _____

   _____

3. Describe the procedure for adjusting the tension of the belt.

   _____

   _____

4. Referring to the service manual, describe the procedure for removing the belt.

   _____

   _____

5. Draw the proper routing for installing a new drive belt.

6. On the vehicle with V-belts, you will find more than one drive belt. List the different belts by their purpose.

_____

_____

_____

_____

7. Carefully inspect the belts, and describe the general condition of each one.

_____

_____

_____

_____

8. Check the tension of the AC generator drive belt. Belt tension should be _____.
   You found _____. Based on these figures, what are your recommendations?

_____

_____

9. Describe the procedure for adjusting the tension of the belt.

_____

_____

_____

_____

10. Describe the procedure for removing an AC generator drive belt.

_____

_____

_____

Problems Encountered

_____

_____

_____

Instructor's Comments

_____

_____

_____

☐ JOB SHEET 19–3

Removing and Replacing an AC Generator

Name _____ Station _____ Date _____

Objective

Upon completion of this job sheet, you will have demonstrated the ability to remove a generator from a vehicle and install one on the vehicle. This task applies to ASE Education Foundation Tasks MLR 6.D.3 (P-2) and AST/MAST 6.D.4 (P-1).

Tools and Materials

Fender cover

Description of Vehicle

Year _____ Make _____ Model _____

VIN _____ Engine Type and Size _____

Mileage _____

PROCEDURE

1. The procedure for removing the generator is as follows: Place fender covers on the fenders. Disconnect the negative (–) battery cable first and then the positive (+) battery cable at the battery.

    WARNING: *Never attempt to remove the generator without isolating the battery first.*

2. Disconnect the wiring leads from the generator.

    Note the condition of the connections. _____

3. Following the appropriate procedures, loosen the adjusting bolts and move the generator to provide sufficient slack to remove the generator drive belt.

4. Remove the bolts that retain the generator.

5. Remove the generator from the vehicle.

    Note the output rating: _____ amps _____volts.

PROCEDURE (INSTALLING GENERATOR)

1. Install the generator onto its mounting bracket with bolts, washers, and nuts. Do not completely tighten.

2. Install the generator drive belt, and tighten to the proper belt tension.

    Specification _____

3. Tighten all bolts to the specified torque.

    Specification _____

4. Install the generator terminal plug and battery leads to the unit.

5. Connect the negative battery cable.

6. Test the generator for specified output.

   Specification _____

   Actual _____

7. If the output is okay, remove the fender covers; if it is not okay, diagnose the charging system.

Problems Encountered

_____

_____

_____

Instructor's Comments

_____

_____

_____

☐ JOB SHEET 19–4

Testing the Charging System

Name _____ Station _____ Date _____

Objective

Upon completion of this job sheet, you will have demonstrated the ability to test the operation of the charging system. This task applies to ASE Education Foundation Tasks MLR 6.D.1 (P-1), 6.D.4 (P-2) AST/MAST 6.D.1 (P-1), 6.D.2 (P-1), and 6.D.5 (P-1).

Tools and Materials

DMM

Starting/charging system tester (VAT-40 or similar)

Description of Vehicle

Year _____ Make _____ Model _____

VIN _____ Engine Type and Size _____

Mileage _____

PROCEDURE

1. Describe the general appearance of the AC generator and the wires that are attached to it.

   _____

   _____

   _____

2. Measure the open circuit voltage of the battery. Your measurement was _____ volts.

3. From the wiring diagram, identify the output, input, and ground wires for the AC generator. Describe these wires by color and location.

   _____

   _____

   _____

4. Start the engine, and allow it to run. Then turn on the headlights in the high-beam mode.

5. Connect the DMM across the charging system's output wire. Measure the voltage drop, and record your readings.

   _____

6. Connect the DMM across the charging system's input wire. Measure the voltage drop, and record your readings.

   _____

7. Connect the DMM across the charging system's ground, normally the generator's case. Measure the voltage drop and record your readings.

   _____

8. What is indicated by the results of the preceding tests?

   _____

   _____

   _____

   _____

9. Identify and record the type and model of AC generator.

   _____

   What are the output specifications for this AC generator? _____ amps and _____ volts at _____ rpm

10. Connect the starting/charging system tester to the vehicle.

11. Start the engine, and run it at the specified engine speed.

12. Observe the output to the battery. The meter readings are _____ amps and _____ volts.

13. Compare readings to specifications, and give recommendations.

    _____

    _____

    _____

14. If readings are outside of the specifications, refer to the service manual to find the proper way to full field the AC generator. Describe the method.

    _____

    _____

    _____

15. If possible, full field the generator, and observe the output to the battery. The meter readings are _____ amps and _____ volts.

    WARNING: *Do not leave the generator at full field output for more than 10 seconds.*

16. Compare readings to specifications and give recommendations:

    _____

    _____

    _____

    _____

Problems Encountered

_____

_____

_____

Instructor's Comments

_____

_____

_____

☐ JOB SHEET 19–5

Testing the Generator Control Circuit

Name _____ Station _____ Date _____

Objective

Upon the completion of this job sheet, you will have demonstrated the ability to obtain and interpret generator output control waveforms using a DSO/GMM. This task applies to ASE Education Foundation Tasks MLR/AST/MAST 6.D.1 (P-1) and AST/MAST 6.D.2 (P-1).

Tools and Materials
DSO/GMM

Description of Vehicle

Year _____ Make _____ Model _____

VIN _____ Engine Type and Size _____

Mileage _____

PROCEDURE

1. Locate a wiring diagram of the charging system of the vehicle being tested.

2. Install a scan tool, and determine whether the tool and vehicle provides for bidirectional control of the generator. _____ YES _____ NO

   Scan tool used _____

3. Attach the DSO/GMM positive lead to the generator control wire and ground the negative lead. Color of control circuit wire _____

   If a two-channel DSO/GMM is being used, attach the positive lead of channel 2 to the battery positive terminal. Set the DSO/GMM to display both channels at the same time.

4. Start the engine, and allow it to idle. Set the DSO/GMM to display a DC voltage waveform. The PCM control circuit is typically a square-wave pattern. Draw the waveform displayed on the DSO/GMM, and note amount of generator on time. Beside this draw or note the battery voltage from channel 2, and compare the two patterns.

5. Turn on the headlights, and note the waveform. How did it change?

   _____

   _____

   _____

6. Turn the headlights off, and using the scan tool, command the generator to its maximum output. Draw the waveform pattern shown when the generator is commanded to full output. Draw or note the battery voltage waveform from channel 2, and compare the two.

   WARNING: *Do not leave the generator at full output for more than 10 seconds.*

7. How did the waveform at full output differ from the previous waveforms?

   _____

   _____

   _____

8. How did the output waveform respond in relation to the battery voltage?

   _____

   _____

   _____

9. Are there any operating conditions in which the PCM may command the generator off?

   _____

   _____

   _____

10. List two possible charging system problems that could cause a low-charge condition even if the PCM is commanding high output.

    _____

    _____

    _____

11. What could cause the generator output to be higher than normal?

    _____

    _____

12. List three possible causes of the generator having no output.

    _____

    _____

Problems Encountered

_____

_____

_____

Instructor's Comments

_____

_____

_____

REVIEW QUESTIONS

Review Chapter 19 of the textbook to answer these questions:

1. Describe how AC voltage is rectified into DC voltage by the generator.

   _____

   _____

   _____

2. Explain how to perform charging system voltage drop tests.

   _____

   _____

   _____

3. Current flow to the rotor is controlled by which of the following?

   a. Diodes

   b. Stator

   c. Voltage regulator

   d. Pole pieces

4. Explain the differences between a wye and a delta stator winding.

   _____

   _____

   _____

5. Explain the voltage regulator A and B circuits.

   _____

   _____

   _____

CHAPTER 20

LIGHTING SYSTEMS

OVERVIEW

Every automobile is equipped with various lighting systems. They include the headlights, taillights, parking lights, stop lights, and turn signal and hazard lights as well as instrument and interior lights. Years ago, there were only a few standard lights used on the vehicle. Today's cars and trucks still use older-style incandescent lights, but they also incorporate vacuum fluorescent lights, light-emitting diodes (LEDs), and high-intensity discharge (HID) lights.

As onboard power consumption has steadily increased, lighting systems, which can consume a large amount of power, have been designed to reduce power usage while improving safety and visibility. Many interior and exterior lights now use LED lights. LEDs use very little power and generate almost no heat, making them ideal for automotive applications, especially in electric or hybrid vehicles.

Many newer vehicles come equipped with HID lights. HID lights operate with high voltage, often needing thousands of volts to start the light but then requiring only 80 volts to remain operating. Because of this high voltage, caution must be used when servicing an HID lighting system.

 Concept Activity 20-1
Lighting Principles

Incandescent lights use a thin wire filament, which, when current passes through it, gets hot enough to glow and radiate light. This type of lightbulb has been in use since Thomas Edison developed it in 1879.

1. How does the material of the filament affect a lightbulb's operation? _____

_____

2. Why is the filament enclosed in a vacuum? _____

_____

3. How will breaking the vacuum seal affect the bulb's operation? _____

_____

Incandescent lights have been in use for well over one hundred years, but they are now being replaced in automotive applications by other types of lighting.

Light emitting diodes (LEDs) are solid-state devices that produce light with very little heat while using very small amounts of electricity.

4. Explain what a standard diode is and its function. _____

_____

5. Describe how light is produced by LEDs. _____

_____

6. Explain why electrical power conservation is important in modern automobiles. _____

_____

High-intensity discharge (HID) lights are available on many vehicles. They provide better illumination than standard bulbs but operate much differently than incandescent or LED bulbs.

7. Explain the basic operation of HID lights.

_____

_____

8. Why do HID lights require high voltages to operate?

_____

_____

9. What are the symptoms of a HID light in the beginning stages of failure?

_____

_____

10. What service and safety precautions must be observed when working in HID lighting systems?

_____

_____

■

Concept Activity 20-2
Lighting Identification

Referring to the following figures, identify the automotive lights.

1.

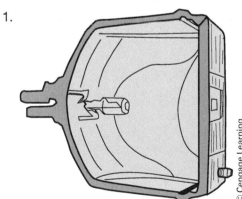

© Cengage Learning

_____

2.

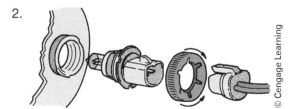

© Cengage Learning

_____

3.

© Cengage Learning 2015

_____

4.

© Cengage Learning 2015

_____

5.

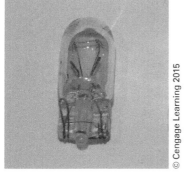

© Cengage Learning 2015

_____

6.

© Cengage Learning 2015

_____

7.

© Cengage Learning 2015

Concept Activity 20-3
Electrical Circuits

To diagnose a malfunctioning lighting circuit, it is necessary to have the ability to read and understand an electrical wiring diagram. Trace the current flow through the low-beam portion of the circuit shown in **Figure 20-1** below.

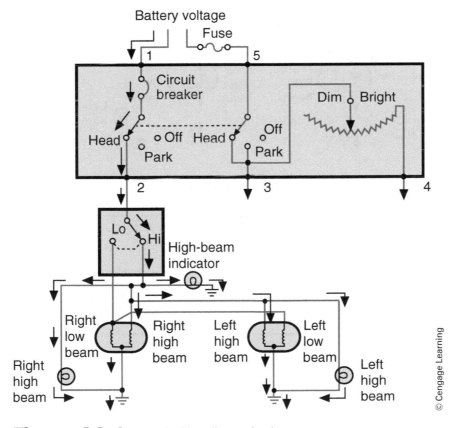

Figure 20-1 A typical headlamp circuit.

© Cengage Learning

Concept Activity 20-4
Electrical Circuit Diagnosis

1. What would be the symptom(s) if there were high resistance at point X in the diagram below?

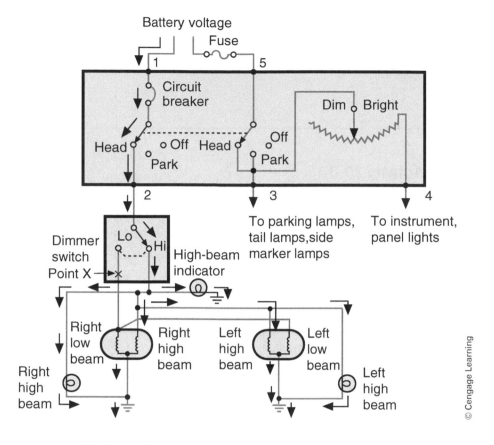

© Cengage Learning

2. List the steps you would take to diagnose this condition.

_____

3. What tools might be needed to diagnose and repair this concern?

_____

_____

LIGHTING SYSTEM DIAGNOSIS AND TESTING

Before performing any lighting system tests, make sure the battery is fully charged and that all battery cable connections are clean and tight. A visual inspection is an important part of diagnosing the lighting system. Check for wires with frayed or damaged insulation, loose connections, or improper harness routing. Refer to the vehicle's wiring diagram.

Any problems found during the visual inspection should be corrected before performing further tests of the lighting system. For example, a loose ground strap may cause intermittent operation of lamps and gauges. If necessary, inspect and tighten the ground strap attaching screws.

Table 20-1 indicates steps to take when trying to isolate lighting system problems. You must know how to perform these tasks in order to pass the ASE Electrical Systems Test.

TABLE 20-1 Lighting System Diagnosis.

Symptom	Possible Cause
Hazard flasher lamps do not flash	Fuse or circuit breaker burned out
	Damaged hazard flasher unit
	Worn or damaged turn-signal switch
	Open circuit in wiring
	Worn or damaged hazard flasher switch
Backup lamps—one lamp does not function	Bulb burned out
	Loose wiring connections
	Open circuit in wiring
Backup lamps—both lamps do not function	Fuse or circuit breaker burned out
	Backup lamp switch out of adjustment
	Worn or damaged backup lamp switch
	Loose wiring connections
	Open wiring or poor ground
	Bulbs burned out
Instrument panel lamp does not light	Bulb burned out
	Fuse burned out
	Open circuit in wiring, rheostat, or printed-circuit board
Dome lamp does not come on when door is opened	Connector loose
	Blown fuse
	Bulb burned out
	Open circuit in wiring
	Worn or damaged door jamb switch
Dome lamp stays on	Worn or damaged door jamb switch
	Worn or damaged main lighting switch
Map lamp does not come on when switch is actuated	Bulb burned out
	Blown fuse
	Open circuit in wiring
	Worn or damaged switch in lamp circuit

© Cengage Learning 2015

(Continued)

TABLE 20-1 *(Continued)*

Symptom	Possible Cause
Map lamp stays on	Worn or damaged switch in lamp circuit
Side or roof marker lamp does not light	Bulb burned out Open circuit or poor ground
Turn-signal lamps do not light	Fuse or circuit breaker burned out Worn or damaged turn-signal flasher Loose wiring connections Open circuit or poor ground Damaged turn-signal switch
Turn-signal lamps light but do not flash	Worn or damaged turn-signal flasher Poor ground
Front turn-signal lamps do not light	Loose wiring connector or open circuit
Rear turn-signal lamps do not light	Loose wiring connector or open circuit
One turn-signal lamp does not light	Bulb burned out Open circuit or poor ground
Headlights do not light	Loose wiring connections Open circuit in wiring Worn or damaged headlight switch
One headlight does not work	Loose wiring connections Sealed-beam bulb burned out Corroded socket
All headlights out; park and taillights are okay	Loose wiring connections Worn or damaged dimmer switch Worn or damaged headlight switch Open circuit in wiring or poor ground
Both low-beam or both high-beam headlights do not work	Loose wiring connections Worn or damaged dimmer switch Open circuit or poor ground
One taillight out	Bulb burned out Open wiring or poor ground Corroded bulb socket
All taillights and marker lamps out; headlights okay	Loose wiring connections Open wiring or poor ground Blown fuse Damaged headlight switch

(Continued)

TABLE 20-1 (*Continued*)

Symptom	Possible Cause
Stop lights do not work	Fuse or circuit breaker burned out
	Worn or damaged turn-signal switch
	Loose wiring connections
	Worn or damaged stop-light switch
	Open circuit or poor ground
Stop lights stay on continuously	Damaged stop-light switch
	Switch out of adjustment
	Short circuit in wiring
One parking lamp out	Bulb burned out
	Open wiring or poor ground
	Corroded bulb socket
All parking lamps out	Loose wiring connections
	Open wiring or poor ground
	Bad switch

The following job sheets will prepare you for lighting system diagnosis and service.

☐ JOB SHEET 20–1

Inspecting and Testing a Headlight Switch

Name _____ Station _____ Date _____

Objective

Upon completion of this job sheet, you will have demonstrated the ability to inspect and test a headlight switch with an ohmmeter. This task applies to ASE Education Foundation Task AST/MAST 6.A.10 (P-1).

Tools and Materials

A DMM

Description of Vehicle

Year _____ Make _____ Model _____

VIN _____ Engine Type and Size _____

Mileage _____

PROCEDURE

1. Put the headlight switch in all possible positions and observe what lights are controlled by each position. List each position and the controlled lights below.

   _____

   _____

   _____

   _____

2. Locate the headlight switch in the wiring diagram and label the lights controlled by each position of the switch.

3. Remove the fuse to the headlights or disconnect the battery's negative cable. Remove the headlight switch by following the procedures outlined in the service manual.

Instructor's Check _____

4. Identify the various terminals of the switch and list the particular terminals that should have continuity in the various switch positions.

_____

_____

_____

_____

5. Connect the ohmmeter across these terminals, one switch position at a time, and record your readings below:

_____

_____

_____

_____

6. Based on the preceding test, what are your conclusions about the switch?

_____

_____

_____

_____

7. Reinstall the switch, and connect the negative battery cable or reinstall the fuse. Then check the operation of the headlights. Note light operation: _____

Problems Encountered

_____

_____

_____

Instructor's Comments

_____

_____

_____

☐ JOB SHEET 20–2

Aiming Headlights

Name _____ Station _____ Date _____

Objective

Upon completion of this job sheet, you will have demonstrated the ability to adjust the aim of headlights using portable headlight aiming equipment. This task applies to ASE Education Foundation Tasks MLR 6.E.2 (P-2), and AST/MAST 6.E.3 (P-2).

Tools and Materials

A vehicle with adjustable headlights

Portable headlight aiming kit

Hand tools

Description of Vehicle

Year _____ Make _____ Model _____

VIN _____ Engine Type and Size _____

Mileage _____

PROCEDURE

1. Describe the type of headlights used on the vehicle.

   _____

   _____

2. Park the vehicle on a level floor. Install the calibrated aiming units to the headlights. Make sure the adapters fit the headlight aiming pads on the lens.

3. Zero the horizontal adjustment dial. Are the split image target lines visible in the view port? _____ If the lines cannot be seen, what should you do?

   _____

   _____

4. Turn the headlight horizontal adjusting screw until the split image target lines are aligned. Then repeat this for the other headlight. List any problems you may have had doing this.

   _____

   _____

5. Turn the vertical adjustment dial on the aiming unit to zero. Turn the vertical adjustment screw until the spirit level bubble is centered. Recheck your horizontal setting after adjusting the vertical. List any problems you had making the vertical adjustment.

_____

_____

_____

_____

6. If the headlight assembly has four lamp assemblies, repeat steps 2 through 5 on the other two lamps. List any problems you may have had doing this.

_____

_____

_____

_____

7. List problems with other aspects of the vehicle that can cause the headlights to not aim correctly.

_____

_____

_____

Problems Encountered

_____

_____

_____

Instructor's Comments

_____

_____

_____

☐ JOB SHEET 20–3

Testing and Adjusting a Brake Light Switch

Name _____ Station _____ Date _____

Objective

Upon completion of this job sheet, you will have demonstrated the ability to test and adjust a stop light switch. This task applies to ASE Education Foundation Tasks MLR 5.F.4 (P-1), AST/MAST 5.F.5 (P-1) and 6.A.10 (P-1).

Tools and Materials

12-volt test light

Jumper wire

Ruler or tape measure

Description of Vehicle

Year _____ Make _____ Model _____

VIN _____ Engine Type and Size _____

Mileage _____

PROCEDURE

1. Locate the brake light switch. Record the color of the wires that supply power to the switch and out to the lights.

   _____

2. Connect a 12-volt test light to ground. Place the test probe into the terminal of the wire supplying power to the switch. Does the test light illuminate?

   _____

3. Next, move the probe end of the test light to the output wire of the switch. Does the light illuminate with the brake pedal not applied?

   _____

4. Slightly depress the brake pedal, and note the test light. Does the light illuminate when the pedal is depressed?

   _____

5. Approximately how far did the pedal travel before the test light illuminated?

   _____

6. Refer to the correct service information, and determine the specification for brake pedal travel for brake light operation.

   _____

7. Is the brake light switch properly adjusted?

_____

8. If the switch needs adjusting, explain the procedure here.

_____

_____

_____

9. If the test light had failed to light in step 2, what steps would you have to take to determine the problem?

_____

10. If the test light illuminated in step 2, but not step 3, what would this indicate?

_____

_____

Problems Encountered

_____

_____

_____

Instructor's Comments

_____

_____

_____

 **REVIEW QUESTIONS**

Review Chapter 20 of the textbook to answer these questions:

1. Explain why you should not touch the glass of a halogen bulb.

   _____

   _____

   _____

2. List three circuits that may be controlled by a multifunction switch.

   _____

   _____

   _____

3. One turn signal indicator blinks normally while the other indicator blinks very rapidly. What is the most likely cause?

   a. Faulty flasher unit

   b. Shorted turn signal switch

   c. Open brake light switch connection

   d. Inoperative bulb

4. Describe the safety precautions you should take when servicing high-intensity discharge lamps.

   _____

   _____

   _____

5. List four types of lighting found on modern vehicles.

   _____

   _____

   _____

INSTRUMENTATION AND INFORMATION DISPLAYS

OVERVIEW

Electrical and electronic instruments include indicator gauges, lights, and audible devices that warn the driver about the engine and the operation of the vehicle systems. The modern vehicle has a vast array of warning lights and symbols as part of the instrument panel. Warning lights, such as the brake warning, air bag, tire pressure, and check engine lights, all illuminate in the event of a problem or malfunction. Drivers are able to monitor coolant temperature and fuel level via their gauges. This is so any potential problem can be caught before it becomes a more severe problem. Audible warnings may alert the driver if the parking brake is applied while moving or a door is not closed completely or about other such conditions that warrant immediate attention.

A switch provides a ground path to illuminate many of the warning lights. The instrument panel (IP) fuse supplies power to the warning lightbulbs. All the IP warning lights should come on when the key is turned to the ON position. This is called a bulb check.

 Concept Activity 21-1
Warning Light Identification

Obtain the keys to a vehicle from your instructor. Note the vehicle information:

Year _____ Make _____ Model _____

1. Turn the key to the ON position, and note the warning lights on the IP. _____

_____

2. Are there any spots on the IP for lights where none illuminates? _____

   Locate a wiring diagram for the IP. Locate the warning lights that are controlled by a switch and those that are computer controlled.

3. Switch-operated warning lights _____

4. Computer-controlled warning lights _____

_____

■

Dash gauges, such as the coolant temperature and fuel level, provide important information to the driver. Although both provide important information, they operate in different ways.

Coolant temperature gauges receive input from a coolant temperature sensor. The sensor, located in a coolant passage in the cylinder head or intake manifold, changes its electrical resistance according to the coolant temperature. When voltage from the IP is supplied to the sensor, the resistance causes a voltage drop, affecting the voltage returning to the dash gauge unit. This voltage drop is displayed as the temperature gauge reading.

Concept Activity 21-2
Coolant Temperature Sensors

Obtain from your instructor a sample of coolant temperature sensors. Measure the resistance of the sensors at room temperature.

1. Sensor 1 resistance, cold _____

2. Sensor 2 resistance, cold _____

3. Sensor 3 resistance, cold _____

Next, carefully heat the sensors and recheck their resistance. Do not use a flame to heat the sensors.

4. Sensor 1 resistance, hot _____

5. Sensor 2 resistance, hot _____

6. Sensor 3 resistance, hot _____

7. How did the increase in temperature affect the sensors' resistance?

_____

_____

_____

Fuel-sending units use variable resistors to change the voltage for the dash fuel gauge. Voltage is supplied to the sending unit. The voltage returned depends on the position of the fuel float level, which is attached to the variable resistor.

Ask your instructor for an old fuel-sending unit.

8. Measure and note the resistance of the sensor with the float at the fully raised position.

_____

9. Measure and note the resistance of the sensor with the float at the fully lowered position.

_____

10. How would unwanted resistance affect the coolant temperature and fuel level circuits?

_____

_____

■

DIAGNOSIS OF ELECTRICAL INSTRUMENTATION

Table 21-1 and **Table 21-2** are basic diagnostic charts for the major electrical instruments and accessories. These diagnostic charts will be beneficial when performing the tasks required for passing the ASE Electrical System Test.

TABLE 21-1 Instrument Panel or Cluster (Analog)

Symptom	Possible Cause
Fuel gauge shows full or partial tank when tank is empty.	Wrong sender installed Sender arm bent or obstructed Improper sender calibration
Fuel gauge reads full at all fuel levels.	Short circuit in wiring Sender arm movement obstructed
Fuel gauge reading fluctuates (erratic).	Loose connection or damaged wiring Loose sender resistor winding
Fuel gauge shows empty at all fuel levels.	Loose or dirty wiring connections or open circuit in wiring Leaking sender float Missing sender float Open circuit in sender Sender arm movement obstructed
Fuel gauge will not read full when tank is full.	Wrong sender installed Sender arm movement obstructed Leaking float Sender calibration is wrong
Temperature gauge does not move, or the gauge is inaccurate.	Inoperative sending unit Open in circuit Inoperative gauge
None of the gauge pointers move.	No voltage at the panel Open instrument voltage regulator
Warning indicator is always on (except LOW FUEL warning indicator and FASTEN BELT indicator).	Inoperative switch
Warning indicator will not light when proper conditions are met (except LOW FUEL warning and FASTEN BELT indicators).	Inoperative switch
Fuel computers, speedometer, and odometer gauge do not switch between English (UCS) and metric (SI) modes.	Open or shorted wiring in circuit Inoperative switch Inoperative instrument panel

TABLE 21-2 Instrument Panel or Cluster (Digital)

Symptom	Possible Cause
Digital displays do not light.	Blown fuse Inoperative power and ground circuit Inoperative instrument panel
Speedometer reads wrong speed.	Faulty odometer Wrong gear on vehicle speed sensor (VSS) Wrong tire size
Fuel gauge displays top and bottom two bars.	Open or short in circuit
Fuel computer displays CS or CO.	Open or short in fuel gauge sender Inoperative instrument panel
Odometer displays error.	Inoperative odometer memory module in instrument panel
Fuel gauge display is erratic.	Vehicle parked on hill Sticky or inoperative fuel gauge sender
Fuel gauge display is erratic.	Faulty wiring Inoperative fuel gauge
Fuel gauge will not display FULL or EMPTY. Fuel economy function of message center is erratic or inoperative.	Sticky or inoperative fuel gauge sender Inoperative fuel flow signal Faulty wiring Inoperative instrument panel
Extra or missing display segments.	Inoperative instrument panel
Speedometer always reads zero.	Faulty wiring Inoperative instrument panel
Temperature gauge displays top and bottom two bars.	Short in circuit Inoperative coolant temperature sender Inoperative instrument panel

© Cengage Learning 2015

The following job sheets will prepare you to diagnose and repair instrumentation systems.

☐ JOB SHEET 21–1

Testing a Dash Warning Light

Name _____ Station _____ Date _____

Objective

Upon completion of this job sheet, you will have demonstrated the ability to test the operation of a warning light circuit. This task applies to ASE Education Foundation Task AST/MAST 6.F.2 (P-2).

Tools and Materials

Jumper wire

Description of Vehicle

Year _____ Make _____ Model _____

VIN _____ Engine Type and Size _____

Mileage _____

PROCEDURE

1. Select a warning light that is controlled by a switch. Often the brake fluid level sensor provides the ground for the red BRAKE warning light.

2. With the key on and the engine running, the red brake warning light should be off.

   Brake warning light status: _____ ON _____ OFF

3. If the light is off, unplug the warning light switch located at the brake fluid reservoir.

   Brake warning light status: _____ ON _____ OFF

4. Use a jumper wire to jump the two wires for the sensor at the connector, and note the status of the warning light.

   Brake warning light status: _____ ON _____ OFF

5. Based on this test, is the brake warning light circuit operating correctly?

6. List two conditions that can cause the red brake warning light to illuminate.

   _____

7. What could be the cause if the red brake warning light does not illuminate during bulb check?

   _____

Problems Encountered

_____

_____

_____

Instructor's Comments

_____

_____

_____

☐ JOB SHEET 21–2

Removing, Checking, and Replacing a Temperature Sending Unit

Name _____ Station _____ Date _____

Objective

Upon completion of this job sheet, you will have demonstrated the ability to remove, check, and replace a temperature sending unit properly. This task applies to ASE Education Foundation Task AST/MAST 6.F.1 (P-2).

Description of Vehicle

Year _____ Make _____ Model _____

VIN _____ Engine Type and Size _____

Mileage _____

PROCEDURE

1. Place fender covers over the fenders. Remove the cap from the radiator to relieve any pressure and then replace it.

   WARNING: *Do not remove the radiator cap if the engine is still hot. Allow the engine to cool before proceeding.*

   Describe the location of the sensor.

   _____

   Disconnect the temperature sending unit wire at the sending unit.

2. Prepare the new temperature sending unit for installation by applying pipe sealant or a small amount of an electrically conductive sealer to the threads.

3. Remove the temperature sending unit and immediately install the new temperature sending unit. Tighten it to the specified torque.

   Torque Specification _____

4. Connect the wire to the temperature sending unit. Refill the cooling system to replace lost coolant.

5. Start the engine, and check the operation of the sending unit.

Problems Encountered

_____

_____

_____

Instructor's Comments

_____

_____

_____

REVIEW QUESTIONS

Review Chapter 21 of the textbook to answer these questions:

1. What is the purpose of the instrument voltage regulator (IVR)?

_____

_____

_____

2. Which two devices trigger the warning and indicator lamps?

_____

_____

_____

3. What items should be inspected when a gauge or indicator system is malfunctioning?

_____

_____

_____

4. Explain how a fuel gauge and sending unit display fuel level.

_____

_____

_____

5. Explain the function of a balancing coil gauge.

_____

_____

CHAPTER

22

BASICS OF ELECTRONICS AND COMPUTER SYSTEMS

OVERVIEW

Semiconductors or solid-state devices have become an important component in today's automotive computer systems. The computers of the various onboard systems interact with each other to allow for seamless operation. It is important to become familiar with the components of these systems and to have a basic understanding of how the onboard systems work.

COMPUTERS

Computers are used to control many systems on today's vehicles. Engine control systems have been used since the early 1980s. Now, electronic control systems are also used to control climate, lighting circuits, cruise control, anti-lock braking, suspension, steering, automatic transmissions, and charging systems. It is important that you know how to properly diagnose problems in electronic control systems.

Electronics is the technology of controlling electricity. Transistors, diodes, semiconductors, integrated circuits, and solid-state devices are all considered to be part of electronics rather than just electrical devices. Keep in mind that all the basic laws of electricity apply to all electronic controls. Equally important is to use care in handling electronic components, which can be easily damaged by electrostatic discharges (ESDs).

Concept Activity 22-1
On-Board Network Data Bus Identification

1. Using a vehicle and a scan tool assigned to you by your instructor, determine what onboard control modules are installed.

   _____

   _____

2. Using the service information, determine the type of computer communication network(s) used by this vehicle.

   _____

3. List the modules and/or functions that utilize the high-speed data bus.

   _____

4. List the modules and/or functions that utilize the medium-speed data bus.

   _____

5. List the modules and/or functions that utilize the low-speed data bus.

_____

_____

_____

■

Below is an example of a computer data network **(Figure 22-1)**.

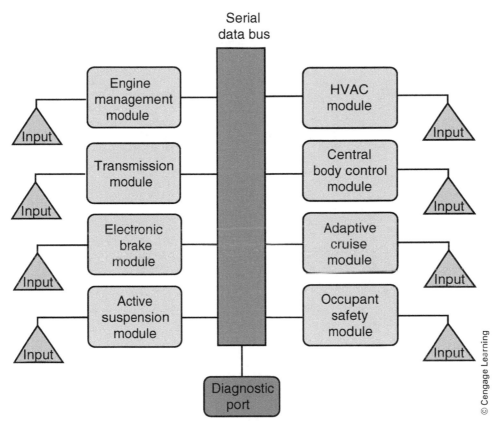

Figure 22-1 A multiplexed system uses a serial data bus to allow communications between the various control modules.

The following job sheets will prepare you to service electronics and computer systems.

☐ JOB SHEET 22–1

Testing a Diode with a DMM

Name _____ Station _____ Date _____

Objective

Upon completion of this job sheet, you will have demonstrated the ability to test a diode for open or shorted condition with a DMM. This task applies to ASE Education Foundation Tasks MLR 6.A.5 (P-1) and AST/MAST 6.A.4 (P-1).

Tools and Materials

DMM   Diode

Description of Vehicle

Year _____ Make _____ Model _____

VIN _____ Engine_____

Mileage _____

PROCEDURE

1. Obtain a diode or a relay containing a diode from your instructor.

2. Set the DMM to test resistance. Connect the leads of the meter to the ends of the diode. What does the meter display?

   _____

3. Reverse the meter leads on the diode, and retest the resistance. What does the meter display?

   _____

4. What does this indicate about the diode?

   _____

   _____

   _____

   If the meter is unable to accurately test the diode on the resistance setting, select the diode test setting.

5. Place the meter leads on the diode and test. What does the meter display?

   _____

6. Reverse the meter leads, and retest the diode. What does the meter display?

   _____

7. What does this indicate about the diode?

   _____

   _____

Problems Encountered

_____

_____

_____

Instructor's Comments

_____

_____

_____

☐ JOB SHEET 22–2

Testing for Module Communication with a Scan Tool and an Oscilloscope

Name _____ Station _____ Date _____

Objective

Upon completion of this task, you will have demonstrated the ability to check for module communication using a scan tool. This task applies to ASE Education Foundation Tasks AST/MAST 6.G.5 (P-2).

Tools and Materials

OBD II vehicle

Scan tool

GMM/DSO

Description of Vehicle

Year _____ Make _____ Model _____

VIN _____ Engine Type and Size _____

Mileage _____

PROCEDURE

1. Obtain a scan tool from your instructor. Scan tool used _____

2. Connect the scan tool to the vehicle's DLC. DLC location _____

3. Turn the ignition to the RUN position and begin to attempt to communicate with the on-board network. Method of communication: _____ Global OBD II _____ SAE

4. Does the scan tool recognize and communicate with the module(s)? _____

5. List the modules that are available. _____

6. "Ping" the modules shown, and record the responses.

   _____

   _____

   _____

7. Connect the DSO/GMM channel 1 to the communication data terminal of the DLC and the negative lead to the signal ground terminal **(Figure 22-2)**. Send a DTC request to a module, and note the waveform pattern on the scope. Draw the pattern here.

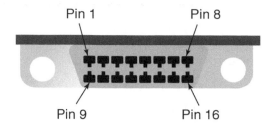

Pin 1: Manufacturer discretionary
Pin 2: J1850 bus positive
Pin 3: Manufacturer discretionary
Pin 4: Chassis ground
Pin 5: Signal ground
Pin 6: Manufacturer discretionary
Pin 7: ISO 1941-2 "K" line
Pin 8: Manufacturer discretionary

Pin 9: Manufacturer discretionary
Pin 10: J1850 bus negative
Pin 11: Manufacturer discretionary
Pin 12: Manufacturer discretionary
Pin 13: Manufacturer discretionary
Pin 14: Manufacturer discretionary
Pin 15: ISO 9141-2 "L" line
Pin 16: Battery power

© Cengage Learning

Figure 22-2   Standard OBD II date link connector.

8. Now request a data stream from a module, and note the waveform pattern on the scope. Draw the pattern here.

9. Compare the two patterns. How are they similar or different? _____

10. Based on the waveform, what type of communication protocol is being used?

_____

Problems Encountered

_____

_____

_____

Instructor's Comments

_____

_____

_____

REVIEW QUESTIONS

Review Chapter 22 of the textbook to answer these questions:

1. Describe the CAN data bus.

_____

_____

_____

2. Compare and contrast Class B and Class C multiplexing.

_____

_____

_____

_____

3. List three advantages of multiplexing.

_____

_____

_____

4. Describe three types of computer memory and their uses.

_____

_____

_____

_____

5. Explain the function of a diode.

_____

_____

CHAPTER 23

ELECTRICAL ACCESSORIES

OVERVIEW

The number of electrical accessories in vehicles has steadily increased over the years. There are accessories that improve driver safety, driver warning and reminder systems, and systems that provide emergency communication. Safety and driver warning accessory systems include windshield wipers and washers, horns, message displays on the instrument panel, windshield heaters, rear-window defrosters, security and antitheft devices, and emergency roadside communication. Other vehicle accessories make driving more convenient, pleasant, and safer for the driver and passengers, including power windows, power seats, power door locks, power mirrors, remote keyless entry, power trunk release, power windows, power sunroofs, cigarette lighters, navigation systems, and entertainment/sound systems. Many of these systems are controlled by the body control module(s).

Concept Activity 23-1
Identify Electrical Accessories

Make a list of five safety-related systems and ten convenience accessory systems found on newer vehicles.

_____

_____

          ■

BODY CONTROL MODULES

A number of the electrical/electronic accessory systems are controlled and monitored by a body control module (BCM), which generates fault codes for the system when it malfunctions. The BCM may also be the gateway module, meaning it is responsible for network communication.

WINDSHIELD WIPER/WASHER SYSTEMS

There are many different types of windshield wiper and washer systems for both front and rear windows, where applicable. The operation of the windshield wiper and washer systems varies depending on the make and model of a vehicle. Many of the systems have a delay or intermittent feature that allows the driver to vary the operation according to the weather conditions. The front and rear wiper systems include a window washer feature for cleaning the windows during driving conditions in which road spray limits driver visibility.

    Table 23-1 lists some symptoms, possible causes, and remedies for problems with windshield wipers and other accessories.

HORNS, CLOCK, AND CIGARETTE LIGHTER SYSTEMS

Horn(s)

The horn is a basic warning system used to alert other drivers and pedestrians to pending danger from the vehicle sounding the horn.

Clock

Many vehicles have a clock to display the current time as a convenience to the driver and passengers of the vehicle. The clock may be separate or incorporated into another accessory such as the radio display.

Power Outlet

The power outlet is a convenience accessory that has been included in vehicles for many years. Some vehicles include a 120 VAC power outlet as well. Both the 12 V and 120 V outlets are designed for low current use, such as cell phone chargers.

CRUISE CONTROL SYSTEMS

Many newer vehicles have what are called "adaptive cruise control" systems. Adaptive cruise control uses either radar or laser ranging to determine the distance to the vehicle ahead. If the distance decreases, the cruise control system can decelerate or even apply the brakes to prevent an accident.

SOUND SYSTEMS

Sound systems are available in a wide variety of models. The complexity of these systems varies significantly, from the basic AM radios to the compact disc (CD) players with high-power amplifiers and multiple speakers. However, the overall operation of the radio itself is electrically and basically the same.

Sound quality depends on the basic system, but especially the quality of the speakers and their placement. A speaker emits sound by the vibration of its diaphragm. The vibration is set up by electrical pulses it receives from the radio, cassette player, or CD player. A permanent magnet and coil of wire change these electrical pulses into motion. This motion causes the speaker's diaphragm to vibrate. The vibration sends out pressure waves that we hear as sound.

Table 23-2 is included to help you diagnose a sound system. This table does not include poor sound quality from the vehicle's speakers. Normally, poor sound is related to the speakers. Rattles and buzzes from the speakers are often caused by loose speakers, speaker mountings, speaker grilles, or trim panels. Distortion can be caused by the speaker, radio chassis, or wiring. If the problem is in the radio chassis, both speakers on the same side of the vehicle will exhibit poor quality. Distortion caused by damaged wiring is most often accompanied by lower than normal sound output.

POWER LOCK SYSTEMS

The basic power lock systems used in vehicles have not changed over the years. What has changed is how and when they lock and unlock. The electronics provides for the personalization of power door locks on many models of vehicles. Doors may automatically lock when a certain speed or gear selection is achieved. The power lock system may have the feature that allows the doors to lock when the last door of the vehicle closes. Because of the many variations in operation of power door locks, it is imperative that the technician be familiar with the exact operation of the door lock system before attempting any repair.

POWER SEATS

Power seats allow the driver or passenger to quickly adjust the seat for his or her comfort. The power seat system may have four to eight adjustments, and certain models of vehicle have a memory-set seat position feature to accommodate two different drivers.

POWER MIRROR SYSTEM/HEATED MIRRORS

Power mirrors allow the driver to control left and right mirrors from one switch located in the driver's side arm rest. Rotating the power mirror switch to the left or right position selects one of the mirrors for adjustment. When the driver moves the joystick control up and down, or right and left, this action moves the mirror to the desired position. Some vehicles have heated outside mirrors. The heated mirror switch is usually located with the power mirror controls.

REAR-WINDOW DEFROSTER SYSTEM

The rear-window defroster, also known as defogger or deicer, heats the rear-window surface to remove moisture or ice depending on climatic conditions. This heating action is accomplished by heating a thermal grid attached to the window.

OTHER ACCESSORIES

Adjustable Pedals

Electrically adjusted pedals help the driver find a comfortable and safe place to sit behind the steering wheel. They may be part of the memory seat system so that the driver can quickly bring the seat, mirrors, and pedals all to the most comfortable position.

Heated Windshields

Heated or self-defrosting front windshield systems are being offered as an option in many vehicles at the upper-end of the price range. This option is of particular interest to drivers operating vehicles in subzero winter climates.

Sunroof System

A power sunroof is an option available on some passenger cars and light trucks. Some sunroofs have a sliding louvered shade panel to block out sun rays and provide additional ventilation.

Night Vision

Night vision is a thermal imaging system that allows drivers to see things in the road ahead of them that they might not otherwise see until too late. This system is extremely helpful in fog conditions.

Navigation Systems

Navigation systems may use global positioning satellites or DVD-based information to help drivers make travel decisions while they are on the road. A navigational system displays traffic and travel information on a screen. It can display a road map marking the exact location of the vehicle, the best route for getting to a specific location, how many miles it is to the destination, and how many miles there are yet to travel. It can also display traffic information regarding traffic backups due to congestion, roadwork, and/or accidents, and then display alternative routes so travel isn't delayed.

Vehicle Tracking Systems

Vehicle tracking systems can monitor the location of a vehicle if it has been stolen or lost. The system can be based on the vehicle's navigational system or the cellular phone in the vehicle. This system is also being used in various commercial applications such as tracking the location of delivery vehicles and dispatching tow trucks more efficiently.

Voice Activation System

The most common application of a voice activation system is for hands-free cellular phone operation, but it can be used on other controls. The voice activation system recognizes the driver's voice and can respond with answers of its own to questions from the driver. An example of this system's use is a voice request to change the radio.

SECURITY AND ANTITHEFT DEVICES

There are three basic types of antitheft devices available: locking devices, disabling devices, and alarm systems. Many of the devices are available as optional equipment from the manufacturers; others are aftermarket installed. Like other electrical systems, computers may control antitheft devices. The sensors and relays can be checked with a digital multimeter and a jumper wire. Read the service manual for operation details.

Remote keyless entry systems are also quite common. These use a handheld transmitter, frequently part of the key fob. With a press of the unlock button on the transmitter, the interior lights turn on, the driver's door unlocks, and the theft security system is disarmed from any direction range from 25 to 50 feet, depending on the type of transmitter. The trunk can also be unlocked. When exiting the car, pressing the lock button locks all doors and arms the security system.

Passkey systems are also quite common. A passkey is a specially designed key that is fitted with a resistor, transistor, or transponder that is selected and programmed just for one particular vehicle. The ignition lock reads the electronic coding on the key. Without the proper key coding, the engine does not start.

The two methods for activating alarm systems are *passive* and *active*. Passive systems switch on automatically when the ignition key is removed or the doors are locked. Active systems are activated manually with a key fob transmitter, keypad, or toggle switch. They allow the driver to leave the car at a service station or parking garage without disclosing the secrets of the system to the attendant.

The following Accessory Diagnosis tables (**Table 23-1** and **Table 23-2**) will help you with diagnosing these systems.

POWER WINDOWS

Power windows have been available on vehicles for many years. The operating characteristics of power windows have changed dramatically since their introduction in the automotive industry. Besides the basic operation of the system, many vehicles have auto down and up functions and may include an obstacle sensing feature. Consulting the service manual for the specific operation of a power window system aids the technician in repairing the system fault the first time.

TABLE 23-1 Accessory Diagnosis

Symptom	Possible Cause
Windshield Wipers	
Wiper will not operate.	Open circuit or blown fuse
	Binding wiper arm, shafts, or linkage
	Link rod loose from drive lever
	Faulty switch
	Faulty motor assembly
Gear teeth are damaged.	Wiper blades striking windshield molding during operation
	Binding connecting links
	Operator stopping blades manually
	Drive arm not held when drive arm nut is tightened
Wiper will not shut off.	Faulty switch
Wiper continuously shuts off.	Binding condition in wiper arm shaft, connecting links, or drive gear and shaft
	Faulty harness connections
	Faulty motor assembly

© Cengage Learning 2015

(Continued)

TABLE 23-1 (*Continued*)

Symptom	Possible Cause
Wiper operates at one speed only.	Faulty switch Faulty connection
Wiper motor speed is excessive under light load but stalls under heavy load. Wiper motor is noisy.	Faulty motor assembly Faulty motor assembly
Interval Windshield/Washer	
Windshield wipers do not operate.	Test circuit breaker. Inspect wire for an open to the interval governor and windshield wiper motor.
Wipers operate in interval mode with wiper switch in OFF position but operate normally in all other positions.	Test wiper switch. Open condition in circuit
Wipers will not park.	Verify voltage at wiper motor switch connector. Inspect ground wire to ground. If this proves acceptable, check wiper motor and switch.
Windshield washer pump motor is inoperative; wipers operate when washer switch is depressed.	Test wire and windshield washer pump motor. Inspect windshield washer pump motor ground wire.
Power Windows	
None of the windows operate.	With ignition switch in "acc" or "run," check for battery voltage at master window/door lock control switch connector. If voltage is not present, check circuit breaker. If circuit breaker is good, repair connecting wire.
One window motor does not operate.	Test switches or motor according to procedures in service manual.
Power Door Locks (without Keyless Entry)	
Power door locks are inoperative from both switches.	Inspect the circuit breakers for the door lock motors. Check voltage at driver's door lock switch. Test continuity of circuit between master window door lock control switch and ground. Test circuit for opens.
Power door locks are inoperative from one switch only.	Verify voltage to inoperative switch. Check continuity of switch.
One door lock motor is inoperative.	Test wiring to motor by connecting test light across wires and operating locks. Inspect door lock motor.
Rear-Window Defroster	
The rear-window defrost grid does not remove frost.	Inspect fuse. Test for voltage at rear-window defrost/control. Verify continuity to ground. With rear-window defrost control on, check voltage across control. If voltage is not present, check rear-window defrost control.
Heated Windshield	

The heated windshield system does not operate.	Inspect fuse. Inspect for battery voltage at heated windshield switch connector.
	Press and hold heated windshield switch on.
	    Test for battery voltage at switch.
	    Verify that the heated windshield switch is grounded.

TABLE 23-1 (*Continued*)

Symptom	Possible Cause
Horns	
Horn does not work.	Check voltage at horn with button pressed.
	Test for voltage at horn relay connector. Ground horn relay wire. If horn does not sound, check horn relay.
	With horn switch depressed, test for continuity to ground. If open, horn switch is inoperative.
Horn will not shut off.	If there is battery voltage at horn relay horn switch may be stuck.
	If no voltage is at horn relay, and horn stops when horn relay is removed, relay contacts are stuck and relay must be replaced.

TABLE 23-2 Radio Troubleshooting

Symptom	Possible Cause
Radio display does not light, but radio plays.	With main light switch in "park" or "head" and knob rotated fully clockwise, verify voltage at connector.
	If voltage is not present, test wire for an open. If voltage is present, remove radio for service.
Radio does not work. No sound from any speaker. Radio display does not light.	With ignition in "acc" or "run," test for battery voltage in wire at radio connector.
	Test the wire for continuity to ground. Test for good connection at radio connectors.
Radio does not appear to work. No sound from any speaker. Radio display lights.	Test wire to amplifier for a short to ground. If okay, check audio fuse.
	Verify battery voltage on wire at amplifier. If not okay, repair as necessary.
	Test for continuity to ground at amplifier connector.
	Disconnect amplifier. Turn radio on and tune to strong station, with balance and fader controls centered and volume control set at maximum. Measure AC voltage with DMM set on low scale between wire and ground. If voltage varies around 1 volt or greater, check amplifier. If not, test radio.
One speaker does not work.	Disconnect suspect connector. Momentarily connect analog ohmmeter of 1.5-volt battery across suspect speaker terminals. If speaker pops, speaker is okay. Inspect wires to amplifier from radio and from suspect speaker for an open or short to ground.
LCD display is erratic.	Inspect connectors to radio and ground. They should be clean and tight.
AM works; FM does not work, or vice versa.	Compare to known good radio for FM operation. If radio does not work as well, remove radio for service.
	Operate radio with a test antenna. If reception improves, check and repair antenna. If reception does not improve, remove radio from service.

© Cengage Learning 2015

(*Continued*)

TABLE 23-2 (*Continued*)

Symptom	Possible Cause
Power antenna does not work properly.	Measure battery voltage at wire with ignition switch in "acc" or "run."
	Test wires for an open ground.
	Momentarily apply battery power to motor in each direction. If motor moves in both directions, it is okay.
	Check for binding or bent mast.
	Check antenna cable.

The following job sheets will prepare you for diagnosing and servicing electrical accessories.

☐ JOB SHEET 23–1

Testing and Repairing a Rear-Window Defogger Grid

Name _____ Station _____ Date _____

Objective

Upon completion of this job sheet, you will have demonstrated the ability to repair a rear-window defogger grid. This task applies to ASE Education Foundation Tasks MLR/AST/MAST 6.A.4 (P-1) and 6.A.5 (P-1).

Tools and Materials

Glass cleaner

Grid repair kit

Description of Vehicle

Year _____ Make _____ Model _____

VIN _____ Engine Type and Size _____

Mileage _____

PROCEDURE

1. Locate the faulty cross wire in the grid. This is usually noticeable when the defogger is used.

2. Visually inspect the cross wire for breaks or burned spots. Note condition.

   _____

3. The broken or faulty spot on the cross wire can normally be found by using a test light. Attach the test light clip to a good ground. Move along the wire while touching the probe to the wire. The test light will fully light at each spot before the break. When the test light no longer lights, the break is between that point and the last point it lit up. Note results.

   _____

   _____

   _____

4. Follow the procedure furnished with the repair kit to repair the open in the circuit. Some kits include small strips of copper tape and epoxy or sealer.

5. Use the sealer supplied to cover and protect the repair.

6. Operate and test the rear-window defogger to determine whether the repair was successful. Note results.

   _____

   _____

Problems Encountered

_____

_____

_____

Instructor's Comments

_____

_____

_____

☐ JOB SHEET 23–2

Identifying the Source of Static on a Radio

Name _____ Station _____ Date _____

Objective

Upon completion of this job sheet, you will have demonstrated the ability to locate the cause of static and/or noise from the speakers for a radio. Before beginning, review the material on the topic in Chapter 22 of *Automotive Technology*. This task applies to ASE Education Foundation Task MAST 6.G.3 (P-3).

Tools and Materials

Jumper wire with alligator clips on both ends

Description of Vehicle

Year _____ Make _____ Model _____

VIN _____ Engine Type and Size _____

Mileage _____

Describe the sound system in the vehicle.

_____

_____

_____

PROCEDURE

1. Turn on the radio, and listen for the noise. Typically, the noise is best heard on the low AM stations. Describe the noise.

_____

_____

Listen to the radio with the engine off and with it running. Describe the difference in the noise level and the reception of the radio when the engine is off and when it is running.

_____

_____

What can you conclude from the above?

_____

_____

2. Operate the radio in AM and FM. Does the noise appear in both bands?

_____

If the noise is only on FM, what could be the problem?

_____

_____

3. If the noise is heard on both AM and FM, continue the test by checking the antenna and its connections. Is the antenna firmly mounted and in good condition?

_____

_____

4. Test the connection of the antenna cable to the antenna. Are the contacts clean and is the cable connector in good condition?

_____

_____

5. Connect a jumper wire from the base of the antenna to a known good ground. Then listen for the noise. Did the noise level change? Describe and explain the results.

_____

_____

6. Refer to the service manual, and identify any noise suppression devices used on this vehicle. List the noise suppression devices here.

_____

_____

Are all of the noise suppression devices present on the vehicle, and are they mounted securely to a clean well-grounded surface?

_____

_____

7. Connect a jumper wire from a known good ground to the grounding tab on each capacitor-type noise suppressor. Listen to the radio, and describe and explain the results of doing this.

_____

_____

8. Turn off the engine, and disconnect the wiring harness from the voltage regulator to the generator. Start the engine, and listen to the radio; describe and explain the results of doing this.

_____

_____

9. Inspect the spark plug wires and spark plugs. Are both of these noise suppressor types?

_____

_____

10. Verify the routing, condition, and connecting points for the spark plug cables. Describe their condition.

_____

_____

11. Connect a jumper wire from a known good ground to the frame of a rear speaker. Listen to the radio, and describe and explain the results of doing this.

_____

_____

12. What are your conclusions and recommendations based on this job sheet?

_____

_____

_____

_____

13. If the noise source has not been identified, what should you do next?

_____

_____

Problems Encountered

_____

_____

_____

Instructor's Comments

_____

_____

_____

 REVIEW QUESTIONS

Review Chapter 23 of the textbook to answer these questions:

1. Explain the operation of the rear-window defroster grid.

_____

_____

_____

2. Describe how express power windows function.

_____

_____

_____

3. Explain the operation of rain-sensing wiper systems.

_____

_____

4. Describe horn circuit basic testing procedures.

_____

_____

5. Explain adaptive and active seating.

_____

_____

ASE PREP TEST

1. Technician A says copper is the most common electrical conductor used in vehicles today. Technician B says if the conductor diameter is doubled, the resistance for any given length is cut in half. Who is correct?

    a. Technician A

    b. Technician B

    c. Both A and B

    d. Neither A nor B

2. Technician A says one characteristic of a series circuit is that all of the current flows through all resistors. Technician B says in a parallel circuit, the current through each leg will be different if the resistance values are different. Who is correct?

    a. Technician A

    b. Technician B

    c. Both A and B

    d. Neither A nor B

3. Technician A says a negative temperature coefficient thermistor is the most commonly used thermistor. Technician B says the resistance of an NTC sensor increases as temperature increases. Who is correct?

    a. Technician A

    b. Technician B

    c. Both A and B

    d. Neither A nor B

4. Technician A says AC voltage may be displayed as rms or average responding voltage. Technician B says rms readings are very close to being average meter readings. Who is correct?

    a. Technician A

    b. Technician B

    c. Both A and B

    d. Neither A nor B

5. Technician A says a parasitic drain test places the ammeter in series with the positive battery cable. Technician B says the maximum permissible current drain is 10 mA (milliamperes). Who is correct?

    a. Technician A

    b. Technician B

    c. Both A and B

    d. Neither A nor B

6. Technician A says an open circuit voltage test reading of 12 volts or better indicates a fully charged battery. Technician B says a battery capacity test indicates how well a battery functions under load. Who is correct?

    a. Technician A

    b. Technician B

    c. Both A and B

    d. Neither A nor B

7. A vehicle has a no-crank condition. Technician A says to check the voltage drop across the starter solenoid motor connections. Technician B says an open in the control circuit may be the cause. Who is correct?

    a. Technician A

    b. Technician B

    c. Both A and B

    d. Neither A nor B

8. High resistance in a circuit can cause which of the following problems?

    a. A component to not operate

    b. A component to operate incorrectly

    c. Damage to connectors and wiring

    d. All of the above

9. Technician A says high voltage will cause lamps to burn out prematurely. Technician B says a shorted field control circuit can cause a high-voltage condition. Who is correct?

   a. Technician A
   b. Technician B
   c. Both A and B
   d. Neither A nor B

10. Technician A says the accumulated mileage (odometer) value of the digital display odometer is stored in a nonvolatile memory. Technician B says if the speedometer has been replaced, there is no need to set the odometer reading of the new to match the old. Who is correct?

    a. Technician A
    b. Technician B
    c. Both A and B
    d. Neither A nor B

11. Technician A says that on a hybrid vehicle, if the "Ready" light is on, the engine is running and the vehicle is ready to drive. Technician B says many hybrid vehicles can be driven on electric power even if the "Ready" light is not illuminated. Who is correct?

    a. Technician A
    b. Technician B
    c. Both A and B
    d. Neither A nor B

12. Technician A says HID headlight require thousands of volts to keep the light in operation. Technician B says HID headlights may require 15 seconds to reach full brightness. Who is correct?

    a. Technician A
    b. Technician B
    c. Both A and B
    d. Neither A nor B

13. In the charging system shown in **Figure 23-1**, the meter is showing:

    a. charging output voltages.
    b. indicator lamp operating voltage.
    c. charging circuit voltage drop.
    d. ignition switch voltage drop.

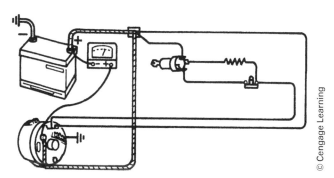

Figure 23-1   Charging system.

© Cengage Learning

14. The oil-pressure light stays on whenever the engine is running. The oil pressure has been checked, and it meets specs. Technician A says that a ground in the circuit between the indicator light and the pressure switch could be the cause. Technician B says that an opening in the pressure switch could be the cause. Who is correct?

    a. Technician A
    b. Technician B
    c. Both A and B
    d. Neither A nor B

15. The voltmeter reads 0.021 volt, as shown in **Figure 23-2**. This indicates that

    a. the ignition switch is shorted.
    b. the safety switch is shorted.
    c. the control circuit is open.
    d. the motor circuit is open.

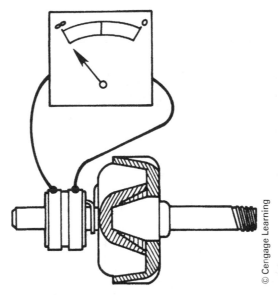

Figure 23-2   Ohmmeter connected across slip rings.

16. The light in the circuit shown in **Figure 23-3** glows dimly. A technician connects a test light as shown. The test light burns at normal brightness. Technician A says that a bad ground at the light could be the cause. Technician B says that high resistance in the circuit from the battery to the light could be the cause. Who is correct?

    a. Technician A
    b. Technician B

    c. Both A and B
    d. Neither A nor B

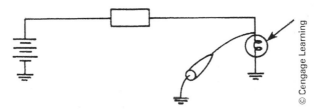

Figure 23-3   A test light connected to a simple circuit.

17. The low-speed position of the windshield wiper system shown in **Figure 23-4** does not work; the high-speed position works normally. Technician A says circuit 58 may be open. Technician B says circuit 63 may be open. Who is correct?

    a. Technician A
    b. Technician B

    c. Both A and B
    d. Neither A nor B

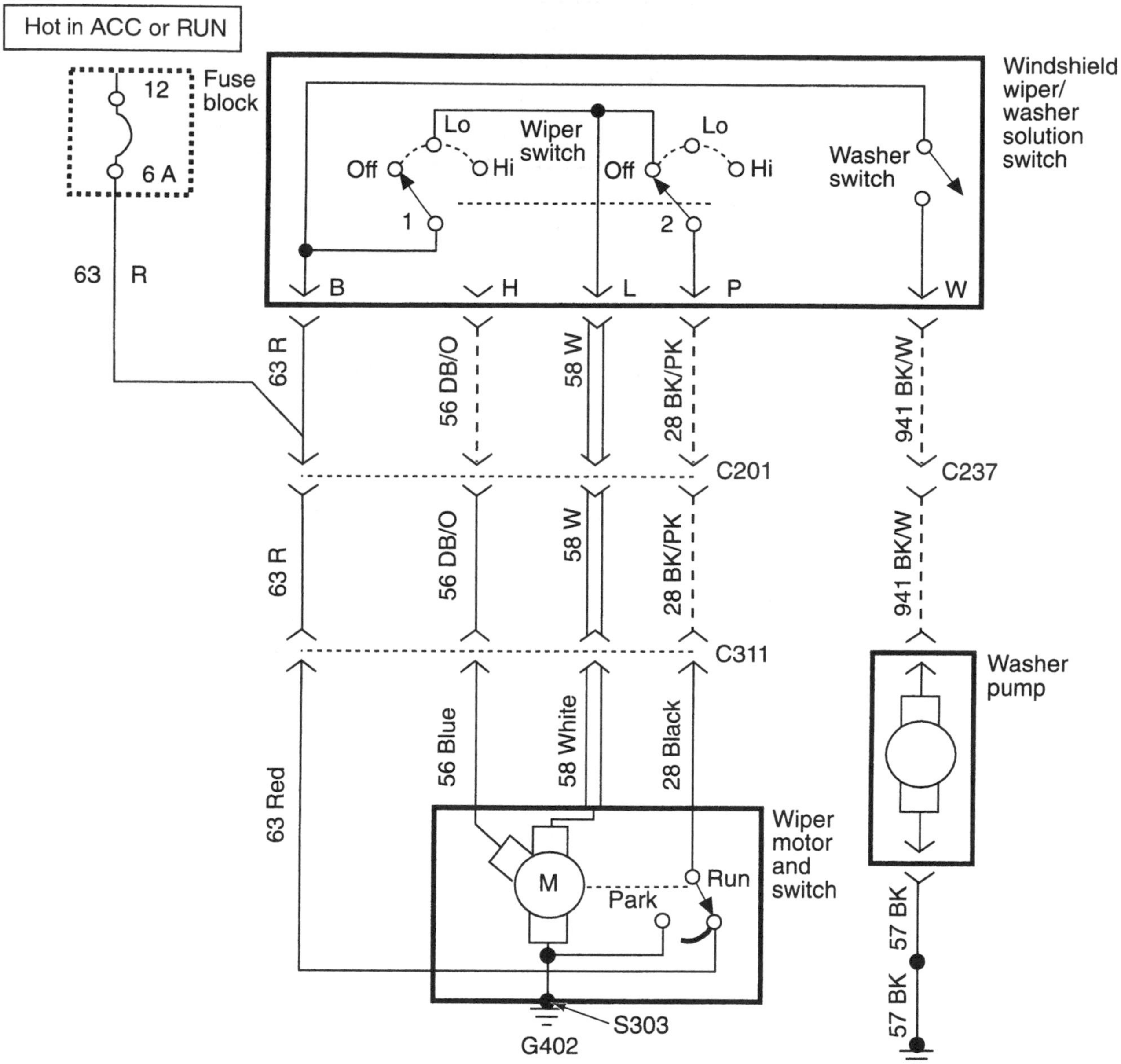

Figure 23-4 Two-speed wiper circuit.

18. A voltmeter that is connected across the input and output terminals of an instrument cluster illumination lamp rheostat (dimmer control) indicates 12.6 volts with the switch in the maximum brightness position and the engine off. Which of the following statements is true?

    a. The voltage available at the lamps will be 12.6 volts.

    b. The voltage available at the lamps will be 0 volts.

    c. The rheostat is working normally.

    d. More information is needed in order to determine whether the lamps will operate correctly.

19. The medium-high speed of the blower motor circuit in **Figure 23-5** is inoperative; the rest of the blower speeds are fine. Technician A says circuit 752 may be open. Technician B says the middle resistor in the blower motor resistor assembly may be open. Who is correct?

    a. Technician A   c. Both A and B

    b. Technician B   d. Neither A nor B

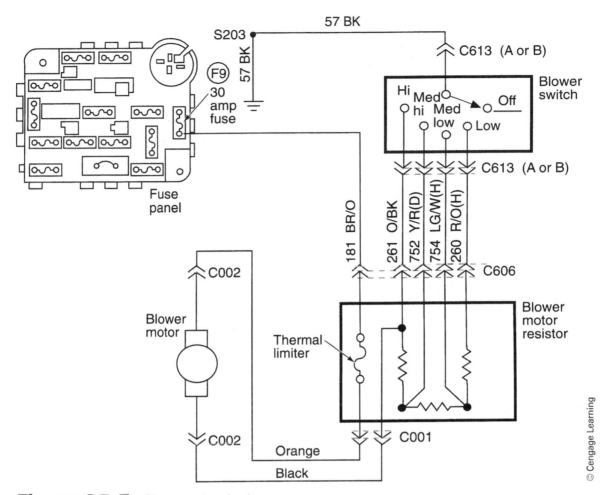

Figure 23-5 Blower motor circuit.

20. The power window motor in **Figure 23-6** is completely inoperative. With the master window switch placed in the "down" position, the following voltages are measured at each terminal:

Terminal No.	Voltage
1	12 V
2	0 V
3	12 V
4	0 V
5	12 V
6	12 V
7	0 V
8	12 V
9	0 V
10	12 V
11	0 V

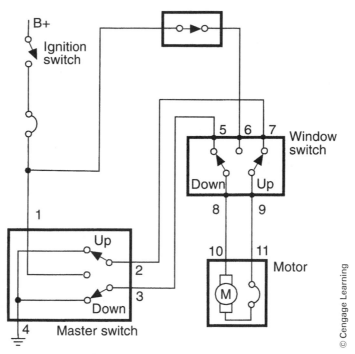

Figure 23-6 Power window circuit.

Which of the following statements represents the cause of this problem?

a. The master switch is faulty.

b. The window switch is faulty.

c. The motor is faulty.

d. There is a poor ground in the circuit.

ENGINE PERFORMANCE SYSTEMS

CHAPTER

24

© Cengage Learning 2015

OVERVIEW

Engine performance systems, such as the ignition, fuel, air, and computer systems, are responsible for how an engine runs. Before electronics became the standard for engines, the ignition and fuel systems were purely mechanical in operation, and there was no computer system. Since the 1970s, engines and vehicles overall have seen a complete transition to electrical/electronic control. Today, computers or modules monitor or control nearly every function of the automobile.

Ignition systems, while performing the same functions, have changed greatly over the decades of vehicle development. Before electronic ignition control became standard in the 1970s, the functions of the ignition system were handled mechanically. The ignition system must supply adequate spark to the correct cylinder at the correct time, while compensating for engine rpm and load. Ignition systems are discussed in detail in Chapters 26 and 27.

Concept Activity 24-1
Ignition System Comparisons

Compare the two ignition systems shown in **Figures 24-1** and **24-2**.

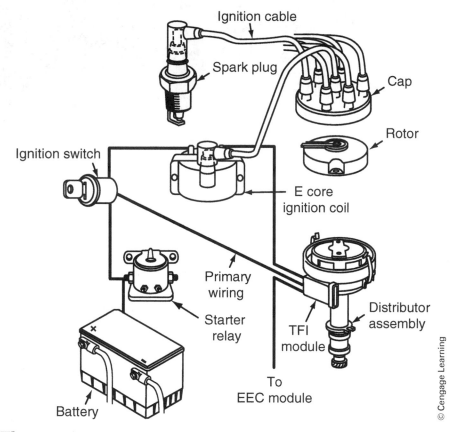

Figure 24-1   One type of electronic distributor type ignition system.

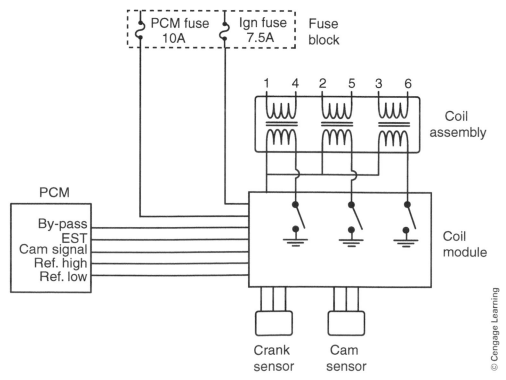

Figure 24-2   This wiring schematic shows the grounding control of the ignition coil module used in the waste spark ignition system.

1. List the locations for mechanical wear in the distributor ignition (DI) system.

_____

2. List the components that need periodic replacement or service with the DI system.

_____

3. List the locations for wear in the distributorless electronic ignition (EI) system.

_____

4. List the components that need periodic replacement or service with the EI system.

_____

5. Explain the benefits of EI systems compared to DI systems.

_____

_____

_____

■

All new vehicles sold in the United States have fuel injection systems. During the 1980s, manufacturers began making the transition from carburetors to fuel injection systems. Carburetors operated on the pressure difference between the atmospheric pressure outside the engine and the lower pressure or vacuum inside the intake manifold. The pressure difference forced air and fuel into the cylinders during the intake stroke. Although this system worked for many decades, it was not very efficient or adaptable for changing operating conditions. Electronic fuel injection systems deliver fuel under pressure to the combustion chamber with far greater efficiency and reliability.

Concept Activity 24-2
Identify the Operating Circuits of a Carburetor

Referring to **Figure 24-3**, fill in the blanks with the names of the operating circuits of a carburetor.

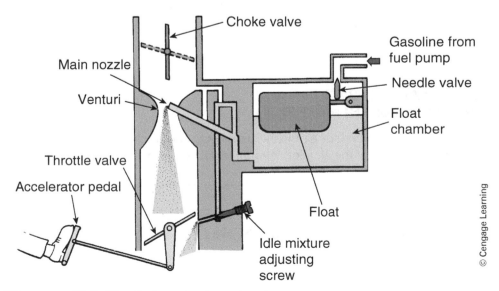

Figure 24-3 Basic parts of a carburetor.

1. _____

2. _____

3. _____

4. _____

5. _____

■

The air induction system has had to evolve with the changes in the fuel system. Air cleaners and housings once sat on top of the engine, covering the carburetor or throttle body injection (TBI) system. Modern vehicles tend to have an aerodynamic shape and lower hood lines than older vehicles; this has necessitated the air induction system to move lower and farther forward to be able to capture incoming air. This can pose a serious problem if a vehicle is driven through water deep enough to be sucked into the air induction system. Water can enter the engine and cause a hydrostatic lock, which can cause severe engine damage. Air filter replacement, once a common tune-up and do-it-yourself item, has also changed. Many cars and light trucks are fitted with an airflow gauge in the air filter box. This gauge measures whether the air filter needs to be replaced based on airflow restriction **(Figure 24-4)**.

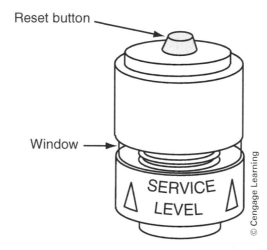

Reset button

Window

SERVICE
LEVEL

© Cengage Learning

Figure 24-4   Airflow restriction indicator.

Concept Activity 24-3
Comparison of TBI and PFI

1. Compare an older TBI or carburetor air cleaner assembly and a newer port fuel injected (PFI) air cleaner assembly. Note the location of the air inlet, the materials, hoses, heat riser, and filter elements.

_____

_____

2. What is the purpose of the heat riser tube from the exhaust manifold to the air cleaner housing?

_____

3. Why are heat risers not used on fuel injection systems?

_____

_____

■

To meet the requirements of environmental legislation, manufacturers have implemented on-board computer systems, which monitor and control the engine performance systems. The ignition, fuel, and emission control systems are managed by a collection of modules communicating on a network.

The first version of this was on-board diagnostics, version one, called OBD I. OBD I had limited monitoring and was implemented differently by each manufacturer. Beginning in 1996, OBD II was phased in to replace OBD I. The main function of OBD II is to monitor emission system operation and alert the driver via the dashboard warning light, called the malfunction indicator light (MIL), of a problem. OBD II also mandated a common communication arrangement, common diagnostic link connector (DLC), and common terminology and codes. OBD III may be implemented in the next several years. OBD III requirements may include remote inspection of the emission system.

Concept Activity 24-4
OBD Systems

1. Compare and contrast five major differences between OBD I and OBD II.

_____

_____

_____

_____

2. Describe the location of the DLC on OBD II vehicles.

_____

_____

3. Explain what is required for the OBD II system to illuminate the MIL.

_____

_____

4. Describe multiplexing.

_____

_____

_____

■

Before electronic ignition systems became standard in the 1970s, the only electronics found on the vehicle were in the radio. Since then, every aspect of vehicle operation has become electronically monitored and controlled.

Modern vehicles use electronic control modules for many functions:

■ Engine management; fuel, ignition, and emissions systems

■ Transmission management; shifting and wear compensation

■ Anti-lock brakes; monitoring wheel speed and controlling brake application during anti-lock brake system (ABS) events

■ Vehicle stability control; monitoring steering angle and body roll and controlling torque output, brake application, adjusting the suspension as needed

■ Climate control; monitoring the in-car, outside air temperature, and sun load to adjust the A/C operation automatically

■ Cruise control; adaptive cruise systems use lasers and radar to adjust speed based on the distance to the vehicle out in front.

This list does not include body, entertainment, navigation, communication, or safety systems, all of which are computer controlled. Many of the systems listed share modules or communicate with other modules on the computer network to ensure proper vehicle operation.

As consumers expect increased efficiency, safety, and comfort, vehicle manufacturers will respond by adding electronic components to satisfy the demand.

The following job sheet will prepare you to service the vehicle's various engine performance systems.

☐ JOB SHEET 24–1

Identifying Engine Performance Systems

Name _____ Station _____ Date _____

Objective

Upon completion of this job sheet, you will have demonstrated the ability to identify the various systems related to engine performance. This task applies to ASE Education Foundation Tasks MLR 8.A.1 (P-1) and AST/MAST 8.A.2 (P-1).

Description of Vehicle

Year _____ _____ Make _____ Model _____

VIN _____ Engine _____

Mileage _____

1. Determine the type of ignition system:

   _____ Distributor (DI)   _____ Distributorless (DIS)

   _____ Distributor points   _____ Coil-on-plug (DIS)

2. Using service information, determine what module controls the ignition timing.

3. Determine the type of fuel system.

   Carburetor _____   TBI _____

   PFI/SFI _____   GDI _____

4. Using service information, determine the normal system fuel pressure.

   _____

5. Remove and inspect the air cleaner. Inspect the housing and related hoses. Note the condition of the components.

   _____

   _____

   _____

6. Determine the computer system type.

   OBD I _____   OBD II _____

7. Is the system CAN 2.0? _____ If the system is not CAN, describe the system used.

   _____

   _____

   _____

8. If the system is OBD I, determine the manufacturer's computer system type. (EEC IV, C^3, and PGM-FI are examples.)

   _____

9. Using the VIN, engine type, and service information, determine whether the engine has variable valve timing (VVT).

_____

10. If the engine has VVT, how is the valve timing changed?

_____

11. Locate the vehicle emission decals.

   Tier _____   Bin _____

   Smog index number _____

12. Briefly summarize the interaction of this vehicle's engine performance systems.

_____

_____

_____

Problems Encountered

_____

_____

_____

Instructor's Comments

_____

_____

_____

REVIEW QUESTIONS

Review Chapter 24 of the textbook to answer these questions:

1. Contrast the purpose and operation of OBD I and OBD II systems.

_____

_____

_____

_____

2. Describe the OBD drive cycle.

_____

_____

3. Describe how freeze-frame data can be important in performing diagnostics.

_____

_____

_____

4. Explain three functions of the ignition system.

_____

_____

_____

5. List four types of fuel injection systems.

_____

_____

_____

CHAPTER
25

DETAILED DIAGNOSIS AND SENSORS

OVERVIEW

All modern vehicles have on-board computer systems to detect faults and provide diagnostic capabilities. These systems contain numerous sensors, actuators, and control modules. A control module contains a microprocessor, memory, and software. Today's cars and trucks can contain over a dozen separate control modules, all linked together in what is called a multiplexed network. *Information from various sensors may be shared between different control modules over the network. Without the correct input from the sensor, the control modules cannot adequately control the operation of a vehicle system.*

Any time there is a vehicle concern that affects the operation of a circuit for which a diagnostic trouble code (DTC) is stored, a thorough visual inspection should be performed. Many problems, such as loose connections, chafed wiring, and broken vacuum hoses, which are often the causes of circuit malfunctions, can be found and corrected during a visual inspection. Job Sheet 25-1 provides a detailed visual inspection procedure.

The on-board computer system relies on the network of sensors to provide data about vehicle operating conditions. Sensors generally fit into one of the following categories:

- Temperature sensors
- Movement sensors
- Pressure sensors
- Speed sensors

Some sensors, such as oxygen sensors, provide information about the exhaust oxygen content by comparing exhaust gas to the outside air. Airflow sensors determine how much air is being drawn into the engine so that the correct amount of fuel can be injected into the cylinder.

Temperature sensors, called thermistors, provide information about the engine coolant temperature, intake air temperature, and on some vehicles, inside and outside air temperature. Temperature sensors can be either positive temperature coefficient (PTC) or negative temperature coefficient (NTC) sensors. As you learned in Concept Activity 15-1, electrical resistance increases as temperature increases. In a PTC sensor, the electrical resistance increases as temperature increases. In a NTC sensor, resistance is high when temperature is low. Many engine coolant and intake air temperature sensors are NTC-type sensors.

Concept Activity 25-1
Measure Engine Coolant Temperature (ECT) Resistance

Obtain from your instructor several coolant and air temperature sensors. Use a digital multimeter (DMM) to test their resistance at room temperature.

Sensor 1 resistance _____ Approximate temperature _____

Sensor 2 resistance _____ Approximate temperature _____

Sensor 3 resistance _____ Approximate temperature _____

Sensor 4 resistance _____ Approximate temperature _____

Next, cool the sensors using ice water or a choke tester and remeasure their resistance and temperature.

Sensor 1 resistance _____ Approximate temperature _____

Sensor 2 resistance _____ Approximate temperature _____

Sensor 3 resistance _____ Approximate temperature _____

Sensor 4 resistance_____ Approximate temperature _____

5. How much did the resistance change compared to the temperature change?

Now, warm the sensors and remeasure their resistance.

Sensor 1 resistance _____ Approximate temperature _____

Sensor 2 resistance _____ Approximate temperature _____

Sensor 3 resistance _____ Approximate temperature _____

Sensor 4 resistance _____ Approximate temperature _____

6. How much did the resistance change compared to the temperature change?

_____

7. Which sensors are PTC and which are NTC sensors?

_____

_____

■

_____

Sensors that detect movement, such as throttle position, EGR position, and fuel level, are variable resistors, also called potentiometers **(Figure 25-1)**. These sensors have three wires: voltage, ground, and signal return. The sensor contains a resistive strip that varies the voltage drop of the circuit returning to a module. By attaching a lever to the signal return connection, the position of a component can be determined.

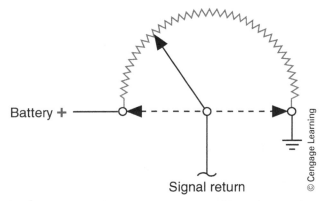

Figure 25-1 In a potentiometer, a movable center contact or wiper senses voltage between ground and its position on a wire-wound resistor.

Concept Activity 25-2
Throttle Position Sensors

1. Obtain several throttle position (TP) sensors and fuel level sensors. Use a DMM to measure the resistance from the voltage to ground terminals.

   Sensor 1 resistance _____

   Sensor 2 resistance _____

   Sensor 3 resistance _____

2. Next, measure the resistance from the voltage to signal return terminals.

   Sensor 1 resistance _____

   Sensor 2 resistance _____

   Sensor 3 resistance _____

3. Now move the sensor position and remeasure the resistance.

   Sensor 1 resistance _____

   Sensor 2 resistance _____

   Sensor 3 resistance _____

4. How do the resistances of the different sensors compare to each other?

   _____

   _____

   _____

5. Why might one type of sensor have a different resistance than another sensor?

   _____

   _____

   _____

6. Describe other circuits or systems that may use variable resistors.

   _____

   _____

   _____

   _____

■

Manifold absolute pressure (MAP) sensors use a flexible electronic capacitor to determine the pressure in the intake manifold. When the engine is running, pressure in the intake is lower than atmospheric pressure. How much lower depends on the engine rpm and load. The MAP sensor detects this pressure difference and reports it back to the powertrain control module (PCM). Job Sheet 25-2 provides detailed testing procedures for a MAP sensor.

Sensors that determine the speed of an object, such as a wheel speed sensor (WSS) or crankshaft position sensor (CKP) are one of two types of sensors; permanent magnet or Hall-effect sensors.

A permanent magnet sensor uses a permanent magnet and a coil of wire. When another magnetic field approaches the sensor, voltage is induced into the coil. This voltage can be used to determine rotational speeds. See Concept Activity 19-1 about induced voltage.

Hall-effect sensors also operate using magnetic fields, but in a slightly different way. A Hall-effect sensor detects the presence of a magnetic field between to points. A shutter wheel alternately opens and closes, interrupting the field **(Figure 25-2)**. This on–off field is converted into a digital on–off signal. This signal can be used to determine a components position and rotational speed. Job Sheets 25-3 and 25-4 provide detailed testing procedures for permanent magnet and Hall-effect sensors.

Exhaust gas oxygen sensors, commonly called oxygen sensors or O_2 sensors, detect the oxygen content of the exhaust gases compared to the surrounding air. Standard O_2 sensors produce a small amount of voltage based on the oxygen content of the sampled exhaust gas. A rich mixture, one that has too little oxygen, produces a high voltage, about 800 mV to 1 V. A lean mixture, one that has excessive oxygen, will produce a lower voltage, around 100 mV to 400 mV.

Many newer engines use an air/fuel ratio sensor. These sensors vary current supplied by the PCM based on the exhaust gas oxygen content. A lean mixture will increase current flow and cause the supplied voltage to increase. A rich mixture decreases current flow and will cause the supplied voltage to decrease. Job Sheet 25-5 provides detailed testing procedures for O_2 sensors.

Several types of airflow sensors are used to calculate the amount of air entering the engine. The most common is the mass airflow (MAF) sensor. This type of sensor uses current to heat a sensing element. The airflow across the sensor cools the sensing element. The amount of current needed to keep the element at the desired temperature is converted into a calculation of incoming air mass. Job Sheet 25-6 provides detailed testing procedures for MAF sensors.

Based on sensor input and programming, a control module determines what the necessary actions are required. Once the output is determined, an output device or actuator is commanded into action. Actuators are typically components such as the engine cooling fan(s), generator output, the fuel pump, fuel injectors, and idle control motors. Many outputs can be tested by using a scan tool.

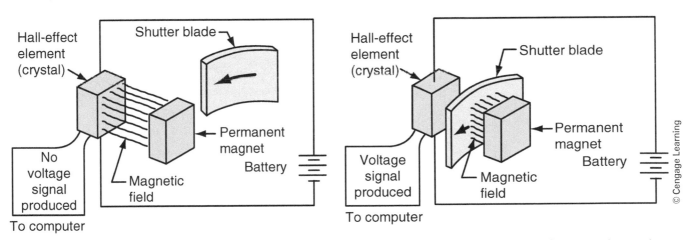

Figure 25-2 As the shutter blade passes in front of the magnet, it reduces the magnetic field, thus increasing the voltage and signal of the Hall-effect element.

Concept Activity 25-3
Computer Output Controls

Obtain a scan tool and lab vehicle from your instructor. Connect the scan tool to the data link connector (DLC), and begin communication. Locate the output control commands.

1. List the available systems or components that can be controlled by the scan tool.

   _____

   _____

2. With guidance from your instructor, attempt to command various systems or components on or off with the scan tool. Note your results.

   _____

   _____

   _____

   ■

The following job sheets will provide you with testing procedures for various computer sensors.

☐ JOB SHEET 25–1

Performing a Visual Inspection

Name _____ Station _____ Date _____

Objective

Upon completion of this job sheet, you will have demonstrated the ability to perform a visual inspection. This task applies to ASE Education Foundation Task AST/MAST 8.D.6 (P-2).

Description of Vehicle

Year _____ Make _____ Model _____

VIN _____ Engine Type and Size _____

Mileage _____

PROCEDURE

1. Check the engine coolant and oil levels. Full _____   Low _____

   Oil condition _____

   Coolant condition _____

2. Inspect the vacuum hoses for positive connection. Check for signs of decay, rot, splits, oil deterioration, and looseness. Note your findings.

   _____

3. With the engine running, listen carefully for vacuum leaks. A vacuum leak will present a hissing sound at the point of failure. Note your findings.

   _____

   _____

   _____

4. With the engine off, inspect the engine compartment wiring and connections. Check connectors for positive fit and for damage. Ensure all ground points are clean and tight. Note your findings.

   _____

   _____

   _____

5. With the engine running, carefully manipulate the wiring harnesses and connectors. If a problem is present with the engine running, the running condition may change, or a DTC may set. Note your findings.

   _____

   _____

   _____

6. With the engine off, open the air filter housing, and inspect the air filter and induction components. Note your findings.

_____

_____

_____

Problems Encountered

_____

_____

_____

Instructor's Comments

_____

_____

_____

☐ JOB SHEET 25–2

Testing a MAP Sensor

Name _____ Station _____ Date _____

Objective

Upon completion of this job sheet, you will have demonstrated the ability to test a frequency modulated MAP sensor using a DMM and a digital storage oscilloscope (DSO) / graphing multimeter (GMM). This task applies to ASE Education Foundation Task MAST 8.B.7 (P-2).

Tools and Materials

DMM   DSO/GMM   Vacuum pump

Description of Vehicle

Year _____ Make _____ Model _____

VIN _____ Engine Type and Size _____

Mileage_____

PROCEDURE

If the sensor being tested produces a frequency modulated signal, attach the DSO/GMM to the signal return connection and ground.

1. Carefully remove the MAP sensor's vacuum hose from the sensor. Note the condition of the vacuum hose and the MAP sensor.

   _____

   _____

   _____

2. Using the appropriate tools, back probe the signal return wire at the MAP sensor connector. Use a wiring diagram to determine which terminal is the signal return. Draw and label the MAP sensor connector and wires in the space below.

3. Connect the positive lead to the signal return connection and the negative lead to a good ground. Turn the key to the ON position. What does the meter/scope read with the key on, engine off (KOEO)?

   _____

   _____

4. Attach the vacuum pump to the vacuum port on the sensor. Apply approximately 20 inches of vacuum to the sensor, and note the voltage/frequency reading.

_____

5. Pump the vacuum tester until the highest vacuum reading is obtained, and note the voltage reading.
Vacuum reading _____ Voltage/Frequency _____

6. Compare your readings to the manufacturer's specifications. Is the sensor's output within specifications?

_____

Problems Encountered

_____

_____

_____

Instructor's Comments

_____

_____

_____

☐ JOB SHEET 25-3

Testing Permanent Magnetic Pickups

Name _____ Station _____ Date _____

Objective

Upon completion of this job sheet, you will have demonstrated the ability to test a permanent magnetic pickup. This task applies to ASE Education Foundation Tasks MAST 8.B.7 (P-2) and 8.C.2 (P-1).

Tools and Materials

DMM   DSO/GMM

Description of Vehicle

Year _____ Make _____ Model _____

VIN _____ Engine Type and Size _____

Mileage _____

PROCEDURE

1. Use a wiring diagram to locate the wiring for a permanent magnet pickup. These sensors are used for crank and camshaft position as well as wheel speed sensors. Note which sensor is being tested, its purpose, and location.

   _____

2. With the sensor unplugged, measure the resistance of the sensor, and compare the reading to the manufacturer's specifications. Spec _____ Reading _____

3. Is the sensor's resistance within specifications?

   _____

4. Attach the test leads to the two terminals of the sensor. With the meter on AC voltage, crank the engine or turn the wheel to produce a signal. How much voltage was generated?

   _____

5. Attach the leads of the scope to the sensor. Set the scope to display AC voltage with a time base of approximately 200 ms. Crank the engine or rotate the wheel. Freeze the screen with the waveform displayed. Draw the waveform in the space below.

6. Using the scope's curser functions, determine the voltage and signal frequency.
   Output Voltage _____ Signal Frequency _____

Problems Encountered

_____

_____

_____

Instructor's Comments

_____

_____

_____

☐ JOB SHEET 25–4

Testing Hall-Effect Pickups

Name _____ Station _____ Date _____

Objective

Upon completion of this job sheet, you will have demonstrated the ability to test a Hall-effect pickup. This task applies to ASE Education Foundation Tasks AST/MAST 8.B.7 (P-2) and 8.C.2 (P-1).

Tools and Materials
DMM   DSO/GMM

Description of Vehicle
Year _____ Make _____ Model _____

VIN _____ Engine Type and Size _____

Mileage _____

PROCEDURE

1. Use a wiring diagram to locate the wiring for a Hall-effect pickup. These sensors are used for crank and camshaft position as well as wheel speed sensors. Note which sensor is being tested, its function, and location.

   _____

2. Attach the positive test lead to the signal return terminal of the sensor and the negative lead to ground. With the meter on DC voltage, crank the engine or turn the wheel to produce a signal. How much voltage was generated?

   _____

3. Attach the leads of the scope to the sensor. Set the scope to display DC voltage with a time base of approximately 200 ms. Crank the engine or rotate the wheel. Freeze the screen with the waveform displayed. Draw the waveform in the space below.

4. Using the scope's curser functions, determine the voltage and signal frequency.
   Output Voltage _____ Signal Frequency _____

5. Why is the voltage displayed on the scope different from the voltage displayed on the DMM?

   _____

Problems Encountered

_____

_____

_____

Instructor's Comments

_____

_____

_____

☐ JOB SHEET 25–5

Testing Oxygen Sensors

Name _____ Station _____ Date _____

Objective

Upon completion of this job sheet, you will have demonstrated the ability to test an oxygen sensor. This task applies to ASE Education Foundation Task MAST 8.B.7 (P-1).

Tools and Materials

DMM   DSO/GMM

Description of Vehicle

Year _____ Make _____ Model _____

VIN _____ Engine Type and Size _____

Mileage _____

PROCEDURE

1. Use a wiring diagram to determine the O_2 sensor output wire, and note its color.

   _____

2. Attach the DMM positive lead to the O_2 signal wire, using an appropriate back probe tool. Attach the DMM negative lead to a good ground.

3. Set the meter to display DC volts.

4. With the KOEO, record the O_2 sensor voltage over approximately 2 minutes. Did the voltage change?

   _____

5. If the voltage decreased, why do you think this occurred?

   _____

   _____

   _____

6. If the voltage did not change, does this indicate a problem?

   _____

7. Start the engine, and note the O_2 voltage after approximately 1 minute. How is the sensor voltage responding?

   _____

8. Using the min/max feature, record the lowest and highest voltage readings over approximately 2 minutes of operation.   High _____   Low _____

9. Do the readings indicate the O_2 sensor is functioning correctly?

_____

10. Attach the positive lead of the DSO/GMM to the O_2 sensor signal wire and ground.

11. Set the scope to display DC voltage over 1 second. Draw the pattern shown on the scope here.

12. Adjust the time index so that at least 10 high/low peaks can be observed. These are called O_2 cross counts. How much time is required for the sensor to achieve 10 high–low peaks or switches?

_____

13. Compare the number of cross counts to the manufacturer's specifications. Is the O_2 sensor responding properly?

_____

14. What could cause the sensor to respond slowly (not enough cross counts)?

_____

_____

_____

15. What could cause the sensor to read a consistently lower than average voltage?

_____

_____

_____

16. What could cause the sensor to read a consistently higher than average voltage?

_____

_____

_____

Problems Encountered

_____

_____

_____

Instructor's Comments

_____

_____

☐ JOB SHEET 25–6

Testing MAF Sensors

Name _____ Station _____ Date _____

Objective

Upon completion of this job sheet, you will have demonstrated the ability to test a MAF sensor. This task applies to ASE Education Foundation Task MAST 8.B.7 (P-1).

Tools and Materials

DMM   DSO/GMM

Description of Vehicle

Year _____ Make _____ Model _____

VIN _____ Engine Type and Size _____

Mileage _____

PROCEDURE

1. Use a wiring diagram to determine the MAF sensor output wire, and note its color.

   _____

2. Attach the scope positive lead to the MAF signal wire, using an appropriate back probe tool. Attach the negative lead to a good ground.

3. Set the scope to display DC voltage waveforms.

4. With the KOEO, record the MAF sensor voltage/frequency.

5. Start the engine and note the voltage/frequency at idle speed. How is the sensor responding?

   _____

   _____

   _____

   Draw the sensor signal here.

6. Increase the engine speed and note the voltage/frequency change.

   _____

7. Locate specifications for MAF sensor output. Compare the readings with the specifications.

8. Do the readings indicate the sensor is functioning correctly?

_____

9. What could cause the sensor to respond incorrectly?

_____

_____

_____

10. What could cause the sensor to read less airflow than is actually present?

_____

_____

_____

11. What could cause the sensor to read more airflow than is normal?

_____

_____

_____

Problems Encountered

_____

_____

_____

Instructor's Comments

_____

_____

_____

☐ JOB SHEET 25–7

Performing a DTC Check on an OBD I Vehicle

Name _____ Station _____ Date _____

Objective

Upon completion of this job sheet, you will have demonstrated the ability to obtain DTCs on an OBD I vehicle. This task applies to ASE Education Foundation Task MLR/AST/MAST 8.B.1 (P-1).

Tools and Materials

Scan tool　　　　　　Jumper wire

On OBD I vehicles, there can be several different methods of obtaining DTCs. This job sheet will introduce you to several of the more common ways to access DTCs.

PROCEDURE

General Motors: Vehicle Information

Year _____ Make _____ Model _____

VIN _____ Engine Type and Size _____

Mileage _____

Most OBD I GM vehicles can provide DTCs either by a scan tool or by counting the flashes of the Check Engine Light (CEL).

1. Locate the DLC. Draw the appearance of the connector.

2. Locate terminals 1 and 2 of the DLC.

Instructor's Check _____

3. Connect a jumper wire or similar tool across terminals 1 and 2.

4. Turn the key to the ON position (engine off). Note the flashes of the CEL. Record any codes.

_____

5. Did the CEL flash a code 12? What does a code 12 indicate?

_____

Ford: Vehicle Information

Year _____ Make _____ Model _____

VIN _____ Engine Type and Size _____

Mileage _____

Most Ford OBD systems require a scan tool to obtain codes. However, on EEC-IV systems, codes can be read by using an analog voltmeter and a jumper wire connected to the DLC.

1. Locate the procedure to read DTCs using an analog voltmeter in the service information.

2. Connect the meter and jumper wire as directed. Draw the connections below.

Instructor's Check _____

3. With the meter and jumper wire connected turn the key to the ON position and note the needle on the meter.

4. Note any codes present.

_____

_____

_____

5. Does the system provide a PASS code?

_____

Chrysler: Vehicle Information

Year _____ Make _____ Model _____

VIN _____ Engine Type and Size _____

Mileage _____

Many OBD I Chrysler vehicles provided DTC information by cycling the ignition key on-off, on-off, on-off, on. This places the PCM into self-test mode and codes are displayed by flashing the CEL.

1. Turn the ignition key on and off three times, ending with the key in the ON position, engine off.

2. Does the CEL flash? _____

3. What DTCs are displayed? _____

4. Does the system flash a PASS code? _____

Import: Vehicle Information

Year _____ Make _____ Model _____

VIN _____ Engine Type and Size _____

Mileage _____

Many imported vehicles display codes by flashing a LED located on the PCM.

1. Refer to the service information for this vehicle and determine how DTCs are retrieved. Note the procedure here.

_____

_____

_____

2. Activate the self-test mode and note any codes.

_____

_____

_____

3. Does the system provide a PASS code?

_____

Problems Encountered

_____

_____

_____

Instructor's Comments

_____

_____

_____

☐ JOB SHEET 25–8

Obtaining DTCs on an OBD II Vehicle

Name _____ Station _____ Date _____

Objective

Upon completion of this job sheet, you will have demonstrated the ability to obtain DTCs on an OBD II-equipped vehicle. This task applies to ASE Education Foundation Task MLR/AST/MAST 8.B.1 (P-1).

Tools and Materials

Scan tool

Description of Vehicle

Year _____ Make _____ Model _____

VIN _____ Engine Type and Size _____

Mileage _____

PROCEDURE

Since the implementation of OBD II, manufacturers have to provide a common diagnostic connection, called the diagnostic link connector, or DLC. The DLC is located on the driver's side of the passenger compartment and has a common connector and basic terminal configuration.

1. Locate the DLC and note its location.

   _____

2. Install the scan tool communication cable and OBD II connector to the DLC.

3. Turn the ignition to the ON position and turn on the scan tool.

4. Select Global OBD II from the menu. Enter vehicle VIN information as necessary. Are any special keys, adaptors, or cables required to access the codes and data?

   _____

5. Navigate to the DTC menu. Record all DTCs present.

   _____

6. Navigate to the freeze-frame failure records menu. List the freeze-frames stored.

   _____

7. Navigate to the history DTCs. Record all history DTCs.

   _____

   If the code(s) stored are for a sensor, navigate to the data menu and load the scan data.

8. Note the KOEO readings for the sensor for which the code is stored. Is the data within the normal parameters for the sensor?

_____

9. Start the engine and note the sensor reading. Is the data within the normal parameters?

_____

10. Perform a visual inspection of the sensor and/or system, if applicable.

Problems Encountered

_____

_____

_____

Instructor's Comments

_____

_____

_____

☐ JOB SHEET 25–9

Testing an ECT Sensor

Name _____ Station _____ Date _____

Objective

Upon completion of this job sheet, you will have demonstrated the ability to check the operation of an engine coolant temperature sensor. This task applies to ASE Education Foundation Task MAST 8.B.7 (P-1).

Tools and Materials

DMM

Description of Vehicle

Year _____ Make _____ Model _____

VIN _____ Engine Type and Size _____

Mileage _____

PROCEDURE

1. Locate the ECT sensor. Describe its location here.

   _____

   _____

   _____

2. What color are the wires that are connected to the sensor?

   _____

   _____

3. Record the resistance specifications for a normal ECT sensor for this vehicle.

   _____

   _____

4. Disconnect the electrical connector to the sensor.

5. Measure the resistance of the sensor.

   It was _____ ohms at approximately _____ °F (_____ °C).

   Note your conclusions from the test here.

   _____

   _____

   _____

   _____

Problems Encountered

_____

_____

_____

Instructor's Comments

_____

_____

_____

☐ JOB SHEET 25–10

Testing the Operation of a TP Sensor

Name _____ Station _____ Date _____

Objective

Upon completion of this job sheet, you will have demonstrated the ability to test the operation of a throttle position sensor with a variety of test instruments. This task applies to ASE Education Foundation Task MAST 8.B.7 (P-1).

Tools and Materials

DMM

Lab scope

Description of Vehicle

Year _____ Make _____ Model _____

VIN _____ Engine Type and Size _____

Mileage _____

PROCEDURE

1. Describe the type of lab scope you are using.

   Model _____

2. Connect the lab scope across the TP sensor.

3. With the ignition on, move the throttle from closed to fully open and then allow it to close slowly.

4. Observe the trace on the scope while moving the throttle. Describe what the trace looked like.

   _____

   _____

   _____

   _____

   Based on the waveform of the TP sensor, what can you tell about the sensor?

   _____

   _____

5. With a voltmeter, measure the reference voltage to the TP sensor. The reading should be _____ volts. It was _____ volts.

6. Measure the output voltage from the sensor when the throttle is closed. It was _____ volts.

7. Measure the output voltage of the sensor when the throttle is opened. It was _____ volts.

8. Move the throttle from closed to fully open and then allow it to close slowly. Describe the action of the voltmeter.

_____

_____

9. Note your conclusions from these tests.

_____

_____

_____

_____

Problems Encountered

_____

_____

_____

Instructor's Comments

_____

_____

_____

☐ JOB SHEET 25–11

Monitoring the Adaptive Fuel Strategy on an OBD II-Equipped Engine

Name _____ Station _____ Date _____

Objective

Upon completion of this job sheet, you will have demonstrated the ability to monitor the short-term and long-term fuel strategies of an OBD II system and determine if the trim is indicative of an abnormal condition. This task applies to ASE Education Foundation Tasks MAST 8.B.5 (P-1) and 8.D.1 (P-2).

Tools and Materials

A vehicle equipped with OBD II

Scan tool

Description of Vehicle

Year _____ Make _____ Model _____

VIN _____ Engine Type and Size _____

Mileage _____

PROCEDURE

1. Describe the scan tool you are using.

   Model _____

2. Connect the scan tool.

3. Start the engine.

4. Pull up the PCM's data stream display. Observe the parameter identification (PID) called short-term fuel trim (STFT), and note and record the value.

   _____

   _____

5. With the engine running, pull off a large vacuum hose, and watch the STFT value. What did it do?

   Now look at the long-term fuel trim (LTFT) value. What did it do?

   _____

   _____

   Note your conclusions from this test.

   _____

   _____

6. Reconnect the large vacuum hose. Observe the STFT and LTFT values. Record what happened.

_____

_____

7. What was happening to the fuel trim, and why did it happen?

_____

_____

_____

_____

Problems Encountered

_____

_____

_____

Instructor's Comments

_____

_____

_____

REVIEW QUESTIONS

Review Chapter 25 of the textbook to answer these questions:

1. Explain how a TP sensor provides input to the PCM.

   _____

   _____

   _____

2. Explain how a MAP sensor operates.

   _____

   _____

   _____

3. Explain how an O_2 sensor operates.

   _____

   _____

   _____

4. What is the purpose of the intake air temperature sensor?

   _____

   _____

   _____

5. List three types of airflow sensors.

   _____

   _____

   _____

CHAPTER

26

IGNITION SYSTEMS

OVERVIEW

The ignition system is responsible for delivering a high-voltage, high-temperature spark to each cylinder at precisely the right moment thousands of times each minute. The ignition process only lasts a couple of thousandths of a second, but it is critical to the correct operation of the gasoline engine.

The ignition system has many different parts that are divided into two categories: those that operate by battery voltage and those that operate at high voltage. Components that operate by battery voltage are part of the primary ignition. The primary ignition components control the operation of the ignition system. Components that operate at high voltage are part of the secondary ignition. The secondary components carry the high-voltage spark to the cylinders. Primary ignition components include the battery, ignition switch, rpm sensors, a switching device, and the primary section of the ignition coil. Secondary components include the distributor cap and rotor, spark plug wires, spark plugs, and the secondary section of the ignition coil.

 Concept Activity 26-1
Identify DI Components

1. Identify the components of the distributor ignition (DI) system in **Figure 26-1**.

    a. _____   b. _____

    c. _____   d. _____

    e. _____   f. _____

    g. _____   h. _____

    i. _____   j. _____

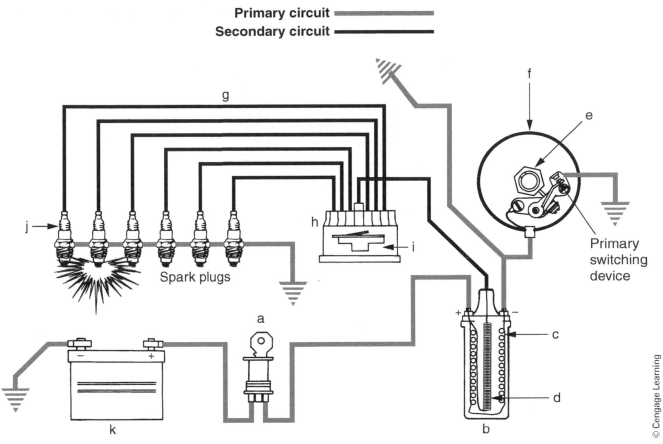

Primary circuit ━━━━━
Secondary circuit ━━━━━

Figure 26-1 The ignition system is two circuits. The primary circuit carries the low voltage, and the secondary circuit carries the high voltage.

2. List the components of the primary ignition from Figure 26-1.

_____

3. List the components of the secondary ignition from Figure 26-1.

_____

4. Which component is part of the primary and secondary ignition?

_____

_____

■

_____

   The ignition system boosts battery voltage to tens of thousands of volts. This voltage boost is required to create a very hot spark. The high heat of the spark is necessary to ignite the air/fuel mixture in the combustion chamber. At the heart of this process to boost the voltage is the ignition coil. An ignition coil is an electrical transformer **(Figure 26-2)**. Inside the coil are two windings of wire. The primary winding is usually a couple hundred turns of heavy wire. The secondary winding is several thousand turns of fine wire. The secondary winding is wound inside of the primary winding. Both windings are wound around an iron core. When current flows through the primary winding, a magnetic field surrounds the wire. When the primary circuit is opened, the magnetic field collapses past the secondary windings. As the field passes the windings of the secondary, voltage is induced in the secondary windings. Because there are more turns of wire in the secondary, the voltage generated in the secondary is increased. The voltage is provided a path to ground by the secondary output terminal, which is ultimately connected to the spark plugs. See Concept Activity 19-1 to review induced voltage **(Figure 26-3)**. **Figure 26-4** shows the relationship of the windings in an ignition coil.

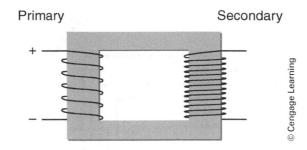

Primary   Secondary

Figure 26-2   Basic ignition coil construction.

Figure 26-3   Coiling the conductor makes a stronger magnetic field.

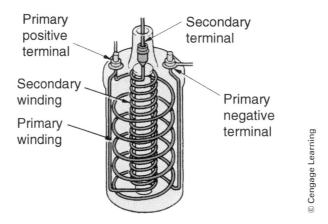

Primary positive terminal — Secondary terminal

Secondary winding — Primary negative terminal

Primary winding

Figure 26-4   A coil has a low-voltage primary winding and a high-voltage secondary winding.

 **Concept Activity 26-2
Ignition Coils**

1. If a coil has 200 primary windings and 20,000 secondary windings, what is the ratio between the primary and secondary?

_____

2. Would you expect the resistance of the primary windings to be high or low?

_____

3. Would you expect the resistance of the secondary windings to be high or low?

_____

4. What is the purpose of the iron core in the coil?

_____

_____

5. An ignition coil is a step-up transformer, meaning it boosts voltage from low to high. What are other common examples of transformers?

_____

_____

_____

■

The primary section of the ignition system controls the ignition coil(s). The primary components are responsible for charging the coil(s), triggering the timing of the spark, and adjusting the spark timing.

Dwell is the time that the coil(s) charge. **Figure 26-5** shows dwell in relation to the number of cylinders of an engine.

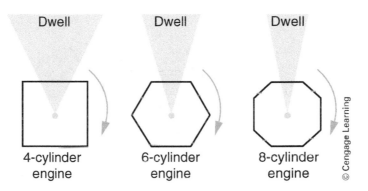

Figure 26-5   Four-cylinder engines have a greater dwell time than eight-cylinder engines.

 Concept Activity 26-3
Dwell

Dwell is the amount of time, in degrees, that the coil has current flowing through the primary windings. This time is important for the coil to fully charge.

1. What two factors determine the amount of dwell time for each cylinder?

_____

2. Describe the effects of insufficient dwell time.

_____

_____

_____

■

A pickup (rpm reference) and switching device are used to control the coil. On vehicles manufactured up until the mid-1970s, contact points were used to control the coil. Contact points are a mechanical switch used to allow current to flow through the coil primary windings. Because points are mechanical, they wear and need periodic adjustment and replacement. Since the 1970s, manufacturers have been using electronic ignition systems. These replaced the points with solid-state transistorized switches, which neither wear nor need adjustment. Instead of contact points, a pickup device and a control module are used. The pickup determines crankshaft position and speed, and supplies that information to the control module. The ignition control module (ICM) interprets the information from the pickup and uses it to determine dwell. **Figure 26-6** shows various types of ignition pickups.

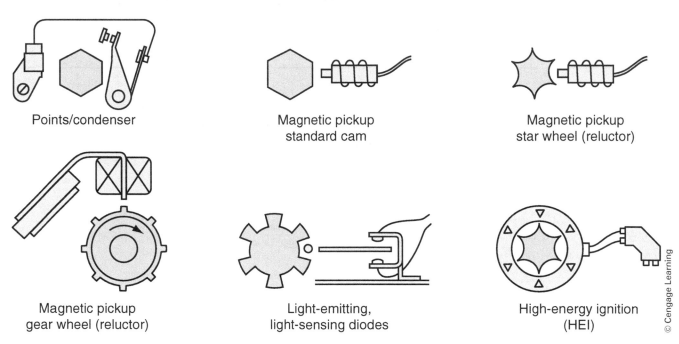

Points/condenser

Magnetic pickup
standard cam

Magnetic pickup
star wheel (reluctor)

Magnetic pickup
gear wheel (reluctor)

Light-emitting,
light-sensing diodes

High-energy ignition
(HEI)

© Cengage Learning

Figure 26-6 Many types of reluctors are used to trigger the transistor on engine ignition systems.

Ignition pickups fall into three categories: permanent magnet generators, Hall-effect generators, and optical or LED pickups.

Figures 26-7 and **26-8** show the operation of a permanent magnet pickup. As the timing disk moves into alignment with the sensor, voltage is produced. This type of sensor is found in ignition systems and as wheel speed sensors for the anti-lock brake system and traction control (ABS/TC).

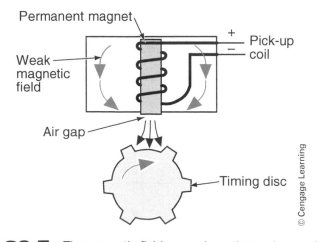

Permanent magnet

Weak
magnetic
field

+
−

Pick-up
coil

Air gap

Timing disc

© Cengage Learning

Figure 26-7 The magnetic field expands as the teeth pass the core.

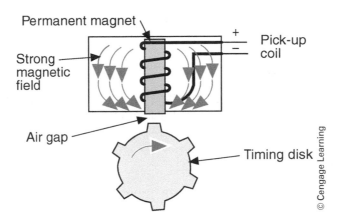

Permanent magnet

Strong magnetic field

Pick-up coil

+
−

Air gap

Timing disk

© Cengage Learning

Figure 26-8 A strong magnetic field is produced in the pick-up coil as the teeth align with the core.

Concept Activity 26-4
Permanent Magnet Sensors

Obtain a selection of permanent magnet type sensors from your instructor and a DMM.

1. Which type of voltage do you think these sensors generate and why?

_____

2. Measure the resistance of the sensor coils, and record your readings.

Resistances _____   _____

_____

3. Why do the sensors have high resistance?

_____

_____

4. What reading would an open coil show? _____

5. What reading would a shorted coil show? _____

6. Are any of the sensors you tested open or shorted? _____

7. Do any of the sensors you tested show excessively high resistance? _____

8. How would excessive resistance affect the performance of the sensor?

_____

_____

Set the DMM to measure AC voltage. Connect the DMM leads to the sensor's terminals. Rotate the sensor-timing disk, or move a magnet back and forth very close to the sensor end of the pickup.

9. Does the sensor produce AC voltage? _____

10. Why is AC voltage produced?

_____

_____

11. Describe the voltage output as you increase the rotation speed of the sensor.

_____

_____

Connect a digital storage oscilloscope or graphing multimeter (DSO/GMM) to the sensor's terminals. Rotate the sensor-timing disk, or move a magnet back and forth very close to the sensor end of the pickup. Obtain a waveform of the sensor output.

12. Draw the waveform pattern here.

13. Describe the sections of the waveform.

_____

_____

14. What happens to the waveform as the sensor rotates faster?

_____

_____

15. What type of signal does this sensor produce? _____

16. Of the three tests you have performed on these sensors—resistance, AC voltage, and waveform observation—which provides the best information for diagnostics and why?

_____

_____

_____

_____

Hall-effect sensors use magnetic fields, but in a slightly different way. **Figures 26-9** and **26-10** show how the movement of a magnetic field produces a signal from a Hall-effect sensor. The Hall effect uses a timing wheel called a shutter wheel. This disk has windows and vanes, which either allow or block the hall field, creating a digital signal called a square wave.

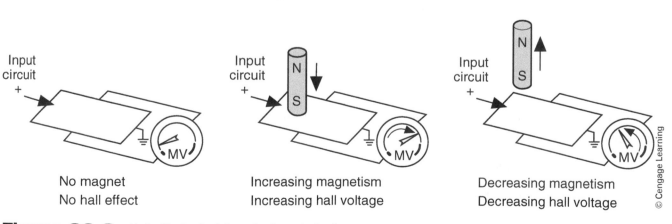

Figure 26-9 Hall-effect principles of voltage induction.

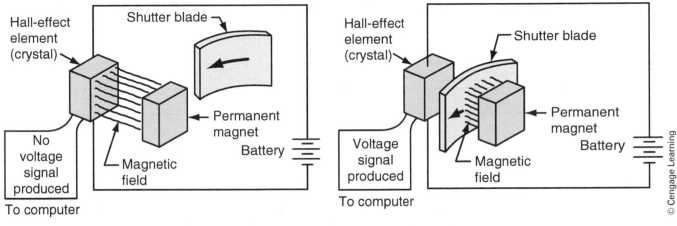

Figure 26-10 As the shutter blade passes in front of the magnet, it reduces the magnetic field, thus increasing the voltage and signal of the Hall-effect element.

Concept Activity 26-5
Hall-Effect Sensors

1. Why do Hall-effect sensors have three wires?

   Connect a DMM to the power and ground terminals of a Hall-effect sensor, and check its resistance.

2. What resistance reading did you obtain from the Hall effect? _____

3. Why do you think you got this reading?

   _____

   Connect a DMM to a Hall-effect sensor on a vehicle.

4. Sensor location and use _____

5. What type of voltage does a Hall effect produce? _____

6. Connect the DMM to the sensor output and ground terminals. Allow the sensor to rotate, and measure its voltage output with the DMM. Describe the reading on the DMM.

7. Why did you get this reading? _____

8. Why are the resistance and voltage output tests not effective for the Hall effect?

   _____

   _____

9. Connect a DSO/GMM to the sensors output wire and ground terminals. Rotate the sensor and obtain a waveform. Draw the waveform below.

10. Label the parts of the waveform you obtained.

11. How will the waveform change as the rpm of the sensor increases?

   _____

12. Why do you think Hall-effect sensors are used in place of permanent magnet sensors in some situations?

_____

_____

_____

_____

■

_____

Optical pickups use two LEDs, a light-emitting diode and a light-sensing diode, to determine rpm and position. Figure 26-6 shows a typical LED configuration. Optical pickups are used for both crankshaft speed and camshaft position. By placing holes in a rotating disc, the light from the LED either passes through the hole or is blocked. The light-sensing diode interprets the light and absence of light as an on–off digital signal similar to a Hall-effect sensor.

Concept Activity 26-6
Optical Pickups

1. Inspect an optical pickup as provided by your instructor. How many output circuits does the pickup have?

_____

2. How many slits does each part of the pickup have? _____

3. How are these slits used?

_____

_____

4. What advantages or disadvantages does the optical pickup have compared to a permanent magnet or Hall-effect sensor?

_____

_____

Connect a DSO/GMM to the pickup signal output wire(s). Rotate the sensor and obtain a waveform pattern.

5. Draw the waveform pattern here.

6. What type of waveform signal does the optical pickup produce?

_____

7. As the rotation speed increases, how does the waveform change?

_____

_____

∎

_____

The ignition pickups supply information to the ICM or the powertrain control module (PCM). The ICM or PCM uses that information to control the ignition coil and ignition timing. **Figure 26-11** shows the primary circuit. The output from the pickup triggers the control module, shown as the transistor, which controls the coil.

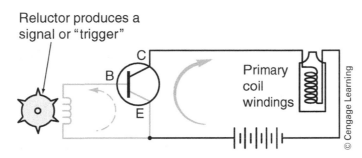

Reluctor produces a signal or "trigger"

C
B
E

Primary coil windings

© Cengage Learning

Figure 26-11 The transistor in a solid-state ignition system is used to turn off and on the primary side of the ignition coil. A small signal voltage triggers the transistor.

Ignition control modules arc usually not serviceable or able to be tested directly. The ICM uses power, ground, and the signal from the pickup to operate. If these are present and no signal from the ICM is present at the coil, the ICM is probably at fault. Before replacing the ICM, always verify its power and ground circuits. Thoroughly test the signal from the pickup to the ICM. If any of these circuits are faulty, the ICM cannot operate properly.

 Concept Activity 26-7
Ignition Control Modules

Locate the ICM, pickups, and related wiring on several vehicles.

Vehicle 1

Year _____ Make _____ Model _____

Engine _____ Ignition System Type _____

ICM location _____

Pickup type and location _____

Wire colors of pickup to ICM _____

Wire colors of ICM power and ground _____

Vehicle 2

Year _____Make _____ Model _____

Engine _____ Ignition System Type _____

ICM location _____

Pickup type and location _____

Wire colors of pickup to ICM _____

Wire colors of ICM power and ground _____

1. What factors do you think affect the location of the ICM?

_____

_____

_____

_____

■

 Concept Activity 26-8
Ignition Timing

The ignition system has to adjust the firing of the coil based on engine speed and load. As the rpm increases, the amount of time between coil firing events gets shorter. This decreases dwell time. If the dwell becomes too short, the coil will not have enough time to fully charge, and the spark output will be decreased. **Figure 26-12** shows how the spark timing must take place sooner as rpm increases. When the ICM advances timing, the process to generate the spark takes place sooner. When the spark generation process takes place later, it is called *timing retard*.

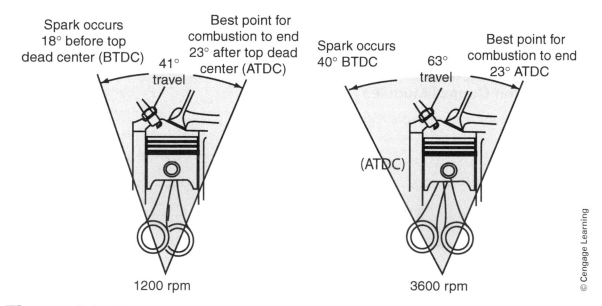

Figure 26-12   As engine speed increases, ignition must begin earlier to end by 23 degrees ATDC.

1. How would the engine performance be affected if the timing did not advance as engine rpm increases?

_____

_____

_____

2. Under what conditions does the engine not need as much timing advance?

_____

_____

3. Under what conditions might the powertrain control module (PCM) command the timing retarded?

_____

_____

In older model vehicles, timing control was achieved by a combination of centrifugal weights and a vacuum diaphragm. The centrifugal weights changed position based on engine speed. The vacuum diaphragm responded to engine load. Between the two, the ignition timing was able to advance as required, but since both were mechanical in nature, they were susceptible to wear and failure. Modern ignition systems use electronic timing control. The PCM controls timing based on input from various engine sensors.

4. List the sensors used by the PCM to determine timing needs.

_____

_____

_____

The secondary side of the ignition system is responsible for getting the spark to the proper cylinder and delivering the spark to the combustion chamber. Modern ignition systems may have a distributor cap and rotor, spark plug wires, and spark plugs. These are called distributor ignition, or DI, systems. Many engines do not have distributors; instead, they use multiple ignition coils, one for each pair of cylinders. These are called distributorless ignition, or DIS, systems. Another version of distributorless ignitions has a coil for every cylinder. These are called coil-on-plug, or COP, systems **(Figures 26-13 and 26-14)**.

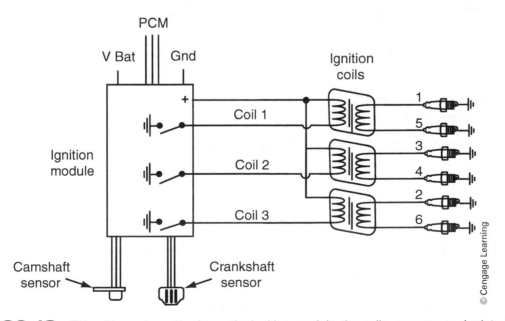

Figure 26-13 This wiring schematic shows the ignition module, the coils, sensors, and related wiring.

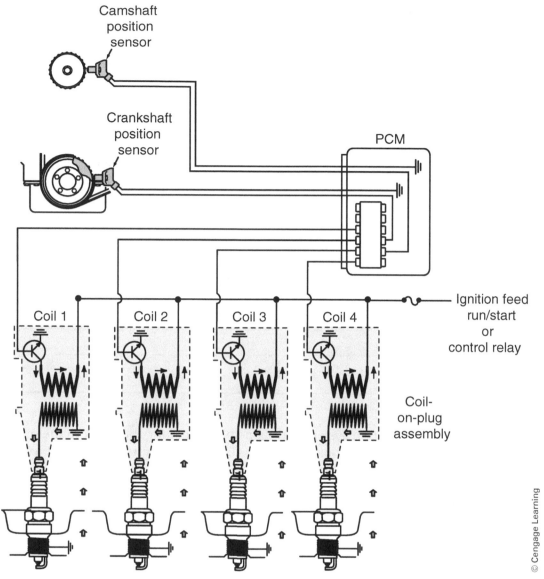

Figure 26-14 Simplified wiring diagram of a coil-on-plug (COP) system.

Concept Activity 26-9
Ignition System Types

1. What are the advantages of DIS and COP systems compared to DI systems?

_____

_____

_____

2. How many ignition coils does a V6 engine equipped with a DIS system have? _____

3. How many ignition coils does a V8 engine with a COP system have? _____

   DIS systems operate using a waste spark system. When a coil fires, it fires both of the secondary termi-nals and both spark plugs connected to that coil. One plug is on the combustion stroke, and the other is on the exhaust stroke. The plug on the combustion stroke requires more energy to ignite the air/fuel mixture than the plug on the exhaust stroke. Because the plugs are in series with each other, one plug fires from positive to negative, and the other fires from negative to positive **(Figure 26-15)**.

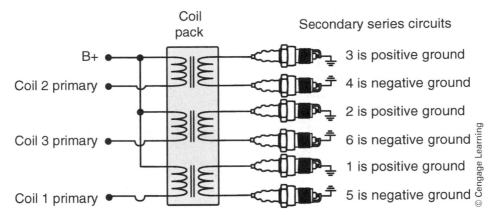

Figure 26-15 Spark plug firing for a six-cylinder engine with electronic ignition (EI).

4. Why does the plug on the combustion stroke require more voltage than the plug on the exhaust stroke?

_____

_____

5. What can happen if one of the spark plugs or plug wires fails on a DIS system?

_____

_____

◼

Spark plugs conduct the spark into the combustion chamber and are designed specifically for their application. Even though most spark plugs are very similar physically, they are not generally interchangeable from one vehicle to another. **Figure 26-16** shows the basic construction of a spark plug. One reason spark plugs are specific to their application is the spark plug heat range. **Figure 26-17** shows two different spark plugs, neither of which is correct for the engine in which it is installed.

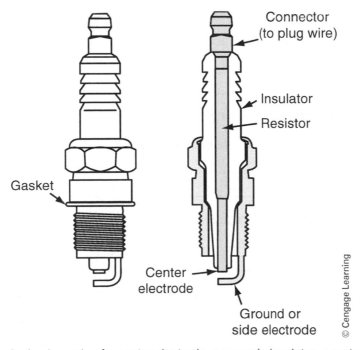

Figure 26-16 A spark plug is made of a center electrode, a ceramic insulator, a metal casing, and the side electrode.

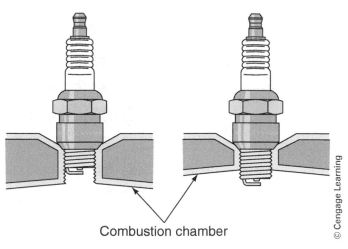

© Cengage Learning

Combustion chamber

Figure 26-17 Incorrect reaches.

 Concept Activity 26-10
Spark Plugs

Obtain a sample of spark plugs and a DMM from your instructor.

1. Examine each plug's physical characteristics. Notice the differences in the threads, fastener size, terminal type, electrode type, and seating area.

2. Measure the resistance of the plugs from the plug cable terminal to the center electrode. Resistances

_____

_____

3. Why do the plugs have high resistance?

_____

_____

Examine the plug insulation for damage.

4. What could happen if the insulation is cracked?

_____

_____

Examine the gap between the center and ground electrodes.

5. How does the spark plug gap affect the plug firing voltage?

_____

_____

6. What effect would decreasing the plug gap have on the plug, firing voltage, and cylinder performance?

_____

_____

7. What effect would increasing the plug gap have on the plug, firing voltage, and cylinder performance?

_____

_____

To carry the spark to the spark plug, a high-voltage or high-tension wire is used, often called a spark plug wire or cable. Spark plug wires are made of several layers of insulation and a flexible core material **(Figure 26-18)**. The ends of the plug wires have specially shaped boots to attach the wire to the spark plug and distributor cap or coil.

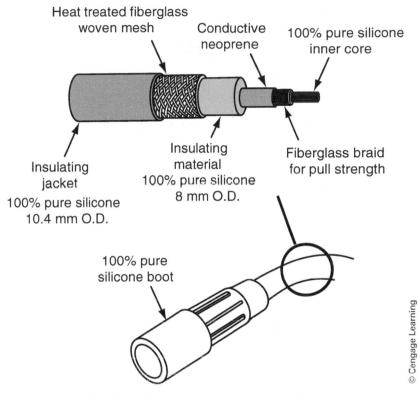

Heat treated fiberglass woven mesh

Conductive neoprene

100% pure silicone inner core

Insulating jacket 100% pure silicone 10.4 mm O.D.

Insulating material 100% pure silicone 8 mm O.D.

Fiberglass braid for pull strength

100% pure silicone boot

© Cengage Learning

Figure 26-18 Construction of a typical ignition wire.

Concept Activity 26-11
Spark Plug Wires

1. Use a DMM to measure the resistance of several spark plug wires. Record your results.

   Wire 1 resistance _____

   Wire 2 resistance _____

   Wire 3 resistance _____

2. Why do spark plug wires have high resistance?

   _____

   _____

3. Inspect the wires for damage to the insulation. Record your findings.

   _____

   _____

4. What problems can result from damaged plug wire insulation?

   _____

   _____

   _____

■

The spark plug wires are attached to the distributor cap or coils and routed to the spark plugs in the firing order. The firing order is the sequence in which each cylinder reaches the combustion cycle and the mixture is ignited. **Figure 26-19** shows examples of common firing orders.

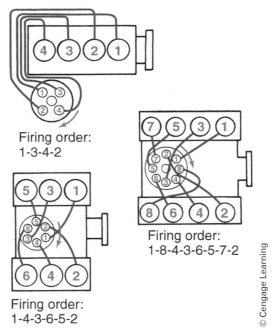

Figure 26-19   Examples of typical firing orders.

 Concept Activity 26-12
Firing Order

1. Locate the firing order diagrams for three lab vehicles with different engine and ignition types. Record the firing order for each.

   Firing order 1 _____

   Firing order 2 _____

   Firing order 3_____

2. What factors do you think apply to selecting a firing order for an engine?

   _____

   _____

3. What could result from the spark plug wires being installed in the wrong firing order?

   _____

   _____

   _____

   _____

To accurately diagnose ignition system concerns, an ignition or lab scope is needed. A scope displays voltages over time **(Figure 26-20)** and usually can be adjusted for different voltages and amounts of time.

Learning how to read and interpret ignition patterns is an important skill necessary for diagnosing both primary and secondary ignition faults.

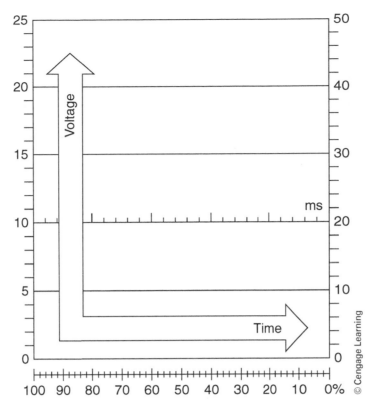

Figure 26-20 A scope displays voltage over time.

 Concept Activity 26-13
Ignition Patterns

A primary pattern displays the events occurring in the primary side of the ignition. **Figure 26-21** show the parts of a typical primary pattern. The firing section displays the voltage being used during the spark plug firing event. The coil section shows unused energy being dissipated to ground. Following the coil section is the dwell section.

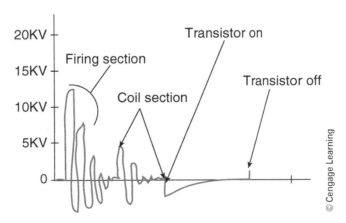

Figure 26-21 A typical primary pattern.

The secondary pattern displays the events of the secondary ignition. The firing section shows the voltage required to reach the spark plug gap and the voltage crossing the gap. The intermediate section displays the unused energy dissipating to ground. The dwell section shows the turning on and off of the primary circuit.

1. Label the parts of the secondary pattern shown in the following figure.

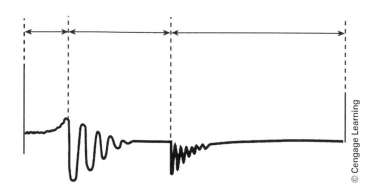

© Cengage Learning

2. What types of faults can be seen in the primary pattern? _____

_____

3. What types of faults can be seen in the secondary pattern?

_____

_____

4. Faults in the spark plugs and plug wires would be visible in what section of the secondary pattern?

_____

_____

_____

■

The following job sheet will prepare you to service and repair the ignition system.

☐ JOB SHEET 26–1

Identify Ignition Systems

Name _____ Station _____ Date _____

Objective

Upon completion of this job sheet, you will have demonstrated the ability to identify different types of ignition systems. This task applies to ASE Education Foundation Tasks MLR 8.A.1 (P-1) and AST/MAST 8.A.2 (P-1).

PROCEDURE

Examine lab vehicles and record the following information.

Distributor Ignition (DI) Systems

Year _____ Make _____ Model _____

VIN _____ Engine Type and Size _____

Mileage _____

1. Location of distributor _____

2. Location of ignition coil _____

3. Location of ICM _____

4. Visual inspection of spark plug wires and distributor cap _____

   _____

5. Type of pickup used _____

Distributorless Ignition (DIS) Systems

Year _____ Make _____ Model _____

VIN _____ Engine Type and Size _____

Mileage _____

1. Location of coils _____

2. Location of ignition module _____

3. Visual inspection spark plug wires and coils _____

   _____

4. Type and number of pickup(s) used _____

   _____

Coil-on-Plug (COP)/Coil-near-Plug (CNP) Systems

Year _____ Make _____ Model _____

VIN _____ Engine Type and Size _____

Mileage _____

1. Location of ICMs _____

2. Visual inspection of coils and boots _____

Problems Encountered

_____

_____

_____

Instructor's Comments

_____

_____

_____

 **REVIEW QUESTIONS**

Review Chapter 26 of the textbook to answer these questions:

1. Explain the construction and operation of the ignition coil.

   _____

   _____

   _____

2. Explain how an optical pickup produces a reference signal.

   _____

   _____

   _____

3. Explain why timing advance is necessary.

   _____

   _____

   _____

   _____

4. Explain why spark plug wires have high resistance.

   _____

   _____

   _____

5. Compare and contrast three types of electronic ignition systems.

   _____

   _____

   _____

   _____

IGNITION SYSTEMS DIAGNOSIS AND SERVICE

CHAPTER 27

Courtesy of Federal-Mogul Corporation

OVERVIEW

The ignition system on modern vehicles needs very little service. Many vehicles need only an occasional spark plug replacement as maintenance. Even though today's systems require very little attention, they are still susceptible to wear, damage, and failure. This chapter looks at the specifics of diagnosing and repairing the ignition system.

 Concept Activity 27-1
Identify Ignition Component Wear

1. Identify the components of the distributor ignition (DI) system that wear and may require service.

a._____ b. _____

c._____ d. _____

e._____ f. _____

■

At the heart of the ignition system is the ignition coil. Inside the coil are two windings of wire. The primary winding is usually a couple hundred turns of heavy wire. The secondary winding is several thousand turns of fine wire. The secondary winding is wound inside of the primary winding. Both are wound around an iron core. When current flows through the primary winding, a magnetic field surrounds the wire. When the primary circuit is opened, the magnetic field collapses past the secondary windings. As the field passes the windings of the secondary, voltage is induced in the secondary windings. Because there are more turns of wire in the secondary, the voltage generated in the secondary is increased. The voltage is provided a path to ground by the secondary output terminal, which is ultimately connected to the spark plugs.

Concept Activity 27-2
Ignition Coils

1. What could result from the primary windings' resistance being too high? _____

_____

2. What could result from the resistance of the secondary windings being too high? _____

_____

3. How can the output of an ignition coil be tested? _____

_____

4. What could be symptoms of low coil voltage output? _____

_____

■

Concept Activity 27-3
Dwell

1. Define what dwell represents._____
   _____

2. How does engine rpm affect dwell?_____
   _____

3. What could result from too short of a dwell period?_____
   _____

4. What kind of faults can cause the dwell to be incorrect?_____
   _____
   _____
   _____

■

 Concept Activity 27-4
Ignition Timing

The ignition system has to adjust the firing of the coil based on engine speed and load. On some engines, base timing must be adjusted after servicing the ignition system. **Figure 27-1** shows an example of how timing is checked.

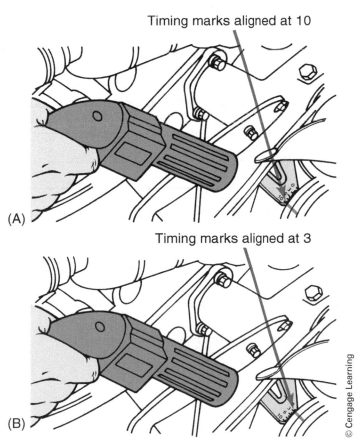

Timing marks aligned at 10

(A)

Timing marks aligned at 3

(B)

© Cengage Learning

Figure 27-1 (A) Timing marks illuminated by a timing light at 10 degrees BTDC and (B) timing marks at 3 degrees BTDC.

1. What engine mechanical components can affect the ignition timing?_____

_____

2. Describe possible symptoms of an inoperative timing control circuit._____

_____

3. What could result from ignition timing that is excessively retarded?_____

_____

4. What could result from ignition timing that is excessively advanced?_____

_____

■

Concept Activity 27-5
Spark Plugs

1. What type of driveability problems can be associated with faulty spark plugs?

_____

_____

2. What could be the result of installing spark plugs with a heat range that is too hot?

_____

_____

3. What could be the result of installing spark plugs with a heat range that is too cold? _____

_____

4. How does a fouled spark plug appear on an ignition scope pattern? _____

_____

5. Why is proper torque on the spark plug necessary? _____

_____

6. Explain why spark plugs should not be removed from hot aluminum cylinder heads. _____

_____

_____

■

Figure 27-2 shows various spark plug concerns.

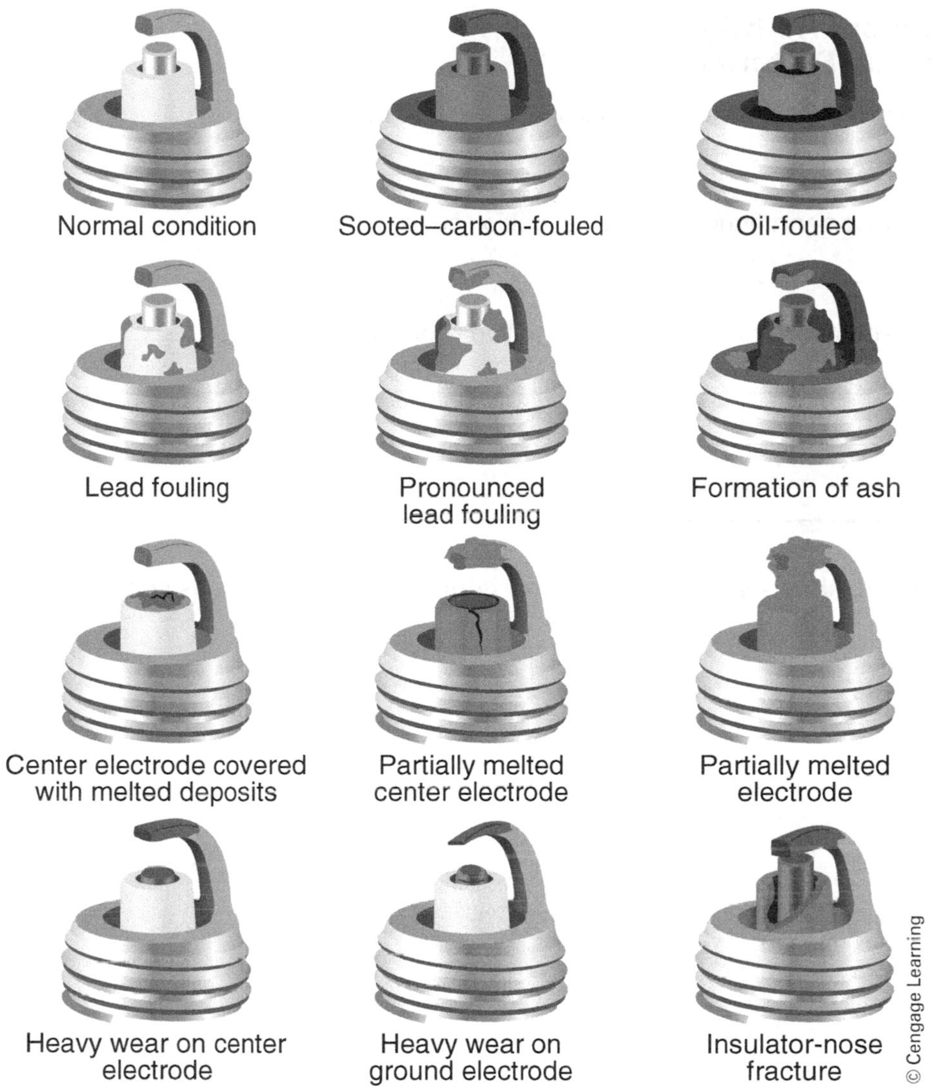

Normal condition	Sooted–carbon-fouled	Oil-fouled
Lead fouling	Pronounced lead fouling	Formation of ash
Center electrode covered with melted deposits	Partially melted center electrode	Partially melted electrode
Heavy wear on center electrode	Heavy wear on ground electrode	Insulator-nose fracture

© Cengage Learning

Figure 27-2 Spark plug conditions tell the story of the condition of the combustion chamber.

Spark plug wires need periodic replacement just like spark plugs. Over time, heat and contact with other components can lead to damage and driveability concerns.

Concept Activity 27-6
Spark Plug Wires

1. What can result from open, or high-resistance, spark plug wires? _____

_____

2. How does an open spark plug wire appear on an ignition scope pattern? _____

_____

3. How does a shorted spark plug wire appear on an ignition scope pattern? _____

_____

4. What precautions should be followed when removing and replacing spark plug wires? _____

_____

_____

■

Concept Activity 27-7
Ignition Patterns

1. What faults cause high firing lines on the ignition scope pattern? _____

_____

2. What faults cause short firing lines on the ignition scope pattern? _____

_____

3. What factors affect the spark lines on the ignition scope pattern? _____

_____

_____

The following job sheets will prepare you for ignition system diagnosis and service.

☐ JOB SHEET 27–1

Testing an Ignition Coil

Name _____ Station _____ Date _____

Objective

Upon completion of this job sheet, you will have demonstrated the ability to test an ignition coil with an ohmmeter. This task applies to ASE Education Foundation Task AST/MAST 8.C.4 (P-1).

Tools and Materials

A vehicle or a separate ignition coil

A DMM

Description of Vehicle

Year _____ Make _____ Model _____

VIN _____ Engine Type and Size _____

Mileage _____

PROCEDURE

1. Describe the general appearance of the coil.

2. Locate the resistance specifications for the ignition coil in the service manual.

   The primary winding should have _____ ohms of resistance.

   The secondary winding should have _____ ohms of resistance.

3. If the coil is still in the vehicle, disconnect all leads connected to it.

4. Connect the ohmmeter from the negative (tach) side of the coil to its container or frame. Observe the reading on the meter. The reading is _____ ohms. What does this indicate?

   _____

   _____

5. Connect the ohmmeter from the center tower of the coil to its container or frame. Observe the reading on the meter. The reading is _____ ohms. What does this indicate?

   _____

   _____

   _____

6. Connect the ohmmeter across the primary winding of the coil. The reading is _____ ohms. Compare this to specifications. What does this reading indicate?

_____

_____

_____

7. Connect the ohmmeter across the secondary winding of the coil. The reading is _____ ohms. Compare this to specifications. What does this reading indicate?

_____

_____

_____

8. Based on the preceding tests, what is your conclusion about the coil?

_____

_____

_____

Problems Encountered

_____

_____

_____

Instructor's Comments

_____

_____

_____

☐ JOB SHEET 27–2

Testing Individual Components

Name _____ Station _____ Date _____

Objective

Upon completion of this job sheet, you will have demonstrated the ability to test components of the primary and secondary ignition system. This task applies to ASE Education Foundation Tasks AST/MAST 8.C.1 (P-2), 8.C.2 (P-1), 8.C.3 (P-3), and 8.C.4 (P-1).

Tools and Materials

12-volt test light

Lab scope digital storage oscilloscope (DSO)

Digital multimeter (DMM)

Description of Vehicle

Year _____ Make _____ Model _____

VIN _____ Engine Type and Size _____

Mileage _____

Type of ignition system _____

PROCEDURE

Check the following components:

Ignition Switch

1. Turn the ignition key off and disconnect the wire connector at the module.

2. Turn the key to the RUN position.

3. With the test light, probe the power wire connection to check for voltage. Was there voltage? _____ YES _____ NO

4. Test for voltage at the bat terminal of the ignition coil. Was there voltage? _____ YES _____ NO

   Conclusions:

   _____

   _____

5. Turn the ignition switch to the OFF position.

6. Backprobe the module's input wire.

7. Connect the digital voltmeter's positive lead to the pin, and ground the negative lead to the distributor base.

8. Turn the ignition to the run position. Your voltage reading is _____.

9. Turn the ignition to the start position. Your voltage reading is _____.

Conclusions:

_____

_____

Magnetic Pickup/CKP

1. Turn the ignition off.

2. Remove the distributor cap (if applicable).

3. Connect the ohmmeter to the pickup coil terminals. Record your readings: _____ ohms.

4. The specified resistance is _____ ohms.

Conclusions:

_____

_____

5. Connect the ohmmeter from one of the pickup leads to ground. Record your readings: _____ ohms

6. The specified resistance is _____ ohms.

Conclusions:

_____

_____

7. Reinstall the distributor cap. Then connect the scope leads to the pickup coil leads.

8. Set the scope on its lowest scale.

9. Spin the distributor shaft by cranking the engine with the ignition disabled.

10. Describe the trace shown on the scope.

_____

_____

Conclusions:

_____

_____

11. Disconnect the scope.

12. Connect a voltmeter set on its low voltage scale. Describe the meter's action.

_____

_____

Conclusions:

_____

_____

Hall-Effect Sensors

1. Connect a 12-volt battery across the plus (+) and minus (–) voltage (supply current) terminals of the Hall layer.

2. Connect a voltmeter across the minus (–) and signal voltage terminals.

3. Insert a steel feeler gauge or knife blade between the Hall layer and magnet. Then remove the feeler gauge.

4. Describe what happens on the voltmeter.

   _____

   _____

   Conclusions:

   _____

   _____

5. Remove the 12-volt power source, and prepare the engine to run.

6. Set a lab scope on its low scale primary pattern position.

7. Connect the primary positive lead to the Hall signal lead; the negative lead should connect to ground or the ground terminal at the sensor's connector.

8. Start the engine, and observe the scope.

9. Record the trace on the scope.

   _____

   _____

   Conclusions:

   _____

   _____

Control Module

1. Connect one lead of the ohmmeter to the ground terminal at the module and the other lead to a good engine ground. Record your readings: _____ ohms.

2. The specified resistance is _____ ohms.

   Conclusions:

   _____

   _____

Secondary Ignition Wires

1. Remove the distributor cap with the spark plug wires attached to the cap but disconnected from the spark plugs.

2. Calibrate an ohmmeter on the ×1,000 scale.

3. Connect the ohmmeter leads from the end of a spark plug wire to the distributor cap terminal inside the cap to which the plug wire is connected. Record your readings: _____ ohms.

4. The specified resistance is _____ ohms.

Conclusions:

_____

_____

Spark Plugs

1. Remove the engine's spark plugs. Place them on a bench, arranged according to the cylinder number.

2. Carefully examine the electrodes and porcelain of each plug. Describe the appearance of each plug.

_____

_____

_____

Conclusions:

_____

_____

3. Measure the gap of each spark plug, and record your findings.

   WARNING: *Be very careful when checking the gap on platinum or iridium search plugs because a steel plug gap gauge can damage these plugs.*

_____

_____

_____

4. What is the specified gap? _____ inches (_____ mm)

Conclusions:

_____

_____

Problems Encountered

_____

_____

_____

Instructor's Comments

_____

_____

_____

☐ JOB SHEET 27–3

Setting Ignition Timing

Name _____ Station _____ Date _____

Objective

Upon completion of this job sheet, you will have demonstrated the ability to check and set the ignition timing on a distributor-type ignition system. This task applies to ASE Education Foundation Task AST/MAST 8.C.1 (P-2).

Tools and Materials

Timing light

Tachometer

Description of Vehicle

Year _____ Make _____ Model _____

VIN _____ Engine Type and Size _____

Mileage _____

PROCEDURE

1. Refer to the service manual or underhood decal and answer the following:

   Ignition timing specs _____

   Source of specifications:

   _____

   What conditions must be met before checking the timing?

   _____

   _____

   _____

   What should the idle speed be? _____ rpm

   What is the idle speed? _____ rpm

   If not within specifications, correct the idle before proceeding.

2. Connect the timing light pickup to the number-one cylinder's spark plug wire. The power supply wires on the light should be connected to the battery terminals with the proper polarity.

3. Start the engine. The engine must be idling at the manufacturer's recommended rpm, and all other timing procedures must be followed.

4. Aim the timing light marks at the timing indicator, and observe the timing marks. Timing found:

   _____ degrees.

5. If the timing mark is not at the specified location, rotate the distributor until the mark is at the specified location. Describe any difficulties you had doing this.

_____

_____

6. Tighten the distributor hold-down bolt to the specified torque. What is the specified torque? _____

7. Connect the vacuum advance hose and any other connectors, hoses, or components that were disconnected so that the timing procedure could be done.

Problems Encountered

_____

_____

_____

Instructor's Comments

_____

_____

_____

☐ JOB SHEET 27–4

Visually Inspecting an EI System

Name _____ Station _____ Date _____

Objective

Upon completion of this job sheet, you will have demonstrated the ability to conduct a preliminary inspection of the components of an electronic ignition system. This task applies to ASE Education Foundation tasks AST/MAST 8.C.2 (P-1), and 8.C.4 (P-1).

Tools and Materials

Clean shop towel

A vehicle with electronic ignition (EI)

Description of Vehicle

Year _____ Make _____ Model _____

VIN _____ Engine Type and Size _____

Mileage _____

Ignition Type   _____

PROCEDURE

1. Examine the EI system's cables.

   Are the spark plug cables pushed tightly into the coil and onto the spark plugs? _____ YES _____ NO

   Do the secondary cables have cracks or signs of worn insulation? _____ YES _____ NO

   Are the boots on the ends of the secondary wires cracked or brittle? _____ YES _____ NO

   Are there any white or grayish powdery deposits on secondary cables? _____ YES _____ NO

   Measure the resistance of the cables and note it below:

Ignition Cable Resistance		Specification
1	5	
2	6	
3	7	
4	8	

2. Examine the ignition coils.

   Do the coils show signs of leakage in the coil towers? _____ YES _____ NO

   Do the coil towers show any signs of burning? _____ YES _____ NO

   Separate the coils and inspect the underside of the coil and the ignition module wires. Are the wires loose or damaged? _____ YES _____ NO

3. Examine the connections of the primary ignition system's wiring.

   Are they tight? _____YES _____NO

4. Inspect the control module mounting and connections.

   Is it tightly mounted to a clean surface? _____ YES _____ NO

   Are the electrical connections to the module corroded? _____ YES _____ NO

   Are the electrical connections to the module loose or damaged? _____ YES _____ NO

5. Record your summary of the visual inspection. Include what looked good as well as what looked bad.

   _____

   _____

   _____

6. Based on the visual inspection, what are your recommendations?

   _____

   _____

   _____

Problems Encountered

_____

_____

_____

Instructor's Comments

_____

_____

_____

☐ JOB SHEET 27–5

Scope Testing an Ignition System

Name _____ Station _____ Date _____

Objective

Upon completion of this job sheet, you will have demonstrated the ability to test the ignition system activity by observing it on a scope. This task applies to ASE Education Foundation Tasks AST/MAST 8.B.7 (P-2), 8.C.1 (P-2), and 8.C.2 (P-1).

Tools and Materials

MM/DSO Scope

Description of Vehicle

Year _____ Make _____ Model _____

VIN _____ Engine Type and Size _____

Mileage _____

Firing order _____

PROCEDURE

1. Connect the scope leads to the vehicle.

   Instructor's Check _____

2. Set the scope to look at the secondary ignition circuit in the display or parade pattern.

3. Start the engine, and observe the height of the firing lines. Are they within 3 kV of each other? _____ If not, which cylinders are the most different from the rest? _____ Are the heights of the firing lines all below 11 kV? _____ Are the heights of the firing lines above 8 kV? _____ Based on these findings, what do you conclude?

   _____

   _____

   _____

   _____

4. Switch the scope to show the patterns in raster. Observe the length, height, and shape of the spark line for each cylinder. Describe them.

   _____

   _____

   _____

   _____

Based on these findings, what do you conclude?

_____

_____

_____

_____

5. Describe the general appearance of the intermediate section of the pattern for each cylinder.

_____

_____

_____

_____

Based on these findings, what do you conclude?

_____

_____

_____

_____

6. Describe the general appearance of the dwell section for each cylinder.

_____

_____

_____

_____

Based on these findings, what do you conclude?

_____

_____

_____

_____

7. Based on the appearance of all sections of the scope pattern for each cylinder, what are your recommendations and conclusions?

_____

_____

_____

_____

Problems Encountered

_____

_____

_____

Instructor's Comments

_____

_____

_____

 REVIEW QUESTIONS

Review Chapter 27 of the textbook to answer these questions:

1. When a spark plug firing voltage test is performed, all the firing lines are the same, but abnormally low. List three possible causes for these results.

_____

_____

_____

_____

2. A vehicle with a DI has repeated ignition coil failures. What components should be checked, and why?

_____

_____

_____

3. Explain how to test permanent magnet pickups.

_____

_____

_____

4. A vehicle with a DI has a crank no-start condition. Which of these is the least likely cause?

   a. Faulty crank position sensor

   b. High secondary resistance

   c. Open coil primary winding

   d. Open electronic spark timing circuit

5. What conditions can be diagnosed by watching the firing and spark lines of an ignition waveform pattern?

_____

_____

_____

OVERVIEW

Fossil fuels have been the main source of power for all types of transportation for over 100 years. However, depletion of the oil reserves and environmental damage are contributing to the search for alternative fuels for future generations.

Gasoline is refined from crude oil. During the refining process, crude oil is heated, and various derivatives are processed from the components of the oil. Once the gasoline is extracted from the crude oil, it undergoes extensive refining and blending with other chemicals until it reaches its final product state.

Concept Activity 28-1
Fuel Additives

Match the following gasoline additives with their functions.

a. MMT   anti-icing

b. Isopropyl alcohol   octane booster (2)

c. Phenols   rust inhibitor

d. Metal deactivators   oxidation inhibitor

e. Ethanol

■

Gasoline is rated by its octane number, typically 87–93. This number represents the fuel's resistance to knock. The higher the number, the more resistant to knock the fuel is. Most vehicle manufacturers recommend 87 octane fuels, though some engines, especially those in high-performance vehicles, need higher-octane fuel.

Concept Activity 28-2
Fuel Octane Requirements

1. List engine and operating conditions that require the use of high-octane fuels.

---

2. List any conditions that allow the use of low-octane fuels.

_____

_____

_____

∎

Diesel fuel is heavier than gasoline and contains more energy per volume. Diesel fuel is rated by its cetane rating. Cetane is a rating of the fuel's volatility. In contrast to gasoline octane, the higher the cetane rating, the more volatile the fuel is.

Since 2007, diesel fuel sold in the United States must be low-sulfur diesel fuel. The low-sulfur fuel is designed to reduce diesel exhaust emissions.

A relatively new and popular alternative is biodiesel. Biodiesel is an organic-based, renewable fuel, such as vegetable oil. Many diesel engines can run on biodiesel without modification.

Concept Activity 28-3
Diesel Fuel

1. List the types of diesel fuel that are available in your area.

_____

_____

_____

2. List the various types of biodiesel that can be used in modern diesel engines.

_____

_____

_____

_____

∎

Concept Activity 28-4
Alternative Energy

1. List some of the alternative energy systems discussed in the textbook under development for automotive use.

_____

_____

_____

2. What are some service and repair differences you foresee for these vehicles?

_____

_____

_____

3. What do think will be the challenges of servicing vehicles in the next 10 years?

_____

_____

_____

4. How do you think the use of high-voltage hybrid drive systems will affect you as a service technician?

_____

_____

_____

_____

■

The following job sheet will prepare you to service alternatively fueled vehicles.

☐ JOB SHEET 28–1

Identifying Fuel System Types

Name _____ Station _____ Date _____

Objective

Upon completion of this job sheet, you will have demonstrated the ability to determine vehicle fuel system types. This task applies to ASE Education Foundation Tasks MLR 8.A.1 (P-1), and AST/MAST 8.A.2 (P-1).

Description of Vehicle

Year _____ Make _____ Model _____

VIN _____ Engine _____

Mileage _____

PROCEDURE

1. Examine the outside of the vehicle for any badges indicating fuel system type. Describe any fuel system identification badges used on this vehicle.

   _____

   _____

2. Explain the differences between the fuel system on this vehicle compared to a gasoline-powered vehicle.

   _____

   _____

3. Are there any special fuel system warning labels on the vehicle? If so, describe them.

   _____

   _____

4. List the warnings and precautions provided in the owner service manual.

   _____

   _____

Problems Encountered

_____

_____

_____

Instructor's Comments

_____

_____

_____

REVIEW QUESTIONS

Review Chapter 28 of the textbook to answer these questions:

1. List 10 products derived from crude oil.

   _____

   _____

   _____

   _____

2. Define octane rating and describe how it is determined.

   _____

   _____

   _____

3. Explain the benefits and disadvantages of E85.

   _____

   _____

   _____

4. Explain why ultra-low sulfur diesel is required in the United States.

   _____

   _____

   _____

5. Explain how diesel combustion differs from gasoline engine combustion.

   _____

   _____

   _____

© Cengage Learning 2015

CHAPTER 29

FUEL DELIVERY SYSTEMS

OVERVIEW

The fuel delivery system contains the components that store, filter, and supply fuel to the engine. Many years ago, the fuel delivery system consisted of a fuel tank, a fuel tank valve, a sediment bowl, and a gravity-fed carburetor. Today's systems still use fuel tanks but share very little else with the systems of the past.

A modern fuel delivery system is comprised of a fuel tank, fuel lines, fuel pump, and filter. Fuel tanks can be either stamped steel or plastic. Plastic is lighter and does not corrode as steel does. Fuel lines can be steel, plastic, and rubber. Most vehicles use a combination of all three types of lines in various locations in the system. The fuel pump draws the fuel from inside the fuel tank and supplies it under pressure to the fuel injection system. As the fuel travels from the pump to the engine, it passes through a filter to remove small contaminates.

Fuel system components should be inspected for signs of leakage, damage, or rust. Steel fuel tanks and lines are susceptible to rust-through and should be carefully inspected.

WARNING: *Do not use a metal-cage trouble light when inspecting the fuel system.*

 Concept Activity 29-1
Fuel Delivery Systems

Examine a vehicle provided by your instructor, and note the following:

Year _____ Make _____ Model _____

1. Fuel system type _____

2. Fuel tank construction and location _____

3. Fuel line construction _____

4. Fuel filter location _____

5. Fuel filter and line connection types _____

■

Since the inception of OBD II in the mid-1990s, vehicles have to have on-board fuel vapor recovery. This system, called the evaporative emission, or EVAP, system, prevents the release of fuel vapors into the atmosphere. Instead, the vapors are collected and streamed into the intake system for burning. However, for the EVAP system to operate correctly, the fuel delivery system must be leak-free. The OBD II system can either pressurize or pull a vacuum on the fuel tank to test for system leaks. One source of EVAP system leaks is the incorrect reinstallation of the gas cap after refueling. Many customers have had their malfunction indicator lamp (MIL) come on after filling their gas tanks. If the gas cap does not seal completely, the EVAP system test will detect a leak, set a DTC, and illuminate the MIL.

Note: *Newer model vehicles have a "check fuel cap" light to warn the driver if the cap is not properly sealed.*

Concept Activity 29-2
Fuel Filler Designs

Using vehicles provided by your instructor, examine several gas cap and filler neck designs.

Year _____ Make _____ Model _____

1. Cap and filler neck type (threaded or turn and lock) _____

Year _____ Make _____ Model _____

2. Cap and filler neck type _____

Year _____ Make _____ Model _____

3. Cap and filler neck type _____

4. Do any of these vehicles have a gas cap warning light on the instrument panel? _____

5. A faulty gas cap can cause what problems? _____

_____

Before any work is performed on the fuel system, the fuel pressure must be relieved. Fuel system pressure can range from 40 psi to over 80 psi on PFI systems and can be over 2,000 psi on GDI systems. If fuel pressure is not relieved and the system opened, pressurized fuel will spray out possibly causing personal injury and creating a fire or explosion hazard.

Concept Activity 29-3
Disable the Fuel System

Using the service information, determine how to disable the fuel system and relieve fuel pressure on two vehicles as provided by your instructor.

Vehicle 1

Year _____ Make _____ Model _____

Fuel system type _____

Fuel system operating pressure _____

Method to relieve fuel pressure _____

Vehicle 2

Year _____ Make _____ Model _____

Fuel system type _____

Fuel system operating pressure _____

Method to relieve fuel pressure _____

_____

The following job sheets will provide you with service and testing procedures for the fuel delivery system.

☐ JOB SHEET 29–1

Relieving Fuel System Pressure in an EFI System

Name _____ Station _____ Date _____

Objective

Upon completion of this job sheet, you will have demonstrated the ability to relieve pressure from the fuel lines and system on a vehicle equipped with electronic fuel injection. This task applies to ASE Education Foundation Tasks AST 8.D.6 (P-2) and MAST 8.D.7 (P-2).

Tools and Materials

Clean shop towels

Pressure gauge with adapters

Bleed hose

Approved gasoline container

Description of Vehicle

Year _____ Make _____ Model _____

VIN _____ Engine Type and Size _____

Mileage _____

PROCEDURE

Note: *Because electronic fuel injection (EFI) systems have a residual fuel pressure, this pressure must be relieved before disconnecting any fuel system component. Failure to relieve the fuel pressure on systems prior to fuel system service may result in gasoline spills, serious personal injury, and expensive property damage.*

1. Loosen the fuel tank filler cap to relieve any fuel tank vapor pressure.

2. Wrap a shop towel around the fuel pressure test port on the fuel rail, and remove the dust cap from this valve. Describe the location of the test port. _____

_____

3. Connect the fuel pressure gauge to the fuel pressure test port on the fuel rail.

4. Install the bleed hose on the gauge in an approved gasoline container, and open the gauge bleed valve to relieve fuel pressure from the system into the gasoline container. Be sure all the fuel in the bleed hose is drained into the gasoline container.

5. Describe any problems encountered while following this procedure.

_____

_____

Note: *On EFI systems that do not have a fuel pressure test port, such as most throttle body injection (TBI) systems, follow these steps for fuel system pressure relief:*

1. Loosen the fuel tank filler cap to relieve any tank vapor pressure.

2. Remove the fuel pump fuse or relay.

Fuse/Relay Location _____

3. Start and run the engine until the fuel is used up in the fuel system and the engine stops.

4. Engage the starter for 3 seconds to relieve any remaining fuel pressure.

5. Disconnect the negative battery terminal to avoid a possible fuel discharge if someone accidentally attempts to start the engine.

Problems Encountered

_____

_____

_____

Instructor's Comments

_____

_____

_____

☐ JOB SHEET 29–2

Testing Fuel Pump Pressure in an EFI System with a Vacuum Regulator

Name _____ Station _____ Date _____

Objective

Upon completion of this job sheet, you will have demonstrated the ability to test fuel pressure on a vehicle equipped with electronic fuel injection. This task applies to ASE Education Foundation Task AST/MAST 8.D.3 (P-1).

Tools and Materials

Clean shop towels

Pressure gauge with adapters

Approved gasoline container

Hand-operated vacuum pump

Description of Vehicle

Year _____ Make _____ Model _____

VIN _____ Engine Type and Size _____

Mileage _____

PROCEDURE

1. Look up the specifications for the fuel pump. Fuel pump pressure specifications: _____ psi.

   _____

   _____

2. Carefully inspect the fuel rail and injectors for signs of leaks. Record findings.

   _____

   _____

3. Connect the fuel pressure tester to the test port on the fuel rail.

4. Connect a hand-operated vacuum pump to the fuel pressure regulator.

5. Turn the ignition on, and observe the fuel pressure readings. Note your readings: _____ psi.

6. Compare the readings to specifications. What is indicated by the readings?

   _____

   _____

7. Create a vacuum at the pressure regulator with the vacuum pump. What happened to the fuel pressure?

_____

_____

8. Conclusions:

_____

_____

Problems Encountered

_____

_____

_____

Instructor's Comments

_____

_____

_____

☐ JOB SHEET 29–3

Testing Fuel Pump Pressure in a Returnless EFI System

Name _____ Station _____ Date _____

Objective

Upon completion of this job sheet, you will have demonstrated the ability to test fuel pressure on a vehicle equipped with a returnless fuel injection system. This task applies to ASE Education Foundation Task AST/MAST 8.D.3 (P-1).

Tools and Materials

Scan tool

Description of Vehicle

Year _____ Make _____ Model _____

VIN _____ Engine Type and Size _____

Mileage _____

PROCEDURE

1. Look up the specifications for the fuel pressure. Fuel pump pressure specifications: _____ psi.

   _____

   _____

2. Locate the fuel pressure testing procedures in the service information. Summarize the procedure.

   _____

   _____

3. Carefully inspect the fuel rail and injectors for signs of leaks. Record findings.

4. Connect the scan tool to the DLC.

5. Turn the ignition on and observe the fuel pressure readings on the scan tool. Note your readings: _____ psi.

6. Compare the readings to specifications. What is indicated by the readings?

   _____

   _____

7. Start the engine and note the fuel pressure: _____ psi.

8. Run the engine at 2000 rpm, and note the fuel pressure: _____ psi.

9. Conclusions:

   _____

   _____

Problems Encountered

_____

_____

_____

Instructor's Comments

_____

_____

_____

☐ JOB SHEET 29–4

Replacing an In-Line Fuel Filter

Name _____ Station _____ Date _____

Objective

Upon completion of this job sheet, you will have demonstrated the ability to locate and replace an in-line fuel filter. This task applies to ASE Education Foundation Tasks MLR 8.C.1 (P-2), and AST/MAST 8.D.4 (P-2).

Tools and Materials

Hose plugs or pinch-off clamps

Description of Vehicle

Year _____ Make _____ Model _____

VIN _____ Engine Type and Size _____

Mileage _____

PROCEDURE

1. Refer to the service manual for the location of the in-line fuel filter of the vehicle to be serviced.

   Location _____

   Note: *On many of the cars, the fuel filter is located under the vehicle near the fuel tank.*

2. Using the recommended procedure, relieve the fuel pressure from the lines.

   Procedure Used _____

3. Place a container under the filter to catch the fuel that will drip out when servicing the fuel filter.

4. Disconnect the lines from the filter.

5. Remove the mounting bracket, and remove the old filter.

6. Install the new filter, and reinstall the hold-down bracket.

   Note: *Pay attention to the direction of fuel flow recommended on the new filter.*

7. Reinstall the fuel lines to the filter.

8. Start the engine. Check the filter and lines for leaks.

Problems Encountered

_____

_____

_____

Instructor's Comments

_____

_____

_____

REVIEW QUESTIONS

Review Chapter 29 of the textbook to answer these questions:

1. How does engine load affect fuel pump pressure?

_____

_____

_____

2. List two causes of higher than normal fuel pressure.

_____

_____

_____

3. What effect can a restricted fuel filter have on engine performance?

_____

_____

_____

4. Explain the function of an inertia switch.

_____

_____

_____

5. Explain how a returnless fuel system controls fuel pressure.

_____

_____

_____

ELECTRONIC FUEL INJECTION

OVERVIEW

All new vehicles sold in the United States are fuel injected. Prior to the early 1990s, cars and trucks used carburetors and/or fuel injection systems. Carburetors provided a mechanical method of delivering air and fuel into the engine. To improve the efficiency of carburetors while fuel injection systems were being introduced, manufacturers added computer control to existing carburetor designs. Even with these improvements, carburetors could not attain the efficiency or reliability offered by fuel injection.

Several types of electronic fuel injection can be found in modern cars and light trucks. Throttle body injection (TBI) was used for many years during the transition to port fuel injection. A TBI system uses one or two injectors centrally mounted with a throttle body. The fuel is injected directly over the throttle plates, where it is carried to the cylinders with the incoming air. Although TBI fuel injection systems were an improvement over carburetors, they are not as efficient as port fuel injection.

Multiport fuel injection (MPI) generally uses one injector per cylinder. Each injector delivers fuel to the intake port, where it is drawn into the cylinder on the intake stroke. MPI systems are bank fired. A bank-fired system triggers all the injectors on the same bank of the engine to open at the same time, regardless of whether the intake valve is open. This can cause the fuel to start to evaporate once it hits the hot intake valve. This can also lead to intake valve deposits, where some of the fuel adheres to the valve, creating a spongelike buildup. Engines today use sequential port fuel injection, or sequential fuel injection (SFI). This means the injectors are fired in the same firing order as the ignition system. Each injector opens only when the intake valve is open, allowing all of the fuel to enter the cylinder.

Centralport injection system, or CPI, is a variation of the TBI and port fuel injection (PFI) systems in use today. Used by General Motors, this system uses a large centrally located injector that feeds fuel to individual nozzles for each cylinder.

Relatively new to gasoline engines, but currently in production on many vehicles, is gasoline direct-injection (GDI) systems. These systems use an injector that injects the fuel straight into the combustion chamber. GDI systems can run at very lean mixtures and therefore save on fuel.

 Concept Activity 30-1
Identify Fuel Injection Types

1. Examine several lab vehicles to determine the type of fuel system they use.

    Vehicle 1:

      Year _____ Make _____ Model _____

      VIN _____ Engine Type and Size _____

   a. Fuel system type (TBI, MPI, SFI, CPI, GDI): _____

   b. How many fuel injectors does the system use? _____

   c. Use the service information and record the fuel system pressure. _____

   d. Is this a returnless fuel system? _____

Vehicle 2:

Year _____ Make _____ Model _____

VIN _____ Engine Type and Size _____

a. Fuel system type (TBI, MPI, SFI, CPI, GDI): _____

b. How many fuel injectors does the system use? _____

c. Use the service information to find the fuel system pressure. _____

d. Is this a returnless fuel system? _____

2. How is correct fuel pressure maintained on a returnless fuel injection system?

_____

_____

3. How is correct fuel pressure maintained on fuel return fuel injection system?

_____

_____

■

An electronic fuel injection system depends on the information provided by the input sensors to supply the correct amount of fuel for each cylinder. Fuel injectors are commanded open by the powertrain control module (PCM) based on engine speed, load, temperature, and other operating conditions. Injector pulse width is the term for the amount of time an injector is open. Injector pulse width varies from about 1 ms to 10 ms, depending on operating conditions. When more fuel is required, the PCM increases the injector pulse width.

Concept Activity 30-2
Injector Pulse-Width

Explain how these operating conditions affect injector pulse width.

A/C request _____

P/S request _____

Deceleration _____

Increased electrical load _____

EVAP purge _____

EGR flow _____

■

Until the recent introduction of electronic throttle control, fuel injection systems relied on an idle air control (IAC) system to provide changes in idle speeds. The IAC typically uses a small bidirectional motor to change the amount of air allowed to bypass the throttle plates. By changing the airflow through the IAC passage, idle speed is adjusted based on engine load demands. Vehicles that have electronic throttle control use small, high-speed motors to open and close the throttle plate based on input from the accelerator pedal. Because the PCM is controlling the throttle plate, a separate IAC is not necessary. The PCM opens and closes the throttle as needed to control the engine idle speed.

Concept Activity 30-3
Idle Control

Connect a scan tool to an OBD II–equipped vehicle. Navigate to the output controls menu. Locate the IAC control.

1. What are the engine's commanded and actual idle speeds? _____

2. What is the throttle position angle or percent? _____

3. Command the idle speed to increase. Does the engine respond? _____

4. What could be the cause if the engine speed did not increase? _____

5. Command the idle speed back to its normal setting. _____

6. What could cause the engine's idle speed to be higher or lower than specified? _____

∎

Table 30-1 provides basic fuel injection troubleshooting information.

TABLE 30-1 Port Fuel Injection System Troubleshooting.

Symptom	Possible Cause
Preliminary checks	Inspect fuel system for fuel leaks.
	Test battery state of charge.
	Inspect wiring and connections.
	Check coolant level.
	Test ignition system.
	Inspect air cleaner and preheat system.
	Check fuel system pressure.
	Inspect fuel lines for restrictions.
	Inspect vacuum hoses for leaks and restrictions.
Hard start when cold or rough idle when cold	Check coolant level
	Fuel pressure bleed-down
	Cold start injector
	Leaking manifold gasket or base gasket
	IAT sensor
	Wrong PCV valve
	Warmup regulator
	Injector
	Pressure regulator
Hesitation or surging— hot or cold	Coolant temperature sensor
	Low fuel system pressure

© Cengage Learning 2015

(Continued)

TABLE 30-1 (*Continued*)

Symptom	Possible Cause
	Restricted air intake system
	TP sensor defective or not adjusted correctly
	Mass airflow sensor
	IAT sensor
	Air leak in air intake system
	Defective oxygen sensor
	Defective computer
Hard start—hot	Bleeding injector
	MAP sensor
	Pressure regulator
Rough idle—hot	MAP sensor and vacuum hose.
	Check coolant level.
	Check TP sensor. Adjust or replace as required.
	Examine injector for variation in spray pattern. Clean or replace injector as required.
	Oxygen sensor
	Defective computer
	Inspect idle speed control device.
	Service or replace as required.
Stalling	Inspect idle speed control device.
	Service or replace as required.
	Check TP sensor. Adjust or replace as required.
	Inspect MAP sensor and vacuum hose.
	Service or replace as required.
	Inspect induction hoses.
Poor power	Observe injector spray pattern.
	Clean or replace injector as required.
	Test fuel pump pressure.
	Examine strainer. Replace as required.
	Inspect fuel filter. Replace as required.
	Check the setting and bleed-down rate of the pressure regulator.
	Service or replace as required.

☐ JOB SHEET 30–1

EFI Visual Inspection

Name _____ Station _____ Date _____

Objective

Upon the completion of this job sheet, you will have demonstrated the ability to perform a visual inspection of the EFI system. This task applies to ASE Education Foundation Tasks AST/MAST 8.D.8 (P-1), and MAST 8.D.1 (P-2).

Description of Vehicle

Year ___ _____ Make _____ Model _____

VIN _____ Engine Type and Size _____

Mileage _____

PROCEDURE

Before more extensive testing takes place, a thorough visual inspection should be performed and any obvious faults corrected.

1. Measure battery voltage with the KOEO and KOER.

   KOEO voltage _____ KOER voltage _____

2. Inspect the battery cables and connections. Note any problems found. _____

   _____

   _____

3. Inspect the underhood wiring harness routing. Is the wiring properly routed? _____

4. Check the electrical connections for positive fit. Are all electrical connections intact? _____

5. Inspect vacuum hoses for proper routing, connection, and condition. Note your findings.

   _____

   _____

6. Is the PCV system operating correctly?_____

7. Are all visible emission control devices correctly hooked up? _____

8. With the engine running, listen carefully for evidence of vacuum or exhaust leaks. Note your findings.

   _____

   _____

   _____

9. Inspect the ignition system components. With the engine off, ensure all spark plug cables are securely fastened and routed properly. Check the cable end terminals for signs of corrosion or arcing. Note your findings. _____

10. Inspect the air induction ductwork. Does the air filter need replaced? _____

11. Are any leaks evident in the induction hoses? _____

12. Inspect the throttle bore for signs of carbon buildup. Describe the condition of the throttle plate(s) and bore. _____

_____

_____

If the throttle needs to be cleaned, complete steps 14–18. _____

13. With the engine running at idle, carefully listen at each injector for its opening and closing. Does each injector click? _____

14. If the throttle shows signs of carbon buildup, locate the manufacturer's procedure for cleaning the throttle.

15. If the throttle is cleaned, the base idle speed should be checked and adjusted if applicable. Describe the procedure to check and adjust the idle speed. _____

_____

16. Idle speed specification _____

17. Actual idle speed _____

18. Method used to measure idle speed _____

Problems Encountered

_____

_____

_____

Instructor's Comments

_____

_____

_____

 **REVIEW QUESTIONS**

Review Chapter 30 of the textbook to answer these questions:

1. Explain the operation of an electronic fuel injector.

_____

_____

_____

2. Describe how the PCM determines how long to keep the fuel injector open.

_____

_____

_____

3. Explain how STFT and LTFT are used for fuel control.

_____

_____

_____

4. Explain the operation of the IAC system.

_____

_____

_____

5. Explain the benefits of GDI.

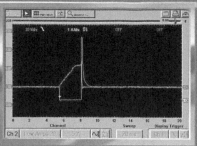

FUEL INJECTION SYSTEM DIAGNOSIS AND SERVICE

OVERVIEW

Although electronic fuel injection systems are more efficient and reliable, periodic service and repairs are still required. Common service procedures include fuel filter replacement and fuel injector cleaning. Fuel injectors are susceptible to heat damage and clogging from fuel deposits.

Concept Activity 31-1
Fuel Filters

Year _____ Make _____ Model _____

Engine Type and Size _____ Fuel System Type _____

1. Locate the fuel filter and describe its location. _____

_____

2. What types of connections are used to connect the filter to the lines? _____

_____

3. What is the manufacturer's recommended replacement interval? _____

_____

4. What driveability symptoms could a restricted fuel filter cause? _____

_____

■

Concept Activity 31-2
Fuel Injectors

Year _____ Make _____ Model _____

Engine Type and Size _____ Fuel System Type _____

1. How many fuel injectors does this system use? _____

2. Does this system use a cold-start injector? _____

3. Explain the operation of a fuel injector. _____

_____

4. What factors determine the length of the injector pulse width? _____

_____

_____

5. What is the purpose of the fuel pressure regulator? _____

_____

6. How is the fuel pressure regulated on a returnless system? _____

_____

7. What can cause fuel pressure to be higher than specified? _____

_____

8. How can higher than specified fuel pressure affect engine performance? _____

_____

9. How can a faulty fuel pressure regulator affect engine performance? _____

_____

10. How can lower than specified fuel pressure affect engine performance? _____

_____

11. Explain how a dirty or restricted fuel filter can affect driveability. _____

_____

12. Describe how to test for dirty or restricted fuel injectors. _____

_____

_____

_____

■

The following job sheets will prepare you for fuel injection system diagnosis and service.

☐ JOB SHEET 31–1

Checking the Operation of the Fuel Injectors on an Engine

Name _____ Station _____ Date _____

Objective

Upon completion of this job sheet, you will have demonstrated the ability to check the operation of the fuel injectors on an engine using a variety of test instruments and techniques. This task applies to ASE Education Foundation Task AST/MAST 8.D.7 (P-2).

Tools and Materials
Current clamp

Lab scope

Description of Vehicle

Year _____ Make _____ Model _____

VIN _____ Engine Type and Size _____

Mileage _____

PROCEDURE

1. Inspect the electrical connectors for the injectors. Note any problems.

   _____

2. Locate a point to install the current clamp to measure injector current draw. Note the testing location.

   _____

3. Connect the current clamp to the scope and configure for measuring injector current. Start the engine and obtain a waveform. Compare the waveform to that shown in the following figure.

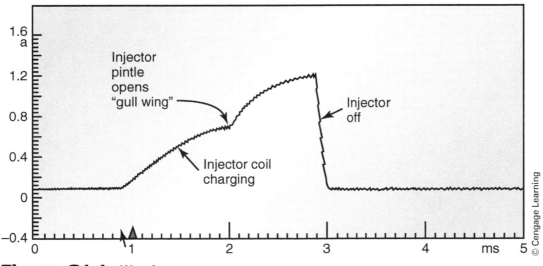

Figure 31-1 Waveform.

4. How do the waveforms compare?

_____

_____

5. Describe how the waveform from a shorted injector would look.

_____

_____

6. Describe how a waveform from an open injector would look.

_____

_____

_____

_____

7. Summarize the results of this test.

_____

_____

_____

_____

Problems Encountered

_____

_____

_____

Instructor's Comments

_____

_____

_____

☐ JOB SHEET 31–2

Conducting an Injector Balance Test

Name _____ Station _____ Date _____

Objective

Upon completion of this job sheet, you will have demonstrated the ability to perform an injector balance test on a port-injected engine. This task applies to ASE Education Foundation Tasks AST/MAST 8.D.7 (P-2), and MAST 8.D.1 (P-2).

Tools and Materials

Fuel pressure gauge

Injector tester

Description of Vehicle

Year _____ Make _____ Model _____

VIN _____ Engine Type and Size _____

Mileage _____

PROCEDURE

1. Carefully inspect the fuel rail and injector for fuel leaks. Do not proceed if fuel is leaking. Note your findings. _____

   _____

   _____

2. Connect the fuel pressure gauge to the test port on the fuel rail.

3. Disconnect the electrical connector to the number-one injector.

4. Connect the injector tester lead to the injector terminals.

5. Connect the injector tester power supply leads to the battery terminals.

6. Cycle the ignition switch several times until the specified pressure appears on the fuel pressure gauge.

   The specified pressure is _____ psi (_____ kPa).

   Your reading was _____ psi (_____ kPa).

   Comparing your reading to the specified pressure indicates:

   _____

   _____

7. Depress the injector tester switch and record the pressure on the pressure gauge. The reading was _____ psi (_____ kPa).

8. Move the injector tester to injector 2.

9. Cycle the ignition switch several times until the specified pressure appears on the fuel pressure gauge.

   Your reading was _____ psi (_____ kPa). Comparing your reading to the specified pressure indicates:

   _____

   _____

10. Depress the injector tester switch and record the pressure on the pressure gauge. The reading was _____ psi (_____ kPa).

11. Continue the same sequence until all injectors have been tested.

    Reading on injector 3 was _____ psi (_____ kPa).

    Reading on injector 4 was _____ psi (_____ kPa).

    Reading on injector 5 was _____ psi (_____ kPa).

    Reading on injector 6 was _____ psi (_____ kPa).

    Reading on injector 7 was _____ psi (_____ kPa).

    Reading on injector 8 was _____ psi (_____ kPa).

12. Summarize the results. _____

    _____

    _____

    _____

Problems Encountered

_____

_____

_____

Instructor's Comments

_____

_____

_____

REVIEW QUESTIONS

Review Chapter 31 of the textbook to answer these questions:

1. What is the purpose of a fuel injector balance test?

   _____

   _____

   _____

2. A signal from an O_2 sensor that stays above 800 mV will cause the PCM to do what?

   _____

   _____

   _____

3. Explain what positive and negative fuel trim numbers represent.

   _____

   _____

   _____

4. Compare and contrast a traditional O_2 sensor and an air/fuel ratio sensor.

   _____

   _____

   _____

5. Explain how to test an injector on a gasoline direct injection (GDI) system.

   _____

   _____

   _____

INTAKE AND EXHAUST SYSTEMS

OVERVIEW

The combustion process depends on the engine cylinders' ability to draw in a fresh supply of air/fuel mixture and their ability to get rid of the exhaust after combustion has occurred. This chapter focuses on these two main engine support systems.

The intake and exhaust systems are critical to the efficient operation of an engine. An engine needs air to operate. This air is drawn into the cylinders through the intake or intake induction system during the intake stroke of the pistons. The air is then mixed with fuel and delivered to the combustion chambers.

After combustion, burned gases are present in the combustion chamber. In order to make room for a fresh supply of intake air and fuel, those exhaust gases must be removed from the combustion chamber during the exhaust stroke of the pistons. That is the function of the exhaust system.

THE AIR INDUCTION SYSTEM

Although there seems to be an endless supply of air outside the engine, often not enough air is available for the engine. Because the actual time of the intake stroke is so short, there is little time to gather, direct, and force air into the cylinders. A vehicle's air induction system is designed to move as much air as possible and as quietly as possible. This statement is more true for high-performance vehicles than normal vehicles, simply because engine performance is directly proportional to intake airflow.

A typical intake system consists of an air cleaner assembly, air filter, air ductwork, and intake manifold. These components are designed to quickly deliver a fresh and clean supply of air to the engine's cylinders. Their design determines the maximum amount of air that can be delivered, and the actual amount is determined by the position of the engine's throttle plates and the vacuum formed on the pistons' intake stroke.

Concept Activity 32-1
Identify Air Induction Components

On a vehicle, locate and identify the components of the air induction system.

1. Examine the air filter element. Describe its condition. _____

2. An excessively dirty air filter can be restrictive to airflow. What can result from restricted or decreased airflow? _____

3. Inspect the air box for debris. Mice often make nests in air filter housings. Note your findings. _____

4. Inspect the induction hoses and clamps. Thoroughly check the hoses for signs of cracks. Describe your findings. _____

5. Describe what problems can result from air induction leaks. _____

_____

■

EXHAUST SYSTEM COMPONENTS

A highly efficient intake system is worthless if the exhaust gases cannot leave the combustion chamber. For the cylinders to form a strong vacuum or low pressure, they need to be empty. The exhaust system has the responsibility of removing the exhaust from the cylinders. As the piston moves up on its exhaust stroke, the exhaust gases are pushed out past the exhaust valve into the exhaust system. To be able to exit the exhaust system, the exhaust gases must remain under pressure and be hotter than the outside air. As the exhaust moves through the exhaust system, it cools some, and the pressure drops. The flow of exhaust will always move from a point of higher pressure to a point of lower pressure. Therefore, as long as the pressure right in front of the moving exhaust gases is lower than the pressure of the exhaust, the exhaust will continue to flow out. An efficient exhaust system ensures that this happens.

A typical exhaust system contains an exhaust manifold and gasket; exhaust pipe, seal, and connector pipe; intermediate pipes; catalytic converter(s); a muffler and resonator; a tailpipe; heat shields; and many clamps, gaskets, and hangers.

 Concept Activity 32-2
Inspect Exhaust System

For the engine to operate correctly, the exhaust system must be sealed and intact.

1. Raise a vehicle, and inspect the exhaust system from the manifold gasket back to the tailpipe. Note your findings. _____

_____

2. Lower the vehicle, and start the engine. Quickly accelerate the engine a few times, and listen for any exhaust leak. Note your findings. _____

_____

3. With the engine running, place the transmission in gear, and listen for any exhaust noises. Note your findings. _____

_____

Small exhaust leaks can be found by having a fellow student block the tailpipe with the engine running. This will cause the exhaust backpressure to increase. The exhaust will then be forced out of any leak. Obtain your instructor's permission to perform this test.

Instructor's Check _____

4. Results of exhaust leak test: _____

_____

5. Exhaust leaks can also be found using a smoke machine. Place the outlet of the smoke machine into the tailpipe, and begin to induce smoke. If a leak is present, smoke should rise from the area of the leak. Results of smoke test: _____

_____

■

Exhaust System Service

Exhaust systems normally are the victims of one of two problems: leaks or restrictions. Leaks not only cause excessive noise but can also decrease the temperature and pressure of the exhaust. Restrictions cause exhaust pressure to build up before the restriction. The difference in pressure determines the rate of flow. When the pressure difference is slight, less exhaust will be able to leave the cylinders, and there will be less room for the fresh intake charge.

TURBOCHARGERS AND SUPERCHARGERS

The power generated by an engine is directly related to the amount of air compressed in the cylinders. Two approaches can be used to increase engine compression. One is to modify the engine to increase the compression ratio. The other and less expensive way is to increase the intake charge. By pressurizing the intake air before it enters the cylinder, more air and fuel molecules can be packed into the combustion chamber. The two processes of artificially increasing the amount of airflow into the engine are known as turbocharging and supercharging.

Turbochargers rely on hot exhaust gases, which are used to spin turbine blades. The turbine is attached to the blades of a compressor. The housing of the turbocharger totally isolates the turbine from the compressor, preventing the exhaust gases from entering into the compressor. As the turbine spins, so does the compressor. The compressor is open to the outside air on one side and sealed to the intake manifold on the other. As the compressor blades spin, intake air is drawn in and thrown into the intake manifold by centrifugal force.

The turbocharger, with proper care and servicing, will provide years of reliable service. Most turbocharger failures are caused by lack of lubricant, ingestion of foreign objects, or contamination of lubricant.

Superchargers are air pumps directly driven by the engine's crankshaft by a belt. They pump extra air into the engine in direct relationship to crankshaft speed. The air is inducted into the supercharger, pressurized by the spinning rotors, and exits the supercharger. If equipped with an intercooler, the cooled air then passes through to the intake manifold. When the intake valves open, the air is forced into the combustion chambers, where it is mixed with fuel delivered by the fuel injectors.

Many of the problems and their remedies given for turbochargers hold true for superchargers. There are also other problems associated specifically with the supercharger.

The following job sheets will prepare you for diagnosing and servicing the intake and exhaust systems.

☐ JOB SHEET 32–1

Inspecting an Exhaust System

Name _____ Station _____ Date _____

Objective

Upon completion of this job sheet, you will have demonstrated the ability to inspect the exhaust system and check it for restrictions. This task applies to ASE Education Foundation tasks MLR 8.C.3 (P-1), AST 8.D.8 (P-1), and MAST 8.D.9 (P-1).

Tools and Materials

Flashlight or trouble light

Tachometer

Hammer or mallet

Description of Vehicle

Year _____ Make _____ Model _____

VIN _____ Engine Type and Size _____

Mileage _____

PROCEDURE

1. Before doing a visual inspection, listen closely for hissing or rumbling that may indicate the beginning of exhaust system failure. With the engine idling, slowly move along the entire system and listen for leaks. Operate the vehicle in well-ventilated area and/or connect to the shop exhaust extraction system.

   WARNING: *Be very careful. Remember that the exhaust system gets very hot. Do not get your face too close when listening for leaks.*

   Did you hear any indications of a leak? ___ YES ___ NO

   If so, where? _____

   _____

   _____

2. Safely raise the vehicle.

3. With a flashlight or trouble light, check for the following:

   ■ Holes and road damage
   　Discoloration and rust
   　Carbon smudges
   　Bulging muffler seams
   　Interfering rattle points
   　Torn or broken hangers and clamps
   　Missing or damaged heat shields

   Did you detect any of these problems? ___ YES ___ NO

If so, where? _____

_____

_____

4. Sound out the system by gently tapping the pipes and muffler with a hammer or mallet. A good part will have a solid metallic sound. A weak or worn-out part will have a dull sound. Listen for falling rust particles on the inside of the muffler. Mufflers usually corrode from the inside out, so the damage may not be visible from the outside. Remember that some rust spots might be only surface rust.

Did you find any weak or worn-out parts? ___ YES ___ NO

If so, which ones? _____

_____

_____

5. Grab the tailpipe (when it is cool), and try to move it up and down and from side to side. There should be only slight movement in any direction. If the system feels wobbly or loose, check the clamps and hangers that fasten the tailpipe to the vehicle.

Did you detect any problems with the clamps or hangers? ___ YES ___ NO

If so, where? _____

_____

_____

6. Inspect all of the pipes for kinks and dents that might restrict the flow of exhaust gases.

Did you find any kinks or dents? ___ YES ___ NO

If so, where? _____

_____

_____

7. Take a close look at each connection, including the one between the exhaust manifold and exhaust pipe.

Did you find any signs or leakage? ___ YES ___ NO

If so, try tightening the bolts or replacing the gasket at that connection.

8. Check for loose connections at the muffler by pushing up on the muffler slightly. If loose, try tightening them.

Problems Encountered

_____

_____

_____

Instructor's Comments

_____

_____

_____

☐ JOB SHEET 32–2

Inspecting the Turbocharger/Supercharger

Name _____ Station _____ Date _____

Objective

Upon completion of this job sheet, you will have demonstrated the ability to inspect components of a turbocharger/supercharger. This task applies to ASE Education Foundation task MAST 8.D.13 (P-2).

Description of Vehicle

Year _____ Make _____ Model _____

VIN _____ Engine Type and Size _____

Mileage _____

PROCEDURE

1. Which device does the vehicle being inspected have?

   _____ Turbocharger   _____ Supercharger

2. Does the vehicle have an intercooler? If so, where is it located? _____

   The following steps are to inspect a turbocharger.

3. Check the engine oil level and condition. Record your findings. _____

   _____

4. Visually inspect the turbocharger housing and the intake and exhaust connections. Record your findings. _____

   _____

5. Inspect the waste gate and its related control devices. Note your findings. _____

   _____

6. How is the waste gate controlled on this vehicle? _____

7. Start the engine, and listen for any noise from the turbocharger. Are any noises present? If yes, describe them. _____

8. Summarize your inspection findings. _____

   _____

   The following steps are for inspecting a supercharger.

9. Inspect the drive belt. Note the belt condition. _____

10. Inspect the engine oil and supercharger oil levels and condition. Record your findings.

   _____

11. Inspect the supercharger for signs of oil leaks, and note your findings. _____

_____

12. If the supercharger is equipped with a clutch, how is the clutch controlled? _____

13. Is the supercharger fitted with a bypass circuit? _____

14. If a bypass circuit is used, when and how is it controlled? _____

Problems Encountered

_____

_____

_____

Instructor's Comments

_____

_____

_____

REVIEW QUESTIONS

Review Chapter 32 of the textbook to answer these questions:

1. Why can a restricted exhaust system reduce engine vacuum at 2000 rpm?

2. How does a restricted air filter affect engine efficiency?

3. Explain the federal catalytic converter warranty.

4. List two causes of excessive boost pressure in a turbocharger.

5. Contrast the operation of a turbocharger and supercharger.

EMISSION CONTROL SYSTEMS

OVERVIEW

Chapter 33 covers the legislation of emissions and the operation, purpose, and components of common emission control systems. At one time, emission control devices were simple add-ons to the engine. Today, they are integral in the design and operation of the vehicle. Failure of one of these systems can not only increase emission levels but can also create drivability problems.

Emission control devices reduce the amount of pollutants and environmentally damaging substances released by vehicles. There are five gases measured as tailpipe emissions: hydrocarbons (HC), carbon monoxide (CO), carbon dioxide (CO_2), oxides of nitrogen (NO_x), and oxygen (O_2). Although CO_2 and O_2 are not pollutants, CO_2 is a major contributor to greenhouse gases and global warming. O_2 is used in the combustion process. If too much O_2 is in the exhaust, a combustion problem is indicated.

Hydrocarbon emissions originate from the fuel burned in the engine. Gasoline and diesel fuels are derived from crude oil. Crude oil can contain between 50% and 97% hydrocarbons, a bonding of hydrogen and carbon atoms. During the combustion process, some hydrocarbons remain unburned. This means that the emission control system must process the remaining HC to reduce the amount of HC emission.

If there is insufficient air to the quantity of fuel during combustion, carbon atoms join with oxygen atoms, forming CO. CO is odorless, colorless, tasteless, and very dangerous. CO poisoning can cause flu-like symptoms or death if not identified and treated.

CO_2 is a byproduct of complete combustion. Although large amounts of CO_2 indicate a well-running engine, it does pose a problem for the atmosphere. CO_2 and other greenhouse gas emissions have been linked to global warming. Automotive engineers are continuing to improve engine design and reduce fuel consumption, which also decreases CO_2 output.

High combustion chamber temperatures, which are good for decreasing CO and HC emissions, create NO_x emissions, a major contributor to smog. Because nitrogen constitutes 78% of our atmosphere, we cannot eliminate it from the combustion process. Engineers try to balance combustion chamber temperatures so that all emissions are as low as possible.

The O_2 that enters into the combustion chamber is used during combustion. There should be little O_2 left from the combustion process.

Because is it not possible to have complete and total combustion every time a cylinder fires, cars and trucks use many types of pre- and post-combustion devices to reduce the amount of hazardous emissions that are emitted.

Concept Activity 33-1
Emissions

Obtain permission from your instructor to perform this experiment.

> **WARNING:** *Torches can be very dangerous. Severe burns can result from improper use or carelessness. Follow all safety guidelines regarding the operation of an oxygen-acetylene torch.*

Instructor's Check _____

1. Turn on the fuel (acetylene) valve and, using a striker, ignite the torch. Note the color of the flame, the amount of smoke generated, and any solids present from the flame. _____

_____

2. Next, turn on the oxygen valve. Note the color of the flame, the amount of smoke generated, and if any solids are present from the flame. _____

3. Why do you think there is such a difference between the acetylene-only flame and the oxygen-acetylene flame? _____

_____

4. Why do you think the oxygen-acetylene flame creates less smoke? _____

_____

5. What opinion can you form about how the extra oxygen affects combustion?

_____

_____

_____

■

Concept Activity 33-2
Emission Control Devices

1. Label the following components as evaporative, pre-combustion, or post-combustion emission control devices.

   a. Exhaust gas recirculation (EGR) valve   _____

   b. Charcoal canister   _____

   c. Gas cap   _____

   d. Positive crankcase ventilation (PCV) valve   _____

   e. AIR pump   _____

   f. Catalytic converter   _____

   g. Improved piston rings   _____

   h. Electronic spark control   _____

   i. Pulse AIR   _____

   j. Diesel exhaust fluid   _____

■

Concept Activity 33-3
Emission Control Device Purpose

Write the exhaust emission (NO_x, CO, CO_2, HC) controlled by the devices listed below.

1. EGR valve   _____

2. Charcoal canister _____

3. PCV valve   _____

4. Air pump   _____

5. Gas cap   _____

6. Catalytic converter   _____

7. Diesel exhaust fluid   _____

■

The following job sheet will prepare you for emission system diagnosis and service.

☐ JOB SHEET 33–1

Identifying Installed Emissions Control Devices

Name _____ Station _____ Date _____

Objective

Upon completion of this job sheet, you will have demonstrated the ability to identify emission control devices installed on a vehicle. This task applies to ASE Education Foundation Tasks MLR 8.A.1 (P-1), and AST/MAST 8.A.2 (P-1).

Description of Vehicle

Year _____ Make _____ Model _____

VIN _____ Engine Type and Size _____

Mileage _____

PROCEDURE

Locate the vehicle emission control label. This is usually under the hood.

1. List the emission control devices installed on this vehicle, according to the emissions decal.

   _____

   _____

   _____

   _____

2. For what emission year is the vehicle certified?

3. If the vehicle has a smog index sticker **(Figure 33-1)**, what is the smog index for this vehicle?

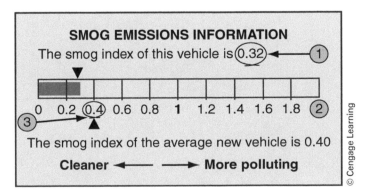

Figure 33-1 Typical smog index sticker.

4. What type of catalytic converter does this vehicle use? _____

   _____

5. How many catalytic converters does this vehicle use? _____

6. What type of EGR valve does this engine use? _____

7. What type of secondary air system does this engine use? _____

_____

Problems Encountered

_____

_____

_____

Instructor's Comments

_____

_____

_____

REVIEW QUESTIONS

Review Chapter 33 of the textbook to answer these questions:

1. Technician A says increases in CO tend to lower NO_x emissions. Technician B says an air leak in the exhaust system may be indicated by a low O_2 reading without a high CO reading. Who is correct?

   a. Technician A

   b. Technician B

   c. Both A and B

   d. Neither A nor B

2. Technician A says EGR valves are postcombustion emission control devices because they used exhaust gas to reduce NO_x. Technician B says EGR valves are precombustion emission control devices because they reduce combustion temperatures. Who is correct?

   a. Technician A

   b. Technician B

   c. Both A and B

   d. Neither A nor B

3. Technician A says an evaporative emission system stores vapors from the fuel tank in a charcoal canister until certain engine operating conditions are present. Technician B says that when the proper conditions are present, fuel vapors are purged from the charcoal canister into the intake manifold. Who is correct?

   a. Technician A

   b. Technician B

   c. Both A and B

   d. Neither A nor B

4. Technician A says that if the engine has excessive blowby or the PCV valve is restricted, crankcase pressure forces crankcase gases through the clean air hose into the air cleaner. Technician B says engines without blowby do not need a PCV system. Who is correct?

   a. Technician A

   b. Technician B

   c. Both A and B

   d. Neither A nor B

5. Technician A says that during an IM240 test, the vehicle is loaded to simulate first a short drive on city streets and then a longer drive on a highway. Technician B says a complete IM240 test cycle includes acceleration, deceleration, and cruising. Who is correct?

   a. Technician A

   b. Technician B

   c. Both A and B

   d. Neither A nor B

CHAPTER 34

EMISSION CONTROL DIAGNOSIS AND SERVICE

OVERVIEW

The emission control system is composed of many different components, both passive and active in their operation. Many of these operate over the lifetime of the vehicle without requiring any service or repair. Some, such as the PCV valve, require periodic inspection and service. This chapter focuses on the servicing of the emission control systems.

The purpose of the on-board diagnostic (OBD II) system is to detect emission-related failures. In fact, the only reason why the malfunction indicator lamp (MIL) comes on is if the emissions exceed 150% of the federal standard for the vehicle. To aid in diagnosing emission-related faults, the OBD II system provides codes and data to help the technician locate and repair the problem. The OBD II system has monitors specific for testing vehicle emission. Some of these monitors run continuously, whereas others run intermittently and only after other monitors have run and passed.

Concept Activity 34-1
OBD II Monitors

1. List the typical OBD II continuous and noncontinuous monitors.

_____

_____

_____

2. Why are some monitors dependent upon the running of another monitor?

_____

_____

_____

_____

The positive crankcase ventilation (PCV) valve was the first emission control device installed in automobiles. The PCV valve requires periodic inspection and service for the system to operate correctly **(Figure 34-1)**.

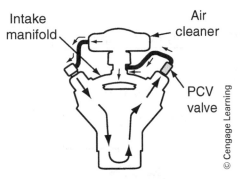

Figure 34-1 Typical PCV system.

Concept Activity 34-2
PCV System

1. What are the main functions of the PCV system?

_____

_____

_____

2. What emission can increase if the PCV system does not operate correctly?

_____

_____

_____

3. What are some other problems that can occur if the PCV system does not operate correctly?

_____

_____

_____

_____

The exhaust gas recirculation (EGR) system is found on nearly all modern engines. EGR valves and their related components draw a small amount of exhaust gas back into the intake manifold for mixing with the incoming air/fuel mixture. Faulty EGR systems can cause a variety of driveability concerns **(Figure 34-2)**.

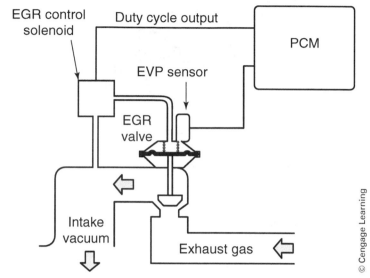

Figure 34-2   Example of a computer-controlled EGR system.

Concept Activity 34-3
EGR Systems

1. What is the primary purpose of the EGR system?

_____

_____

_____

2. What problems can result from a malfunctioning EGR system?

_____

_____

_____

3. Describe three types of EGR valves found on modern engines.

_____

_____

_____

_____

_____

_____

Catalytic converters have been installed on gasoline engine since the mid-1970s, with the introduction of unleaded gas. Lead, used for many years in gasoline as an octane booster, poisons the materials in the catalytic converter. This is why all catalytic converter-equipped vehicles require unleaded gas **(Figure 34-3)**.

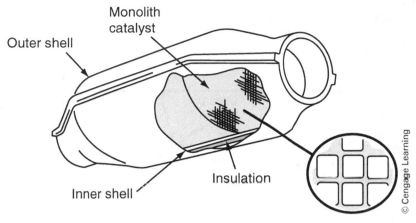

Figure 34-3 Cutaway view of a monolithic catalytic converter.

Concept Activity 34-4
Catalytic Converters

1. Describe in detail the function of the catalytic converter. _____

_____

2. Describe three types of catalytic converters. _____

_____

_____

3. What materials are used inside of the catalytic converter? _____

_____

4. How long is the federal catalytic converter warranty? _____

5. How long must a catalytic converter be kept by a shop after replacement? _____

_____

■

To aid in the function of the catalytic converter, some vehicles use secondary air injection reactor (AIR) systems. These systems can use a belt-driven air pump or an electric air pump, or rely on exhaust pressure to pump air into the catalytic converter **(Figure 34-4)**.

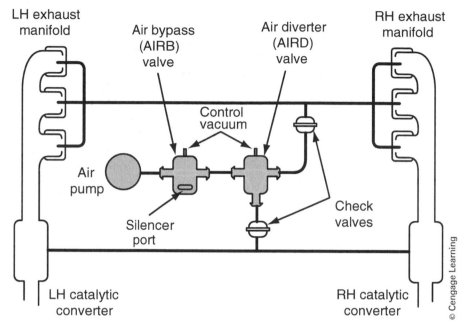

Figure 34-4 Typical components of the air injection reactor (AIR) system.

 **Concept Activity 34-5**
Secondary Air Injection

1. How does the secondary air system aid catalytic converter operation?

_____

_____

_____

2. What are some problems from an inoperative secondary air system?

_____

_____

_____

■

The evaporative (EVAP) emission control system is used to trap and burn fuel vapors. Under certain conditions, the PCM will command that stored fuel vapors be routed to the intake manifold for recirculation back into the intake stream **(Figure 34-5)**.

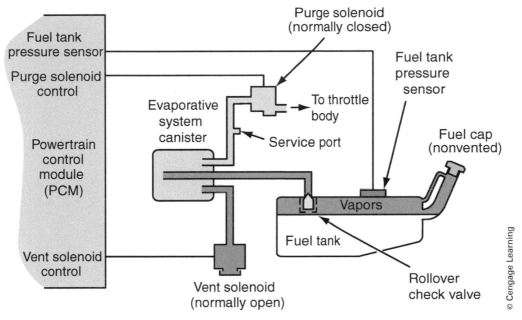

Figure 34-5 Typical components of the air injection reactor (AIR) system.

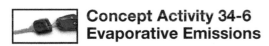 **Concept Activity 34-6**
Evaporative Emissions

1. Which emission is reduced by the EVAP system?

_____

2. How does inadequate fuel cap tightening affect the EVAP system?

_____

_____

3. Why does the PCM need to accurately know the fuel level in the fuel tank?

_____

_____

_____

_____

◼

The following job sheets will prepare you for emission system diagnosis and service.

☐ JOB SHEET 34–1

Testing the Operation of a PCV System

Name _____ Station _____ Date _____

Objective

Upon completion of this job sheet, you will have demonstrated the ability to test the PCV valve and check the hoses to determine if they are working properly. This task applies to ASE Education Foundation Tasks MLR 8.D.1 (P-2), and AST/MAST 8.E.2 (P-2).

Tools and Materials

Exhaust gas analyzer

Clean engine oil

Description of Vehicle

Year _____ Make _____ Model _____

VIN _____ Engine Type and Size _____

Mileage _____

PROCEDURE

1. Locate the PCV valve. Describe its location.

   _____

   _____

2. Run the engine until normal operating temperature is reached. Record the CO reading from the exhaust analyzer: _____ %

3. Remove the PCV valve from the valve or camshaft cover. Record the CO reading now: _____%

   Explain why there was a change in CO.

   _____

   _____

   _____

4. Place your thumb over the end of the PCV valve. Record the CO reading now: _____%

   Explain why there was a change in CO.

   _____

   _____

   _____

5. Remove the valve from its hose and check for vacuum. Record your results.

_____

_____

_____

6. Hold and shake the PCV valve. Record your results.

_____

_____

7. State your conclusions about the PCV system on this engine.

_____

_____

_____

Problems Encountered

_____

_____

_____

Instructor's Comments

_____

_____

_____

☐ JOB SHEET 34–2

Testing the Operation of an EGR Valve

Name _____ Station _____ Date _____

Objective

Upon completion of this job sheet, you will have demonstrated the ability to test the operation of an EGR valve and associated circuits. This task applies to ASE Education Foundation Task AST/MAST 8.E.3 (P-2).

Tools and Materials
Hand-operated vacuum pump

Vacuum gauge

Exhaust gas analyzer

Scan tool for assigned vehicle

A vehicle with an electronic EGR transducer and solenoid

Description of Vehicle
Year _____ Make _____ Model _____

VIN _____ Engine Type and Size _____

Mileage _____

PROCEDURE

1. Examine the EGR system found on the vehicle. Describe it here.

   _____

   _____

2. Inspect the hoses and electrical connections within the EGR system. Record the results.

   _____

   _____

3. If the vehicle is equipped with a vacuum-operated EGR valve, pull the hose to the EGR transducer, and install a vacuum gauge.

4. Run the engine, and check for vacuum at the hose.

   Vacuum Reading _____

5. Turn off the engine, and connect the scan tool to the diagnostic link connector (DLC).

   Scan Tool Used _____

6. Insert a tee fitting into the vacuum hose, and reconnect the supply line to the transducer. Connect the vacuum gauge to the tee fitting.

7. Start the engine.

   Record the vacuum reading: _____ inches of mercury (_____ mm Hg).

   There should be a minimum of 15 inches of mercury (381 mm Hg). Summarize your readings.

   _____

   _____

8. Actuate the solenoid with the scan tool.

   Did the solenoid click? ___ YES ___ NO

   Did the vacuum fluctuate with the cycling of the solenoid? ___ YES ___ NO

9. Disconnect the vacuum hose and back-pressure hose from the transducer.

10. Disconnect the electrical connector from the solenoid.

11. Plug the transducer output port.

12. Apply 1 (6.89 kPa) to 2 (13.79 kPa) psi of air pressure to the back-pressure port of the transducer.

13. With the vacuum pump, apply at least 12 inches of mercury (305 mm Hg) to the other side of the transducer. How did the transducer react?

    _____

    _____

    What did this test tell you?

    _____

    _____

14. Run the engine at fast idle.

15. Insert the probe of the gas analyzer in the vehicle's tailpipe.

    Record these exhaust levels.

    _____ HC _____ CO _____ CO_2 _____ O_2 _____ NO_x

    What was the injector pulse width at this time? _____ ms

16. Remove the vacuum hose at the EGR valve, and attach the vacuum pump to the valve.

17. Apply just enough vacuum to cause the engine to run rough. Keep the vacuum at that point for a few minutes, and record the readings on the gas analyzer while the engine is running rough.

    _____ HC _____ CO _____ CO_2 _____ O_2 _____ NO_x

    What was the injector pulse width at this time? _____ ms

    Describe what happened to the emissions levels, and explain why.

    _____

    _____

    _____

    _____

18. Apply the vehicle's parking brake.

19. Disconnect the vacuum pump, and plug the vacuum hose to the EGR valve.

20. Firmly depress and hold the brake pedal with your left foot, and place the transmission in drive.

21. Raise engine speed to about 1,800 rpm.

    Record the readings on the gas analyzer.

    _____ HC _____ CO _____ CO_2 _____ O_2 _____ NO_x

    What was the injector pulse width at this time? _____ ms

22. Return the engine to an idle speed; put the transmission in park; and then shut off the engine.

23. Describe what happened to the emissions levels and explain why.

    _____

    _____

    _____

    _____

24. Reconnect the EGR valve.

25. Apply the vehicle's parking brake. Start the engine.

26. Firmly depress and hold the brake pedal with your left foot, and put the transmission in drive.

27. Raise the engine speed to about 1,800 rpm.

    Record the readings on the gas analyzer.

    _____ HC _____ CO _____ CO_2 _____ O_2 _____ NO_x

    What was the injector pulse width at this time? _____ ms

28. Return the engine to an idle speed; put the transmission in park; and then shut off the engine.

29. Describe what happened to the emissions levels, and explain why.

    _____

    _____

    _____

    _____

30. Use the scan tool and retrieve any DTCs. Record any that are displayed.

    _____

31. Clear the codes.

    The following steps are applicable if the engine uses an electrical EGR valve.

32. Connect a scan tool to the DLC.

    Scan Tool Used _____

33. With the KOEO, record the EGR position.

    _____

34. Start the engine, and note the EGR position.

   _____

35. With the exhaust analyzer installed in the tail pipe, record the exhaust gases at idle and at 2,500 rpm.

   Idle: _____ HC _____ CO _____ CO_2 _____ O_2 _____ NO_x

   2,500 rpm: _____ HC _____ CO _____ CO_2 _____ O_2 _____ NO_x

36. Command the EGR to open.

   EGR%: _____

   rpm change: _____

   Idle: _____ HC _____ CO _____ CO_2 _____ O_2 _____ NO_x

   2,500 rpm: _____ HC _____ CO _____ CO_2 _____ O_2 _____ NO_x

37. What conclusion can you make about the operation of the EGR system?

   _____

   _____

   _____

Problems Encountered

_____

_____

_____

Instructor's Comments

_____

_____

_____

☐ JOB SHEET 34–3

Testing a Catalytic Converter for Efficiency

Name _____ Station _____ Date _____

Objective

Upon completion of this job sheet, you will have demonstrated the ability to check the condition and efficiency of a catalytic converter. This task applies to ASE Education Foundation Tasks AST 8.D.10 (P-2), and MAST 8.D.11 (P-2).

Tools and Materials

Rubber mallet   Propane enrichment tool

Pyrometer   Exhaust gas analyzer

Pressure gauge

Description of Vehicle

Year _____ Make _____ Model _____

VIN _____ Engine Type and Size _____

Mileage _____

PROCEDURE

1. Securely raise the vehicle on a hoist. Make sure you have easy access to the catalytic converter and that it is not HOT.

2. Smack the exhaust pipe, by the converter, with a rubber mallet. Did it rattle? ___ YES ___ NO

   If it did, it needs to be replaced and there is no need to do any more testing. A rattle indicates loose substrate, which will soon rattle into small pieces. If the converter passed this test, it does not mean it is in good shape. It should be checked for plugging or restrictions.

3. Remove the precatalytic converter O_2 sensor.

4. Install a pressure gauge into the sensor's bore.

   Sensor Location _____

5. After the gauge is in place, start the engine, and hold the engine's speed at 2,000 rpm. Record the reading on the pressure gauge. _____ psi (_____ kPa).

   Exhaust pressure should be under 1.25 psi (8.27 kPa). A very bad restriction could be over 2.75 psi (18.96 kPa). What does your reading tell you?

   _____

   _____

   Newer cars should have pressures well under 1.25 psi (8.27 kPa). Some older ones can be as high as 1.75 psi (12.06 kPa) and still be good. You will notice that if you quickly rev up the engine, the pressure goes up. This is normal. Remember, do this test at 2,000 rpm, not with the throttle wide open.

6. Remove the pressure gauge; turn off the engine; allow the exhaust to cool; and then install the O_2 sensor.

7. If the converter passed this test, you can now check its efficiency. There are three ways to do this. The first way is the delta temperature test. Start the engine, and allow it to warm up. With the engine running, carefully raise the vehicle, using the hoist or lift.

8. With a pyrometer, measure and record the inlet temperature of the converter. The reading is _____ .

9. Now measure the temperature of the converter's outlet. The reading is _____ . What is the percentage increase of the temperature at the outlet compared to the temperature of the inlet? What are your conclusions based on this test?

_____

_____

How does temperature show the efficiency of a catalytic converter?

_____

_____

10. Now you test O_2 storage. Begin by disabling the air injection system.

Procedure to disable the AIR system _____

11. Turn on the gas analyzer, and allow it to warm up. Start the engine and warm the catalytic converter up as well.

12. When everything is ready, hold the engine at 2,000 rpm. Watch the exhaust readings. Record the readings.

_____ HC _____ CO _____ CO_2 _____ O_2 _____ NO_x

If the converter was cold, the readings should continue to drop until the converter reaches light-off temperature.

13. When the numbers stop dropping, check the oxygen levels. Check and record the oxygen level; the reading is _____ . O_2 should be about 0.5% to 1%. This shows the converter is using most of the available oxygen. There is one exception to this: If there is no CO left, there can be more oxygen in the exhaust. However, it still should be less than 2.5%. It is important that you get your O_2 reading as soon as the CO begins to drop. Otherwise, good converters will fail this test.

14. If there is too much oxygen left and no CO in the exhaust, stop the test and make sure the system has control of the air/fuel mixture. If the system is in control, use your propane enrichment tool to bring the CO level up to about 0.5%. Now the O_2 level should drop to zero.

15. Once you have a solid oxygen reading, snap the throttle open, and then let it drop back to idle. Check the oxygen. The reading is _____ . It should not rise above 1.2%. If the converter passes these tests, it is working properly. If the converter fails the tests, chances are that it is working poorly or not at all. The final converter test uses a principle that checks the converter as it is doing its actual job, converting CO and HC into CO_2 and water.

16. Allow the converter to warm up by running the engine.

17. Calibrate the gas analyzer, and insert its probe into the exhaust pipe. If the vehicle has dual exhaust with a crossover, plug the side that the probe is not in. If the vehicle has a true dual exhaust system, check both sides separately.

18. Turn off the engine, and disable the ignition.

19. Crank the engine for 9 seconds as you pump the throttle. Look at the gas analyzer and record the CO_2 reading. _____

    The CO_2 for injected cars should be over 11%. If you are cranking the engine and the HC goes above 1,500 rpm, stop cranking; the converter is not working. Also stop cranking once you hit your 10 or 11% CO_2 mark; the converter is good. If the converter is bad, you should see high HC and low CO_2 at the tailpipe. What are your conclusions from this test?

    _____

    _____

    Do not perform this test more than *one* time without running the engine in between tests.

20. Reconnect the ignition, and start the engine. Do this as quickly as possible to cool off the converter.

Problems Encountered

_____

_____

_____

Instructor's Comments

_____

_____

_____

REVIEW QUESTIONS

Review Chapter 34 of the textbook to answer these questions:

1. Which exhaust emission increases from complete combustion?

   _____

   _____

2. Which of these components is used to control NO_x emissions?

   a. Charcoal canister   c. Air pump

   b. EGR valve   d. PCV valve

3. Describe how the PCM determines when to purge the EVAP system.

   _____

   _____

4. Technician A says tailpipe O_2 readings should be about 15%. Technician B says O_2 readings should be 0%. Who is correct?

   a. Technician A   c. Both A and B

   b. Technician B   d. Neither A nor B

5. Technician A says an inoperative EGR valve will cause a lean air/fuel mixture. Technician B says an inoperative EGR valve can cause a misfire. Who is correct?

   a. Technician A   c. Both A and B

   b. Technician B   d. Neither A nor B

HYBRID VEHICLES

CHAPTER 35

© Cengage Learning 2015

OVERVIEW

Hybrid-electric vehicles (HEVs) continue to gain acceptance and market share due to their "green" factor and fuel economy. However, not all hybrids are designed solely for high gas mileage, though most offer at least some improvement in fuel economy.

Hybrid vehicles can take many forms, all related to how the internal combustion engine (ICE) and electric drive system manage vehicle propulsion. Some HEVs are full, or strong, hybrids, meaning they can be driven by the ICE, the battery/motor, or a combination of the two. A mild hybrid cannot operate solely on its battery/motor alone, but can be used to assist the ICE during periods of heavy load. An assist hybrid uses an integrated starter/alternator for start–stop operation, also known as idle–stop capability. All hybrids utilize idle–stop to maximize fuel economy.

Concept Activity 35-1
Hybrid Vehicles

Of the following new model hybrids, determine whether each is a mild, assist, or full hybrid.

a. Toyota Prius   _____

b. Toyota Highlander   _____

c. Toyota Camry   _____

d. Honda Civic   _____

e. Ford Fusion   _____

f. Ford C-Max Energi   _____

g. Lexus ES Hybrid   _____

h. Chevrolet Malibu   _____

i. Chevrolet Volt   _____

j. Porsche Cayenne S Hybrid _____

■

The near future can expect to bring new types of energy-saving concepts and propulsion systems to the automotive industry.

Concept Activity 35-2
Types of Hybrid Drive Systems

1. Describe the construction and operation of a full hybrid vehicle, such as the Toyota Prius.

2. Describe the construction and operation of an assist hybrid vehicle, such as the Chevrolet Malibu eAssist.

3. Describe the construction and operation of a plug-in hybrid vehicle, such as the Honda Accord Plug-In.

■

Concept Activity 35-3
Hybrid Batteries

1. What color is used on full hybrid vehicles to indicate high-voltage wiring?

_____

2. What color is use on mild hybrid vehicles to indicated high-voltage wiring?

_____

3. Why must high-voltage gloves be worn when servicing a hybrid?

_____

■

The following job sheets will prepare you to service alternatively fueled vehicles.

☐ JOB SHEET 35–1

Identifying Hybrid High-Voltage Circuits, Service Disconnect, and Service Precautions

Name _____ Station _____ Date _____

Objective

At the completion of this job sheet, you will have demonstrated the ability to identify hybrid high-voltage circuits, locate the high-voltage service disconnect, and identify hybrid service safety precautions. This task applies to ASE Education Foundation Task MLR/AST/MAST 6.B.7 (P-2).

Description of Vehicle

Year _____ Make _____ Model _____

VIN _____ Engine Type and Size _____

Mileage _____

PROCEDURE

1. Hybrid vehicle type: _____ Full hybrid _____ Mild hybrid _____ Assist hybrid.

2. Describe the major components of the hybrid system used.

   _____

3. What operating voltages are used by this vehicle?

   _____

4. Describe the location(s) of the high- and low-voltage batteries.

   _____

   _____

5. What is the color of the hybrid electrical wiring? _____

6. Where are the warning and precaution labels located? _____

   _____

7. Using service information and/or the vehicle owner's manual, determine the location of the high-voltage service disconnect, and explain in detail the steps required to disconnect the high-voltage system.

   _____

   _____

   _____

8. List the warnings or precautions for disabling the high-voltage system.

   _____

9. What is the length of time specified after the high-voltage system is disabled before service can begin?

_____

10. Why do you think the manufacturer includes a wait time between power disconnect and starting to service the vehicle?

_____

_____

_____

11. List the procedures to reconnect the high-voltage system.

_____

_____

Problems Encountered

_____

_____

_____

Instructor's Comments

_____

_____

_____

☐ JOB SHEET 35–2

Identifying Hybrid Vehicle High-Voltage Components

Name _____ Station _____ Date _____

Objective

Upon completion of this job sheet, you will have demonstrated the ability to locate hybrid vehicle safety and service warnings and precautions. This task applies to ASE Education Foundation Task MLR/AST/MAST 6.B.7 (P-2).

Description of Vehicle

Year _____ Make _____ Model _____

VIN _____ Engine Type and Size _____

Mileage _____

PROCEDURE

1. Describe the type of hybrid system used on this vehicle.

   _____

   _____

2. Explain the operation of the HVAC system on this vehicle.

   _____

   _____

3. Where is the high-voltage system disconnect located? _____

   _____

4. List the procedures for powering down the high-voltage system. _____

   _____

5. Examine the engine compartment, and note any visible components of the high-voltage system.

   _____

   _____

6. Raise the vehicle on a hoist, and note any visible components of the high-voltage system.

   _____

   _____

Problems Encountered

_____

_____

_____

Instructor's Comments

_____

_____

_____

 REVIEW QUESTIONS

Review Chapter 35 of the textbook to answer these questions:

1. Define the term *hybrid-electric vehicle.*

   _____

   _____

   _____

2. Explain the differences between a hybrid and a plug-in hybrid vehicle.

   _____

   _____

   _____

3. Describe how a General Motors belt alternator starter (BAS) system operates.

   _____

   _____

   _____

4. List several safety precautions for working on hybrid vehicle high-voltage systems.

   _____

   _____

   _____

5. Explain the operational differences between series and parallel hybrids.

   _____

   _____

   _____

ELECTRIC VEHICLES

OVERVIEW

Recently, pure electric vehicles have made a resurgence in the automotive market. Although few models are available, their increasing popularity means more models can be expected from more manufacturers in the future. Currently, there are three types of electric vehicles for sale in the United States: the fuel-cell electric vehicle, the extended range electric vehicle, and the pure battery electric vehicle.

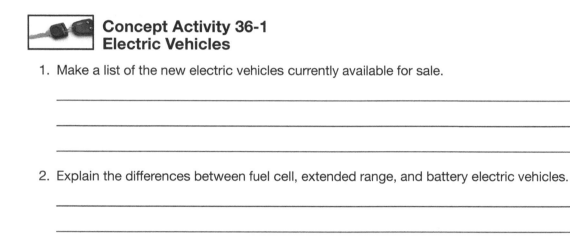

Concept Activity 36-1
Electric Vehicles

1. Make a list of the new electric vehicles currently available for sale.

   _____

   _____

   _____

2. Explain the differences between fuel cell, extended range, and battery electric vehicles.

   _____

   _____

   _____

   ■

Concept Activity 36-2
Battery Charging

1. Choose an electric vehicle, and research the various charging options available for the vehicle.

   _____

   _____

   _____

2. Compare and contrast the advantages and disadvantages of the different charging systems.

   _____

   _____

   _____

3. Using the average cost of a kilowatt-hour (kWh) for your location, determine the cost of recharging a fully discharged battery. _____

   ■

☐ JOB SHEET 36–1

Identifying Electric Vehicle Safety and Service Precautions

Name _____ Station _____ Date _____

Objective

Upon completion of this job sheet, you will have demonstrated the ability to locate hybrid vehicle safety and service warnings and precautions. This task applies to ASE Education Foundation Task MLR/AST/MAST 6.B.7 (P-2).

Description of Vehicle

Year _____ Make _____ Model _____

VIN _____ Engine Type and Size _____

Mileage _____

PROCEDURE

1. Describe the type of electric drive system used on this vehicle.

   _____

   _____

2. Explain the major components of this electric drive system and their functions.

   _____

   _____

3. List the warnings and precautions provided on the vehicle decals and owner/service manual.

   _____

   _____

4. List the procedures for powering down the high-voltage system.

   _____

   _____

5. List any special tools or equipment required for servicing the electrical system.

   _____

   _____

6. Describe the location of the high-voltage service plug or switch.

   _____

   _____

Problems Encountered

_____

_____

_____

Instructor's Comments

_____

_____

_____

REVIEW QUESTIONS

Review Chapter 36 of the textbook to answer these questions:

1. Explain why traditional transmissions are not necessary in most electric vehicles (EVs).

_____

_____

_____

_____

2. Describe the differences between inductive and conductive charging systems.

_____

_____

_____

3. Describe how a fuel cell generates electricity.

_____

_____

_____

4. How has the federal government provided an incentive for purchasing an electric vehicle?

_____

_____

_____

5. Describe the precautions for charging an EV.

 ASE PREP TEST

1. While testing a waste spark ignition system, a spark plug wire is disconnected and a spark plug tester is inserted. The plug is grounded and the engine is cranked, but no spark is produced. Technician A says a faulty coil may be the cause. Technician B says to inspect the spark plug wires, distributor, and rotor. Who is correct?

   a. Technician A
   b. Technician B
   c. Both A and B
   d. Neither A nor B

2. Which of the following is least likely to cause a misfire DTC on an OBD II system?

   a. Defective thermostat
   b. EVAP system faults
   c. EGR system faults
   d. Restricted exhaust

3. Technician A says a crankshaft sensor failure with the engine running causes the SFI engine to continue to run, but the PCM reverts to multiport fuel injection. Technician B says a crankshaft sensor failure will cause the engine to stall and not restart. Who is correct?

   a. Technician A
   b. Technician B
   c. Both A and B
   d. Neither A nor B

4. In discussing a waste spark system, Technician A says this system may have two spark plugs and a single coil. Technician B says this system gives a more efficient combustion and cleaner emissions than distributor ignition systems. Who is correct?

   a. Technician A
   b. Technician B
   c. Both A and B
   d. Neither A nor B

5. In discussing coil-on-plug (COP) ignition, Technician A says this system reduces radio frequency interference and electromagnetic interference. Technician B says when performing a manual cylinder power balance test, the coil needs to be fitted with an adapter or plug wire. Who is correct?

   a. Technician A
   b. Technician B
   c. Both A and B
   d. Neither A nor B

6. Technician A says most COP systems can be tested using the inductive pickup on the test scopes and engine analyzers. Technician B says an advantage of this system is that no adapters are required to gain access to secondary ignition data. Who is correct?

   a. Technician A
   b. Technician B
   c. Both A and B
   d. Neither A nor B

7. Technician A says a ruptured turbocharger wastegate diaphragm causes reduced boost pressure. Technician B says excessive carbon buildup in the wastegate can keep the valve open and reduce boost. Who is correct?

   a. Technician A
   b. Technician B
   c. Both A and B
   d. Neither A nor B

8. Technician A says a large amount of back pressure in the exhaust is used to allow a slower passage of exhaust gases through the catalytic converter. Technician B says excessive back pressure reduces an engine's volumetric efficiency. Who is correct?

   a. Technician A
   b. Technician B
   c. Both A and B
   d. Neither A nor B

9. Technician A says that on-the-car injector cleaners are the most popular and practical method for cleaning injectors. Technician B says one way to clean injectors is to soak them in cleaning solvent. Who is correct?

   a. Technician A
   b. Technician B
   c. Both A and B
   d. Neither A nor B

10. In a discussion of catalytic converters, Technician A says diesel engines use on oxidation catalyst to reduce CO emissions. Technician B says that modern gasoline engines use three-way catalytic converters to control CO, HC, and NO_x. Who is correct?

    a. Technician A
    b. Technician B
    c. Both A and B
    d. Neither A nor B

11. Technician A says a digital EGR may have two electric solenoids operated by the PCM. Technician B says a linear EGR valve contains three electric solenoids that are operated by the PCM with a pulse-width modulated signal. Who is correct?

    a. Technician A
    b. Technician B
    c. Both A and B
    d. Neither A nor B

12. An engine backfires on deceleration. Technician A says the air pump diverter valve may be defective. Technician B says the EGR valve may be sticking open. Who is correct?

    a. Technician A
    b. Technician B
    c. Both A and B
    d. Neither A nor B

13. Technician A says ignition components can be tested for intermittent failure by stress-testing them. Technician B says a higher than normal firing line can be caused by a faulty spark plug wire. Who is correct?

    a. Technician A
    b. Technician B
    c. Both A and B
    d. Neither A nor B

14. Technician A says two spark plug leads should only be placed side by side when these wires fire one after another. Technician B says EMI in spark plug leads may cause detonation and reduce engine power. Who is correct?

    a. Technician A
    b. Technician B
    c. Both A and B
    d. Neither A nor B

15. Technician A says that on a waste spark system half the plugs fire on reverse polarity. Technician B says the spark plugs, ignition wires, and ignition coil are in a series circuit in waste spark systems. Who is correct?

    a. Technician A
    b. Technician B
    c. Both A and B
    d. Neither A nor B

16. Technician A says many technicians conclude that an EGR valve is working if the engine stalls or idles very rough when the EGR valve is opened. Technician B says a restricted exhaust passage of only ⅛ inch (3.18 mm) will still cause the engine to run rough or stall at idle, but not control combustion temperatures at higher engine speeds. Who is correct?

    a. Technician A
    b. Technician B
    c. Both A and B
    d. Neither A nor B

17. In discussing catalytic converter diagnosis, Technician A says if the converter rattles, it needs to be replaced. Technician B says a plugged converter can cause popping or backfiring in the throttle body. Who is correct?

    a. Technician A   c. Both A and B
    b. Technician B   d. Neither A nor B

18. A low voltage on a sensor's reference wire could be caused by any of the following except:

    a. a poor computer ground.   c. high resistance in the sensor.
    b. a shorted reference wire.   d. excessive voltage drop on the sensor circuit.

19. Which of the following would most likely cause a high voltage on a sensor's reference wire?

    a. Poor computer ground   c. High resistance in the circuit from the computer
    b. Shorted reference wire   d. Excessive voltage drop in the wires to the sensor

20. Technician A says one of the best ways to inspect for injector leaks is to conduct an injector resistance test. Technician B says the best way to find an injector leak is to perform an injector balance test. Who is correct?

    a. Technician A   c. Both A and B
    b. Technician B   d. Neither A nor B

21. Technician A says OBD II standards define a warm-up cycle as a period where the coolant rises 40°F (4°C) and reaches at least 160°F (71°C). Technician B says an OBD II drive cycle allows for all monitoring sequences to complete their tests. Who is correct?

    a. Technician A   c. Both A and B
    b. Technician B   d. Neither A nor B

22. A technician connects a scan tool to an OBD II vehicle with a CAN network, but the scan tool will not turn on. Technician A says the fuse powering the DLC connector may be blown. Technician B says one of the two 60-ohm resistors may be shorted. Who is correct?

    a. Technician A   c. Both A and B
    b. Technician B   d. Neither A nor B

23. Technician A says when reading stored DTCs, you should also check for pending DTCs. Technician B says pending DTCs are stored for faults that require two trips before a code to set. Who is correct?

    a. Technician A   c. Both A and B
    b. Technician B   d. Neither A nor B

24. In discussing regenerative braking, Technician A says the regenerative system is designed to completely stop the vehicle in order to generate a maximum amount of electricity. Technician B says only full or strong hybrid vehicles use regenerative braking. Who is correct?

    a. Technician A   c. Both A and B
    b. Technician B   d. Neither A nor B

25. In discussing fuel trim, Technician A says short-term fuel trim is based on post-converter HO_2S activity. Technician B says fuel trim numbers that are positive indicate a lean exhaust condition. Who is correct?

    a. Technician A   c. Both A and B
    b. Technician B   d. Neither A nor B

CHAPTER

37

CLUTCHES

Courtesy of LuK Automotive Systems

OVERVIEW

A clutch has two primary purposes: to mechanically connect the engine's crankshaft to the transmission and to disconnect the crankshaft from the transmission. The clutch assembly is mounted to the flywheel and is engaged and disengaged by the clutch pedal. To transmit the action of the clutch pedal to the clutch, a linkage system is used. The clutch linkage can be a rod and lever, a cable, or it can be hydraulically operated.

Before we examine the manual transmission clutch, you should be familiar with the different types of clutches. You may have used a cordless screwdriver or drill in the lab. Those that have a selectable torque setting use a type of clutch that stops transmitting torque through the drill chuck once the torque value is reached. This type of clutch allows for slip. Slip occurs when there is a different amount of movement between the input and output.

Used in automatic transmissions, one-way clutches allow a shaft to rotate in one direction, but not the other. If you have ever ridden a bicycle with hand brakes, you know that the rear axle rotates freely when pedaled backward but applies torque to the wheel when pedaled forward. This is an example of a one-way clutch **(Figure 37-1)**. This type of clutch does not slip; it is either applied or freewheeling.

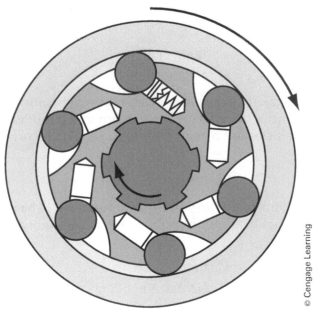

© Cengage Learning

Figure 37-1   This type of one-way clutch rotates freely clockwise but locks when turned counterclockwise.

In manual transmission arrangements, the clutch allows for a connection between the engine and transmission that can be disengaged and engaged as required by the driver, while allowing for slip. When taking off from a stop, the engine is spinning, but the transmission is not. The clutch allows for the gradual coupling of the two. Without this gradual synchronizing of speeds, shifts would be much harsher.

By pressing the clutch pedal, the driver disengages the clutch, allowing the engine and transmission to spin separately. With the engine and transmission free from each other, the engine can run at idle speeds when the vehicle is stopped. Using the clutch and allowing for some slip allows for easier shifting of the gears in manual transmissions.

Concept Activity 37-1
Clutch Theory

Obtain from your instructor three sheets of paper and a thumbtack. With the paper in a stack, push the tack through all three sheets in the center. With the papers lying flat, grasp the top sheet and spin it on the tack's axis.

1. Did the other two sheets spin with the first sheet?

_____

2. If not, why not?

_____

Pick up the papers and tack so that the tack is horizontal and the papers are vertical; spin the middle sheet.

3. Did all three sheets rotate?

_____

4. If not, why not?

_____

5. What would need to be done to the papers for them to all spin together?

_____

_____

■

For a clutch to operate correctly, it must make good contact with the engine and transmission. If the clamping force on the clutch is not sufficient, the clutch will slip. Slipping increases heat buildup and accelerates wear of the clutch components.

To understand how a clutch works, you must know something about friction. Both the brake system and clutch use friction. Brakes use friction to slow and stop the vehicle; clutches use friction to connect the engine to the transmission. Just as poor friction in the brake system will cause poor brake performance, poor friction in the clutch will cause shifting problems. See Concept Activity 48-3 in Chapter 48 for activities about friction.

 **Concept Activity 37-2**
Clutch Components

Label the parts of the clutch system shown below **(Figure 37-2)**.

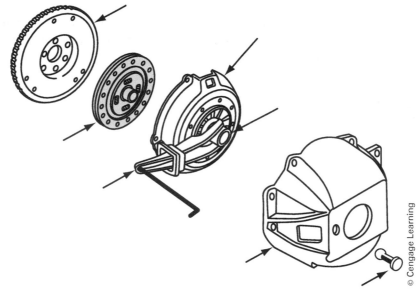

© Cengage Learning

Figure 37-2   Identify the clutch system components.

■

 Concept Activity 37-3
Clutch Linkage

Clutches are operated by rod and lever, cable, or hydraulic linkage. Rod and lever linkage is similar to the transmission linkage found on many automatic and manual transmission vehicles. Cable systems are similar to parking brake systems. Refer to Chapter 48 for additional information about cable-operated devices. Hydraulic clutch linkage operates the same way as the hydraulic brake system. Refer to Chapters 8 and 48 for additional information.

Examine several vehicles with manual transmission, and note the type of clutch linkage system used.

Vehicle 1

Year _____ Make _____ Model _____

Linkage type: _____ Rod and lever _____ Cable _____ Hydraulic

Vehicle 2

Year _____ Make _____ Model _____

Linkage type: _____ Rod and lever _____ Cable _____ Hydraulic

Vehicle 3

Year _____ Make _____ Model _____

Linkage type: _____ Rod and lever _____ Cable _____ Hydraulic

4. Why do you think different vehicles uses different types of clutch linkage? _____

5. What benefits or drawbacks do you see with one linkage type versus another? _____

_____

■

Concept Activity 37-4
Clutch Service

Clutch service is typically straightforward. The transmission/transaxle is separated from the engine to gain access to the clutch assembly. The pressure plate and clutch disc are bolted together as an assembly to the flywheel. The release bearing can be attached to the pilot bearing shaft on the transmission/transaxle or be part of the clutch slave cylinder.

1. Clutch friction material can contain asbestos. Describe what precautions must be taken when servicing the clutch system to avoid exposure to asbestos dust.

_____

_____

2. Describe why a clutch alignment tool is used.

_____

_____

3. Describe what problems can be caused by a misaligned clutch.

_____

_____

4. Why is the clutch pilot bearing or bushing replaced during clutch replacement?

_____

_____

_____

■

Table 37-1 can be used as a guide to help you diagnose clutch system concerns.

TABLE 37-1 Clutch Troubleshooting Chart

Symptom	Possible Cause
Pedal fails to release	Improper linkage adjustment
	Improper pedal
	Loose linkage or worn cable
	Faulty pilot bearing
	Faulty clutch disc
	Fork off ball stud
	Clutch disc hub binding or input shaft spline
	Clutch disc warped or bent
	Pivot rings loose, broken, or worn

TABLE 37-1 (Continued)

Symptom	Possible Cause
Slipping	Improper adjustment
	Oil-soaked clutch disc
	Worn facing or facing torn from clutch disc
	Warped pressure plate or flywheel
	Weak diaphragm spring
	Clutch disc not seated
	Clutch disc overheated
Grabbing or chattering	Oil on facing; burned or glazed facings
	Worn splines on input shaft
	Loose engine mountings
	Warped pressure plate or flywheel
	Burned or smeared resin on flywheel or pressure plate
Rattling or transmission click	Weak retracting springs
	Clutch fork loose on ball stud or in bearing groove
	Oil in clutch plate damper
	Clutch disc damper spring failure
Release bearing noise with clutch fully engaged	Improper adjustment
	Release bearing on transmission bearing retainer
	Insufficient tension between clutch fork spring and ball stud
	Fork improperly installed
	Weak linkage return spring
Noisy bearing	Worn clutch bearing
	Springs weak in pressure plate
	Pilot bearing loose in crankshaft
Pedal stays on floor when disengaged	Bind in cable or clutch bearing
	Springs weak in pressure plate
	Springs being overtraveled
Pedal hard to depress	Bind in linkage
	Clutch disc worn
	Friction in cable

The following job sheets will provide you with testing and service procedures for clutches and their linkages.

☐ JOB SHEET 37-1

Checking and Adjusting Clutch Pedal Free Travel

Name _____ Station _____ Date _____

Objective

Upon completion of this job sheet, you will have demonstrated the ability to check and adjust clutch pedal free travel. This task applies to ASE Education Foundation Task AST/MAST 3.B.2 (P-1).

Tools and Materials
Ruler or tape measure

Description of Vehicle

Year _____ Make _____ Model _____

VIN _____ Engine Type and Size _____

Mileage _____

PROCEDURE

1. Perform a visual inspection of the clutch pedal, linkage, and clutch housing. Note your findings.

   _____

2. If free travel is present, what is the specification for free travel given by the manufacturer?

   _____

3. Use a ruler or tape measure to measure the pedal free travel. Move the pedal just enough to take up the play in the pedal and linkage **(Figure 37-3)**. Record your reading.

   _____

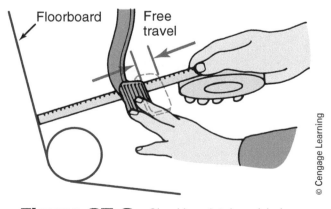

Figure 37-3 Checking clutch pedal play.

4. Is the free travel within specification?

_____

5. If the reading is not within specifications, adjust the linkage until the desired free travel is achieved, and remeasure the free travel.

Problems Encountered

_____

_____

_____

Instructor's Comments

_____

_____

_____

☐ JOB SHEET 37–2

Troubleshooting a Clutch Assembly

Name _____ Station _____ Date _____

Objective

Upon completion of this job sheet, you will have demonstrated the ability to troubleshoot a clutch assembly. This task applies to ASE Education Foundation Task AST/MAST 3.B.1 (P-1).

Tools and Materials

Droplight	Socket set
Pry bar	Wheel chocks
Ruler	Wrenches

Description of Vehicle

Year _____ Make _____ Model _____

VIN _____ Engine Type and Size _____

Mileage _____

WARNING: *Be sure that the wheels are chocked properly and that the brake system is in good operating condition before beginning this task. Do not allow anyone to stand in front of or behind the vehicle during this test. Do not take any longer than necessary to determine whether a problem exists.*

PROCEDURE (CHECK CLUTCH CHATTER)

1. Start the engine; set the parking brake; depress the clutch pedal fully; and shift the transmission into first gear. Increase the engine speed to about 1,500 rpm, and slowly release the clutch pedal. When the pressure plate first makes contact with the clutch disc, notice the clutch operation. Depress the clutch pedal, and reduce the engine speed. Record your results.

   _____

2. Shift the transmission into reverse, and repeat step 1. Record your results.

   _____

3. If clutch chatter does not occur, increase the engine speed to about 2,000 rpm, and repeat steps 1 and 2. Record your results.

   _____

4. If chatter occurs during the tests, raise the vehicle on a hoist. Check for loose or broken engine mounts, loose or missing bell-housing bolts, and damaged linkage. Record your results.

   _____

5. Lower the vehicle, and repeat steps 1–3.

PROCEDURE (CHECK CLUTCH SLIPPAGE)

1. Block the front wheels with wheel chocks, and set the parking brake. Start the engine, and run it for 15 minutes or until it reaches normal operating temperature.

2. Shift the transmission into high gear, and increase the engine speed to about 2,000 rpm. Release the clutch pedal slowly until the clutch is fully engaged. Record your results.

CAUTION: *Do not keep the clutch engaged for more than five seconds at a time. The clutch parts could become overheated and be damaged.*

3. If the engine does not stall, raise the vehicle on a hoist and check the clutch linkage. Correct any problems found during the inspection. Record your results.

---

4. If any linkage problems were found and corrected, repeat steps 1 and 2.

PROCEDURE (CHECK CLUTCH DRAG)

Note: *The clutch disc and input shaft require about 3 to 5 seconds to come to a complete stop after engagement. This is known as* clutch spin-down time. *This is normal and should not be mistaken for clutch drag.*

1. Start the engine; depress the clutch pedal fully; and shift the transmission into first gear.

2. Shift the transmission into neutral, but do not release the clutch pedal.

3. Wait 10 seconds. Shift the transmission into reverse.

4. If the shift into reverse causes gear clash, raise the vehicle on a hoist.

5. Inspect the clutch linkage. Record your results.

---

6. If any linkage problems were found and corrected, repeat steps 1–3.

PROCEDURE (CHECK PEDAL PULSATION)

1. Start the engine. Slowly depress the clutch pedal until the clutch just begins to disengage.

Note: *A minor pulsation is normal.*

Depress the clutch pedal further, and check for pulsation as the clutch pedal is depressed to a full stop. Record your results.

---

2. Check for bell housing misalignment, and record your results.

Pulsation can be caused by a damaged flywheel, pressure plate, or clutch disc.

Problems Encountered

_____

_____

_____

Instructor's Comments

_____

_____

_____

☐ JOB SHEET 37-3

Removing the Clutch

Name _____ Station _____ Date _____

Objective

Upon completion of this job sheet, you will have demonstrated the ability to remove a clutch from a vehicle safely and properly. This task applies to ASE Education Foundation Task AST/MAST 3.B.3 (P-1).

Note: *Before the clutch can be removed, both the transmissions/transaxle (Chapter 39) and the drive line (Chapter 40) must be removed from the vehicle. Refer to these chapters for the appropriate job sheets for driveline and transmission/transaxle removal.*

Tools and Materials

Clutch alignment tool   Special vacuum cleaner made for removal of asbestos fibers

Flywheel turning tool

Transmission jack

Description of Vehicle

Year _____ Make _____ Model _____

VIN _____ Engine Type and Size _____

Mileage _____

PROCEDURE (REMOVE CLUTCH ASSEMBLY)

1. Place fender covers over the fenders. Remove the battery negative cable.

   WARNING: *When disconnecting the battery, disconnect the grounded terminal (usually the negative terminal) first. Disconnecting the positive terminal first can create a spark, which can cause the battery to explode and result in serious injury.*

2. Raise the vehicle on a hoist. Remove the starter and driveline.

3. Remove the transmission following the appropriate procedures.

4. Remove the clutch fork and cross-shaft linkage assemblies.

5. Remove the bell housing. There are three basic bell-housing configurations commonly used:

   a. The bell housing and transmission are removed as a unit. The bell housing may be part of the transmission or bolted to it.

   b. The transmission is removed first. The bell housing is unbolted and removed from the back of the engine to expose the clutch assembly.

   c. The transmission is removed, but the bell housing remains bolted to the engine. An access opening in the bell housing allows the technician to work on the clutch assembly.

6. Remove any dust from the clutch assembly using an OSHA- and EPA-approved vacuum cleaner.

WARNING: *The dust inside the bell housing and on the clutch assembly contains asbestos fibers that are a health hazard. Do not blow this dust off with compressed air. Any dust should be removed only with a special vacuum cleaner designed for removal of asbestos fibers.*

7. To ensure proper reassembly, mark the flywheel and pressure plate clutch cover using a hammer and punch **(Figure 37-4)**. Insert the clutch alignment tool through the clutch disc **(Figure 37-5)**. Loosen and remove the bolts holding the clutch assembly to the flywheel.

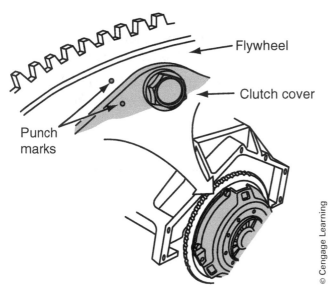

Figure 37-4 Clutch cover and flywheel marked to ensure proper reassembly.

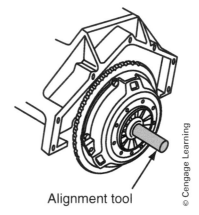

Figure 37-5 A clutch alignment tool is used to keep the disc centered and to prevent it from falling while loosening the pressure plate mounting bolts.

CAUTION: *The pressure plate springs will push the pressure plate away from the flywheel. The bolts must be loosened a little at a time. If the bolts are loosened completely and one at a time, the pressure plate springs will push against the flywheel. This causes the clutch cover to become distorted or bent, ruining it.*

WARNING: *The pressure plate assembly is heavy. Be sure the clutch alignment tool remains seated in the pilot bearing. Loosening the mounting bolts may unseat the alignment tool. The alignment tool may have to be tapped in frequently with a soft-faced mallet.*

8. Remove the pressure plate assembly from the vehicle.

9. Remove the clutch alignment tool and clutch disc from the flywheel.

Problems Encountered

_____

_____

_____

Instructor's Comments

_____

_____

_____

☐ JOB SHEET 37–4

Inspecting and Servicing the Clutch

Name _____ Station _____ Date _____

Objective

Upon completion of this job sheet, you will have demonstrated the ability to inspect and repair the major components of a clutch assembly. This task applies to ASE Education Foundation Tasks AST/MAST 3.B.3 (P-1), and 3.B.7 (P-1).

Tools and Materials

Bell-housing alignment tool post assembly

Clutch alignment tool

Dial indicator

Feeler gauges

Micrometer

Pilot bearing puller

Small mirror

Straightedge

Surface plate

Torque wrench

Vernier calipers with depth gauge

Description of Vehicle

Year _____ Make _____ Model _____

VIN _____ Engine Type and Size _____

Mileage _____

PROCEDURE (FOR GENERAL INSPECTION)

1. Perform a visual inspection of the components listed on the accompanying "Report Sheet for Clutch Inspection and Servicing." Check for damage, wear, and warpage. Record your results on the report sheet.

2. Locate the specifications for flywheel lateral and radial runout in the service manual, and note them on the report sheet.

3. Mount the dial indicator on the bell housing or engine block. Place the tip of the indicator's plunger against the clutch surface of the flywheel **(Figure 37-6)**. Check the flywheel for excessive lateral and radial runout. Record your results on the report sheet.

4. Locate the specification for flywheel warpage in the service manual, and note it on the report sheet.

5. Using a straightedge and feeler gauge set, check the amount of flywheel warpage and taper. Record your results on the report sheet.

6. If the flywheel has excessive runout, warpage, taper, light scoring, or light checking, remove the flywheel and resurface it.

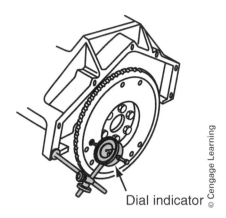

Figure 37-6 Checking a flywheel for excessive lateral runout.

WARNING: *Do not remove more than 10% of the flywheel's original thickness. A flywheel that is too thin can explode, causing serious injury or death.*

PROCEDURE (CHECK FACE RUNOUT)

1. Locate the specification for face runout in the service manual, and note it on the accompanying "Report Sheet for Clutch Inspection and Servicing."

2. Attach the clutch assembly to the flywheel.

   WARNING: *The clutch assembly is heavy. If it falls, it can be damaged or cause physical injury. Have a helper assist in holding and positioning the clutch assembly during installation.*

3. Install the bell-housing alignment tool post assembly through the bell-housing bore and into the clutch disc. Tighten the nut on the end of the post assembly until the clamp inside the clutch grips the clutch hub tightly.

4. Install the dial indicator on the tool post assembly, and position it so the indicator's plunger contacts the face of the housing in a circular pattern just outside the housing bore **(Figure 37-7)**.

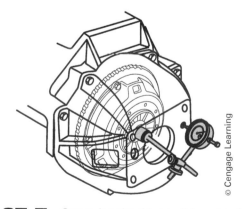

Figure 37-7 Setup for checking bell-housing face runout.

5. Gently pry the crankshaft rearward to eliminate all end play of the crankshaft. Hold the crankshaft in a rearward position. Set the dial indicator to zero.

6. Rotate the crankshaft through one complete revolution, and check to see that the dial indicator returns to zero position after one revolution.

7. If the dial indicator did not return to zero, repeat steps 2–4.

8. If the dial indicator returned to zero, rotate the crankshaft through two revolutions, making note of the highest reading obtained during revolutions. Record your results on the report sheet.

PROCEDURE (CHECK BORE RUNOUT)

1. Position the bell-housing alignment tool post assembly and dial indicator to check the bore's runout.

   Note: *The rubber band should be installed just snugly enough to provide a light pressure on the lever tip in the bore of the housing. If the rubber band is too tight, it may bind the dial indicator or distort the readings.*

2. Zero the dial indicator, and rotate the crankshaft through one revolution, as in steps 5 and 6 of the procedure for checking face runout.

3. Make a note of the highest indicator reading through two complete crankshaft revolutions. Record your results on the accompanying "Report Sheet for Clutch Inspection and Servicing."

Problems Encountered

_____

_____

_____

Instructor's Comments

_____

_____

_____

Name _____   Station _____   Date _____

REPORT SHEET FOR CLUTCH INSPECTION AND SERVICING

	Serviceable	Nonserviceable
1. Visual inspection		
Oil leaks		
Flywheel		
Pilot bearing		
Clutch disc		
Pressure plate		
Finger height		
Release bearing		
Clutch fork		
Pressure plate cap screws		
2. Flywheel lateral runout		
Specifications		
Actual		
3. Flywheel radial runout		
Specifications		
Actual		
4. Flywheel warpage		
Specifications		
Actual		
5. Face runout		
Specifications		
Actual		
6. Bore runout		
Specifications		
Actual		

Conclusions and Recommendations _____

_____

_____

☐ JOB SHEET 37–5

Reassembling and Installing a Clutch

Name _____ Station _____ Date _____

Objective

Upon completion of this job sheet, you will have demonstrated the ability to assemble and install a clutch assembly safely and properly. This task applies to ASE Education Foundation Task AST/MAST 3.B.3 (P-1).

Tools and Materials

Bearing support fixtures   Soft-faced mallet

Clutch alignment tool   Torque wrench

Dowel installation tool   Transmission jack

Press   Vise

Description of Vehicle

Year _____ Make _____ Model _____

VIN _____ Engine Type and Size _____

Mileage _____

PROCEDURE

1. Using the service manual, locate the bolt torque specifications, and note them on the accompanying "Report Sheet for Clutch Reassembly and Installation."

2. Lubricate the pilot bearing or bushing according to the manufacturer's recommendation. Install the pilot bearing or bushing into the crankshaft flange.

3. Install any flywheel dowels.

   CAUTION: *Do not damage the surface of the flywheel around the dowel holes during installation.*

4. Position the flywheel over the crankshaft mounting flange, and screw in, by hand, two mounting bolts to hold the flywheel in position.

5. Insert the remaining crankshaft mounting bolts, and finger-tighten them. Torque the flywheel mounting bolts in a crisscross pattern to the manufacturer's specifications.

6. Place the clutch disc in its position against the flywheel. Line up the match-marks on the flywheel and clutch cover, and finger-tighten two bolts through the pressure plate assembly to hold it in position.

   WARNING: *The clutch assembly is heavy. If it falls, it can be damaged or cause physical injury. Have a helper assist in supporting and positioning the clutch assembly during reassembly.*

7. Select the correct-size clutch alignment tool, and insert it through the center of the pressure plate and into the clutch disc. Make sure the splines of the disc line up with the splines on the tool if the tool has them. Then carefully push the tool through the disc and into the pilot bearing or bushing.

8. Thread the rest of the pressure-plate mounting bolts, and finger-tighten them. Torque the bolts in a crisscross pattern. Then remove the alignment tool.

    CAUTION: *The clutch assembly-to-flywheel mounting bolts are specially hardened bolts. If it is necessary to replace any of these bolts because of damage, use only the recommended hardened bolts. Do not use ordinary bolts.*

9. Clean the front and rear mounting surfaces of the bell housing and mounting faces on the rear of engine and front of the transmission.

10. Install the bell-housing mounting bolts finger-tight. Torque them to the manufacturer's specifications.

11. Press the old throwout bearing out of the clutch fork. Press a new throwout bearing into the clutch fork.

12. Slip or clip the new throwout bearing to the clutch fork.

13. Put a small amount of the recommended lubricant inside the hub of the release bearing.

14. Install the dust cover or seal.

15. At this point, the transmission/transaxle and driveline are reinstalled in the vehicle. See the job sheets given in the following chapters that pertain to these procedures. After completing all of the reassembly procedures, road test the vehicle and verify the proper clutch operation.

Problems Encountered

_____

_____

_____

Instructor's Comments

_____

_____

_____

Name _____   Station _____   Date _____

REPORT SHEET FOR CLUTCH REASSEMBLY AND INSTALLATION

1. Torque specifications	Acceptable	Not Acceptable
Flywheel to crankshaft		
Pressure plate to flywheel		
Bell housing to block		
Transmission to bell housing		
Transmission to mounts		

2. Final road test	Acceptable	Not Acceptable
Pedal free travel		
Pedal effort		
Pedal pulsation		
Clutch chatter		
Clutch slippage		
Clutch drag		
Vibrations		

Conclusions and Recommendations _____

_____

_____

REVIEW QUESTIONS

Review Chapter 37 of the textbook to answer these questions:

1. List three types of clutch linkages found in modern vehicles.

   _____

   _____

   _____

2. Why should the pressure plate retaining bolts be loosened a little at a time in a rotating manner?

   _____

   _____

3. List four possible causes of clutch slippage.

   _____

   _____

   _____

   _____

4. What safety precautions must be taken regarding the dust in the bell housing?

   _____

   _____

   _____

5. Describe the procedure to checking clutch pedal free travel.

   _____

   _____

   _____

MANUAL TRANSMISSIONS AND TRANSAXLES

CHAPTER 38

© Cengage Learning 2015

OVERVIEW

Both Chapters 38 and 39 deal with the operation and servicing of manual transmissions and transaxles. General transmission/transaxle work includes in-vehicle inspections and linkage adjustments. Shops that specialize in this type of repair perform the majority of internal transmission/transaxle work.

To know how a manual or automatic transmission operates, an understanding of gears and gear ratio is required.

A gear is a circular component that transmits rotational force to another gear or component. Gears have teeth so that they can mesh with other gears without slipping. The amount of leverage or mechanical advantage gained by the gear, and ultimately the gear ratio also, is determined by calculating the distance from the center of the gear to the end of a tooth **(Figure 38-1)**.

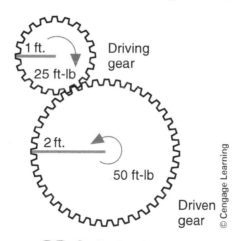

Figure 38-1   Radius determines gear ratio.

In Figure 38-1, the driving gear has a radius of 1 foot. As the driving gear turns clockwise, the driven gear rotates counterclockwise. The driven gear can apply twice the torque of the driving gear because the driven gear has a radius of 2 feet. By using a gear that is larger, more force can be applied because the distance from the center of the gear to the teeth creates leverage. This advantage allows gears to transmit the large amount of torque through the drivetrain, to propel the vehicle.

Because the driven gear has twice the number of teeth of the driving gear, it will turn at one-half of the speed of the driving gear. The relationship between gear radius and the number of gear teeth is called gear ratio. The driving gear will turn twice for every one rotation of the driven gear, so the gear ratio is 2:1 **(Figure 38-2)**.

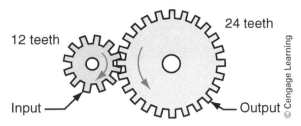

Gear ratio = 2 to 1
(2 revolutions input to
1 revolution output)

12 teeth

24 teeth

Input

Output

© Cengage Learning

Figure 38-2 Gear ratio is based on the number of teeth on the driven gear divided by the number of teeth on the driving gear.

Concept Activity 38-1
Gear Ratios

Determine the gear ratios of the following sets of gears **(Figure 38-3)**.

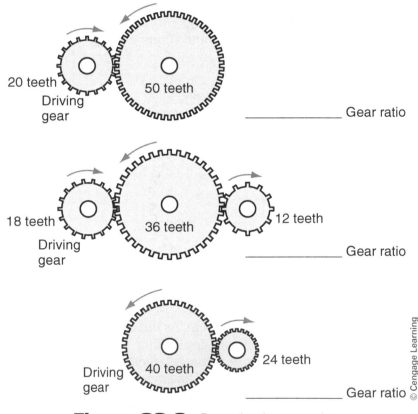

20 teeth
Driving
gear

50 teeth

_____ Gear ratio

18 teeth
Driving
gear

36 teeth

12 teeth

_____ Gear ratio

Driving
gear 40 teeth

24 teeth

_____ Gear ratio

© Cengage Learning

Figure 38-3 Determine the gear ratios.

Concept Activity 38-2
Gear Operation

Construct a simple gear train, using assorted container lids, such as from a coffee can, a gallon container of coolant, and from a quart bottle of oil. Place a rubber band around the circumference of each lid to increase the coefficient of friction of each. Place the lids against each other and secure the lids in their centers with a tack. Place a mark on each lid at points of intersection. Rotate one lid and note effect on the other lids **(Figure 38-4)**.

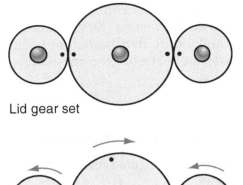

Lid gear set

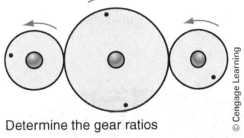

Determine the gear ratios

Figure 38-4 Determine the gear ratios.

© Cengage Learning

1. Record the diameter of the lids.

   Lid 1 _____ Lid 2 _____ Lid 3 _____

2. Based on the diameter of each lid, determine the gear ratios between the three. _____

   Reconfigure the lids so that a different lid is the driving gear, and repeat the activity.

3. How did changing the gear positions change the gear ratios? _____

4. In an automobile, where might a small gear drive a bigger gear? _____

5. For what reasons would a small gear be used to drive a larger gear? _____

6. Where might a large gear be used to drive a smaller gear? _____

7. For what reasons would a large gear be used to drive a smaller gear? _____

   When a small gear drives a larger gear, it is called gear reduction. The larger driven gear will turn at a slower speed than the input gear, but it will be able to apply more torque due to its larger size. When a large gear drives a small gear, it is called overdrive. The smaller gear will turn faster and with less torque than the input. In the engine, a crankshaft gear drives a camshaft gear with twice the number of teeth. This causes the camshaft to rotate at one-half of the crankshaft speed. In a transmission/transaxle, gear reduction ratios are used for torque and acceleration, whereas overdrive ratios are used for high-speed driving and fuel economy.

8. In a five-speed transmission/transaxle, first, second, and third gears would be which type of gear ratio?

   _____

9. In a five-speed transmission/transaxle, fifth gear would be which type of gear ratio? _____

   ■

Manual transmissions and transaxles gears are constant mesh. This means that all the gears, except for reverse, are always meshed together. Only when the transmission/transaxle is shifted and a set of gears is locked together is motion and torque transmitted through the gears. Inside of a typical five-speed manual transmission/transaxle, there are 10 gears meshed together to produce the five forward speeds. Three gears are used for reverse.

Concept Activity 38-3
Manual Transmission Component Identification

Most rear-wheel drive (RWD) manual transmissions are similar in construction. Identify the main components of the transmission shown in **Figure 38-5**.

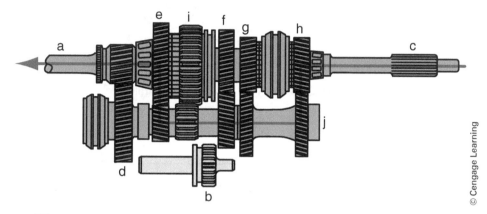

© Cengage Learning

Figure 38-5   Identify the components of the manual transmission.

_____ Input shaft

_____ Output shaft

_____ Counter shaft

_____ Input gears

_____ First gear

_____ Second gear

_____ Third gear

_____ Fourth gear

_____ Fifth gear

_____ Reverse gear

_____ Reverse idler

Concept Activity 38-4
Power Flow through a Manual Transmission

1. Use **Figure 38-6** below to trace the power flow through the transmission for each gear.

   First gear

   Second gear

   Third gear

   Fourth gear

   Fifth gear

   Reverse

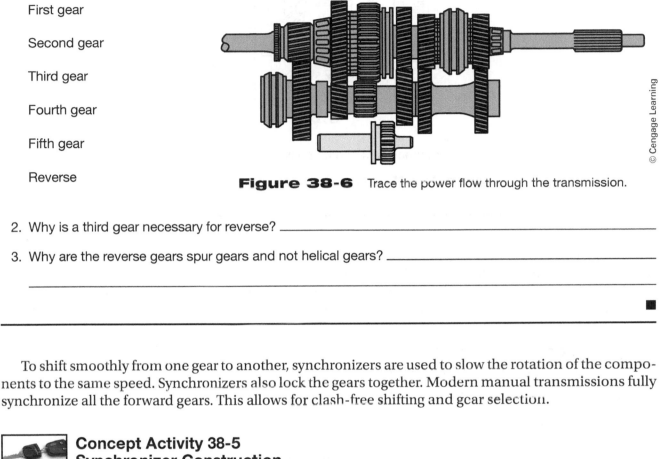

Figure 38-6 Trace the power flow through the transmission.

© Cengage Learning

2. Why is a third gear necessary for reverse? _____

3. Why are the reverse gears spur gears and not helical gears? _____

_____

■

To shift smoothly from one gear to another, synchronizers are used to slow the rotation of the components to the same speed. Synchronizers also lock the gears together. Modern manual transmissions fully synchronize all the forward gears. This allows for clash-free shifting and gear selection.

Concept Activity 38-5
Synchronizer Construction

1. Identify the parts of the synchronizer shown below **(Figure 38-7)**.

2. Why is reverse not synchronized? _____

3. Why are brass and bronze used for synchronizer blocking rings? _____

_____

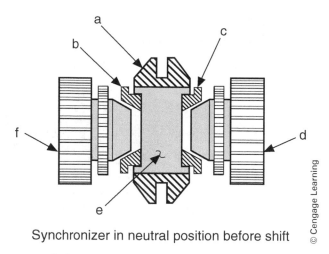

Synchronizer in neutral position before shift

© Cengage Learning

Figure 38-7 A block synchronizer in its neutral position.

Manual transaxles are similar in design and operation to manual transmissions, with the exception that the transaxle also contains the differential gearing and drive axle connections.

Many modern vehicles are now equipped with dual-clutch transmissions (DCTs). A dual clutch transmission is essentially a fully automated manual transmission with a computer-controlled clutch.

Concept Activity 38-6
Dual Clutch Transmissions

1. Compare the use of dry- and wet-clutch DCT transmission applications.

2. Explain the basic operation of a DCT.

3. Explain three benefits of using dual-clutch transmissions.

REVIEW QUESTIONS

Review Chapter 38 of the textbook to answer these questions:

1. Define the term *constant mesh*. _____

_____

_____

2. Explain the operation of the synchronizer assembly. _____

_____

_____

3. Explain why the synchronizer components are made of brass. _____

_____

_____

4. Explain the operation of the reverse idler gear. _____

_____

_____

5. Technician A says during torque multiplication, a small gear drives a large gear. Technician B says during overdrive a small gear drives a large gear. Who is correct?

    a. Technician A   c. Both A and B
    b. Technician B   d. Neither A or B

CHAPTER 39

MANUAL TRANSMISSION/ TRANSAXLE SERVICE

OVERVIEW

Servicing a manual transmission or transaxle is similar because they share the same basic construction and configuration. The major difference is that a transaxle contains the differential gearing, whereas a transmission does not.

Before attempting to diagnose and service a manual transmission/transaxle, verify that the cause of the complaint is not because of components or systems separate from the transmission/transaxle.

 Concept Activity 39-1
Preliminary Transmission/Transaxle Inspection

1. Inspect all powertrain mounts for wear and excessive movement. How can worn or loose powertrain mounts affect transmission/transaxle operation?

   _____

   _____

2. Inspect the transmission/transaxle linkage for looseness, binding, or damage. Problems with the linkage could cause what problems?

   _____

   _____

3. Ensure the engine is running properly and smoothly.

4. Verify proper operation of the clutch components. Explain how faulty clutch operation could affect the operation of the transmission/transaxle.

   _____

   _____

5. Inspect FWD half-shafts for wear, looseness, and damage. Check RWD driveshafts for missing balance weights. Drive shafts are covered more thoroughly in Chapter 40.

   _____

   _____

   ■

The following job sheets will prepare you for manual transmission/transaxle service and repair.

☐ JOB SHEET 39–1

Checking the Fluid Level in a Manual Transmission and Transaxle

Name _____ Station _____ Date _____

Objective

Upon completion of this job sheet, you will have demonstrated the ability to check the fluid level in a manual transmission and transaxle. This task applies to ASE Education Foundation Task MLR/AST/MAST 3.A.3 (P-1).

Description of Vehicle

Year _____ Make _____ Model _____

VIN _____ Engine Type and Size _____

Mileage _____

PROCEDURE

1. Refer to the service manual to determine the fluid level checkpoint on the specific vehicle you are checking.

   Location _____

2. Refer to the service manual to determine the type of fluid for the specific vehicle you are checking.

   Fluid Type and Amount _____

3. The transmission/transaxle gear oil level should be checked at the intervals specified in the service manual. Normally, these range from every 7,500 to 30,000 miles (12,070–48,280 km). For service convenience, many units are now designed with a dipstick and filler tube accessible from beneath the hood. Check the oil with the engine off and the vehicle resting on level grade. If the engine has been running, wait 2 to 3 minutes before checking the gear oil level.

   Fluid Level _____

4. Some vehicles have no dipstick. Instead, the vehicle must be placed on a lift and the oil level checked through the fill plug opening on the side of the unit. Clean the area around the plug before loosening and removing it. Insert a finger or bent rod into the hole to check the level. The oil may be hot. Lubricant should be level with, or not more than 1/2 inch (13 mm) below the fill hole. Add the proper grade lubricant as needed, using a filler pump.

   Fluid Level Prior to Filling _____

   Note: *Manual transmission/transaxle lubricants in use today include single and multiple viscosity gear oils, engine oils, synchromesh fluid, and automatic transmission fluid. Always refer to the service manual to determine the correct lubricant and viscosity range for the vehicle and operation conditions.*

Problems Encountered

_____

_____

_____

Instructor's Comments

_____

_____

_____

☐ JOB SHEET 39–2

Inspecting and Adjusting Shift Linkage

Name _____ Station _____ Date _____

Objective

Upon completion of this job sheet, you will have demonstrated the ability to adjust manual transmission and transaxle shift linkage. This task applies to ASE Education Foundation Task AST/MAST 3.C.1 (P-2).

Tools and Materials

Shifter alignment tool

Solvent

Description of Vehicle

Year _____ Make _____ Model _____

VIN _____ Engine Type and Size _____

Mileage _____

PROCEDURE (PREPARATION)

1. Place the shift lever in the NEUTRAL position. Raise the vehicle on a hoist. Wipe off all linkage parts with a shop towel.

2. Inspect the linkage for damage or wear. Check the shift-linkage rods for looseness and for worn or missing bushings. Record your results.

   _____

3. Insert the shifter alignment tool to hold the linkage in its neutral position.

PROCEDURE (LINKAGE RODS WITH SLOTTED ENDS)

1. Loosen the adjustment nut with an open-end or box-end wrench until the rod is free to move **(Figure 39-1)**. Hold the slotted end, using a screwdriver inserted into the slot. Loosen the adjustment nut with an open-end or box wrench.

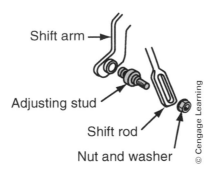

Shift arm →

Adjusting stud

Shift rod

Nut and washer

© Cengage Learning

Figure 39-1 Slotted shift-linkage rod adjustment.

2. Grasp the shifting arms, and move them back and forth by hand, checking for smooth and positive movement.

3. Place the shifting arms in their neutral position. When reconnecting the shift-linkage rods, adjust the rods to match the distance between the levers and arms. After the rods have been adjusted for length, tighten them in place.

4. Tighten the adjustment nut by hand until the nut begins to contact the arm. Hold the slotted end of the rod by using a screwdriver inserted into the slot. Tighten the adjustment nut with an open-end or box-end wrench.

Adjustment nut torque spec _____

PROCEDURE (LINKAGE RODS WITH THREADED ENDS)

1. Remove the holding pins from the swivels. Pull the swivels from the shift levers (**Figure 39-2**).

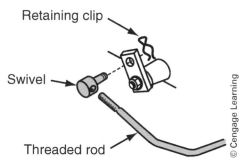

Retaining clip

Swivel

Threaded rod

© Cengage Learning

Figure 39-2   Threaded shift-linkage rod adjustment.

2. Grasp the shifting arms, and move them back and forth by hand, checking for smooth and positive movement.

3. Place the shifting arms in their neutral position. When reconnecting the shift-linkage rods, the rods must be adjusted to match the distance between the levers and the arms. After the rods have been adjusted for length, tighten them in place.

4. Lubricate the bushings with white grease. Rotate the adjustment swivels until the pin ends of the swivels slip easily into the bushings in the shift levers. Install new cotter pins.

Note: *Always use new cotter pins. Old cotter pins will break if they are reused.*

PROCEDURE (LINKAGE RODS WITH CLAMPS)

1. Loosen the bolts that hold the clamps to the transmission arms. The bolt should be loosened until the rods are free to move **(Figure 39-3)**.

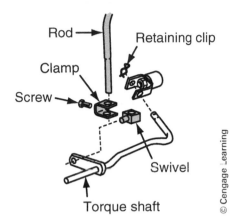

Rod → Retaining clip

Clamp

Screw →

Swivel

Torque shaft

© Cengage Learning

Figure 39-3 Clamped-end shift-linkage rod adjustment.

2. Grasp the shifting arms, and move them back and forth by hand, checking for smooth and positive movement.

3. Place the shifting arms in their neutral position. When reconnecting the shift-linkage rods, adjust them to match the distance between the levers and the arms. After the rods have been adjusted for length, tighten them in place.

4. Tighten the clamp bolts with an open-end or box-end wrench.

PROCEDURE (FINAL CHECKS—ALL TYPES)

1. Remove the alignment tool.

2. Lubricate the shift linkage with the proper lubricant.

   Lubricant used _____

3. Lower the vehicle.

4. Test the shift lever through all gears. The operation of the shift lever should be smooth. Note any shifting noise or hard shifting problems. Record your results.

   _____

5. Adjust the back-drive rod.

Problems Encountered

_____

_____

_____

Instructor's Comments

_____

_____

_____

☐ JOB SHEET 39–3

Road Testing a Vehicle for Transmission Problems

Name _____ Station _____ Date _____

Objective

Upon completion of this job sheet, you will have demonstrated the ability to road test a vehicle for transmission problems. This task applies to ASE Education Foundation Task AST/MAST 3.A.1 (P-1).

Description of Vehicle

Year _____ Make _____ Model _____

VIN _____ Engine Type and Size _____

Mileage _____

PROCEDURE

Note: *Have a fellow student accompany you on the road test to record your findings.*

1. Before beginning the road test, check the feel of the clutch pedal. Does it feel normal? Describe the feel and action of the clutch pedal.

   _____

   _____

   _____

Note: *While driving the vehicle, obey traffic laws.*

2. Begin driving the vehicle. Shift the transmission through the gears, and identify any problem you might feel.

   a. Did it easily shift into first? _____ YES _____ NO

   b. Did the gear change feel smooth? _____ YES _____ NO

   c. Did it easily shift into second? _____ YES _____ NO

   d. Did the gear change feel smooth? _____ YES _____ NO

   e. Did it easily shift into third? _____ YES _____ NO

   f. Did the gear change feel smooth? _____ YES _____ NO

   g. Did it easily shift into fourth? _____ YES _____ NO

   h. Did the gear change feel smooth? _____ YES _____ NO

   i. Did it easily shift into fifth? _____ YES _____ NO

   j. Did the gear change feel smooth? _____ YES _____ NO

   k. Did it easily shift into reverse? _____ YES _____ NO

   l. Did the gear change feel smooth? _____ YES _____ NO

3. Describe what you think could be the cause of any shifting problem.

_____

_____

4. Now drive the vehicle in one gear at a time. Accelerate, and then back off the throttle.

   a. Did the transmission stay in first gear? _____ YES _____ NO

   b. Did the transmission stay in second gear? _____ YES _____ NO

   c. Did the transmission stay in third gear? _____ YES _____ NO

   d. Did the transmission stay in fourth gear? _____ YES _____ NO

   e. Did the transmission stay in fifth gear? _____ YES _____ NO

   f. Did the transmission stay in reverse gear? _____ YES _____ NO

5. Describe what you think could be the cause of any problem involving jumping out of gear.

_____

_____

_____

6. Describe the general condition of the transmission.

_____

_____

_____

Problems Encountered

_____

_____

_____

Instructor's Comments

_____

_____

_____

 REVIEW QUESTIONS

Review Chapter 39 of the textbook to answer these questions:

1. List four common causes of hard shifting.

   _____

   _____

   _____

2. List three possible causes of manual transmission/transaxle noise in neutral.

   _____

   _____

   _____

3. List three possible causes of a transmission/transaxle sticking in gear.

   _____

   _____

   _____

4. Explain the possible causes of a clicking noise from the transmission/transaxle.

   _____

   _____

   _____

5. Explain what problems can result from using the improper lubricant in a manual transmission.

   _____

   _____

   _____

CHAPTER 40

Courtesy of Federal-Mogul Corporation

DRIVE AXLES AND DIFFERENTIALS

OVERVIEW

The output of the transmission or transaxle is transferred to the vehicle's driving wheels through several major components. In a rear-wheel-drive (RWD) vehicle, these include the drive shaft and rear axle assembly, which consists of the differential and rear drive axles. Front-wheel-drive (FWD) vehicles use a differential contained in the transaxle, along with half-shaft drive axles using inner and outer constant velocity joints. Four-wheel-drive (4WD) vehicles use the elements of both front- and rear-wheel-drive systems, plus special transfer case gearing, to split the driving power between the front and rear drive axles.

To transfer power effectively, drive shafts must be balanced and installed at proper operating angles. Drive shaft universal joints and drive axle constant-velocity (CV) joints must be properly installed, lubricated, and protected against dirt penetration. Finally, differential gearing must be maintained in good working order, with particular attention given to proper lubrication.

Concept Activity 40-1
Identifying CV Joints

Identify the six basic types of constant-velocity joints shown in **Figure 40-1**.

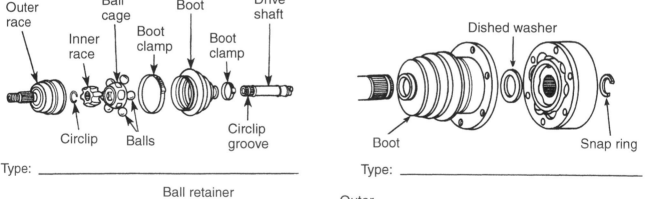

Type: _____   Type: _____

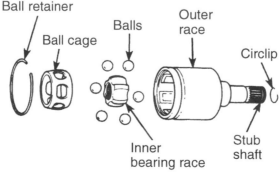

Type: _____

Figure 40-1   Six types of inboard and outboard constant velocity joints.

© Cengage Learning

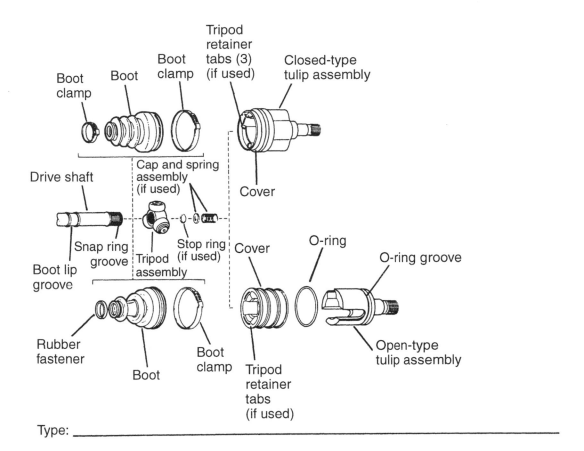

Boot clamp

Boot

Boot clamp

Tripod retainer tabs (3) (if used)

Closed-type tulip assembly

Drive shaft

Cap and spring assembly (if used)

Cover

Snap ring groove

Tripod assembly

Stop ring (if used)

Cover

O-ring

O-ring groove

Boot lip groove

Rubber fastener

Boot

Boot clamp

Tripod retainer tabs (if used)

Open-type tulip assembly

Type: _____

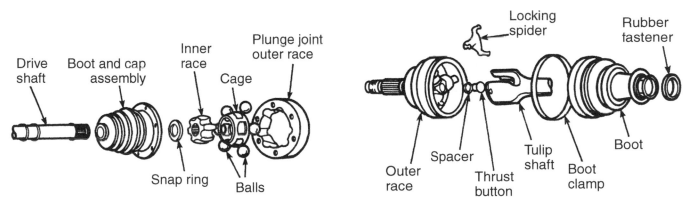

Drive shaft

Boot and cap assembly

Inner race

Cage

Plunge joint outer race

Locking spider

Rubber fastener

Snap ring

Balls

Outer race

Spacer

Thrust button

Tulip shaft

Boot clamp

Boot

Type: _____

Figure 40-1 Continued.

☐ JOB SHEET 40–1

Inspecting and Diagnosing a FWD Drive Axle

Name _____ Station _____ Date _____

Objective

Upon completion of this job sheet, you will have demonstrated the ability to inspect and diagnose the front drive axles and joints on a vehicle. This task applies to ASE Education Foundation Tasks AST/MAST 3.D.1 (P-1), and 3.D.4 (P-1).

Tools and Materials

A FWD vehicle

Description of Vehicle

Year _____ Make _____ Model _____

VIN _____ Engine Type and Size _____

Mileage _____

PROCEDURE

1. Raise the vehicle on a hoist.

2. Carefully look at the CV joint boots on both drive axles, and describe their condition and your recommendation.

   _____

   _____

   _____

3. Inspect the tightness of the boot clamps. Record your findings.

   _____

   _____

   _____

4. Visually inspect the shafts, and describe their condition.

   _____

5. If there are no obvious problems with the boots, clamps, and shafts, lower the vehicle, and prepare it for a road or parking lot test. Do a quick safety check of the vehicle; this should include the tires and lights. Note any problems found.

   _____

   _____

   _____

CAUTION: *Be sure to have the permission of the owner to drive the vehicle. While driving it, make sure you obey all traffic laws.*

6. Drive the vehicle straight ahead. Pay attention to any unusual noises or handling problems. Describe your findings (see Step 10).

_____

_____

_____

7. While moving straight at a low speed, accelerate hard, and then let off the throttle. Pay attention to any unusual noises or handling problems. Describe your findings (see Step 10).

_____

_____

_____

8. Now turn the vehicle to the right. Pay attention to any unusual noises or handling problems. Describe your findings (see Step 10).

_____

_____

_____

9. Now turn the vehicle to the left. Pay attention to any unusual noises or handling problems. Describe your findings (see Step 10).

_____

_____

10. Return to the shop, and record your conclusions from the road test.

_____

_____

_____

11. If problems were found during the road test, describe what service needs to be performed.

_____

_____

Problems Encountered

_____

_____

_____

Instructor's Comments

_____

_____

_____

☐ JOB SHEET 40–2

Removing and Replacing a FWD Drive Axle

Name _____ Station _____ Date _____

Objective

Upon completion of this job sheet, you will have demonstrated the ability to remove and replace a FWD drive shaft. This task applies to ASE Education Foundation Tasks MLR 2.D.2 (P-2), and AST/MAST 3.D.4 (P-1).

Description of Vehicle

Year _____ Make _____ Model _____

VIN _____ Engine Type and Size _____

Mileage _____

Transaxle Model _____ Specified Transaxle Fluid _____

PROCEDURE

1. Remove the cotter pin (if applicable), retaining the drive axle nut, and discard.

   Note: *Never reuse an old cotter pin; always replace with a new one.*

2. Locate the tightening specifications you will need to remove and replace the axle.

   Axle Nut Torque _____   Lower Ball Joint Torque _____

   Axle-to-Transaxle Bolts _____   Other _____

3. Remove the axle nut and washer. Remove the lower ball joint to lower control arm fastener. Inspect the threads of the ball joint stud or retaining bolt. Note your findings.

   _____

   _____

   _____

4. Pry down on the lower control arm to separate it from the ball joint.

5. On some vehicles, the axle can be pulled from the hub by hand. Other vehicles need a puller to remove the axle from the hub. Was a puller needed? _____ YES _____ NO

6. Some vehicles use bolts to secure the inner CV joint and axle to the transaxle. Remove the bolts as needed. On many vehicles, the axle is held in the transaxle with a ring on the axle shaft inside the transaxle. To remove this type, place a pry bar between the axle and the transaxle and push the axle out. Place a drain pan below this area, as the transaxle may lose fluid when the axle is removed.

7. Carefully slide the axle out of the vehicle, and place on a bench.

8. Inspect the axle for CV boot damage and deterioration. Note your findings.

   _____

   _____

9. Move the inner and outer CV joints, and check for looseness, binding, or noise. Note your findings.

_____

10. Compare the replacement axle to the original. Check overall length, shaft-end diameter, and spline count. Note your findings.

_____

11. Place a small amount of grease on the end of the inner CV joint retaining ring, and center the ring in the groove. This will allow easier reinstallation of the CV joint. Generally, wheel-bearing grease can be used on manual transaxles. Assembly lubricant can be used on automatic transaxles. Refer to the service information for the correct type of lubricant.

Lubricant used _____

12. Line the axle as straight as possible, and carefully insert the inner CV joint shaft back into the transaxle. Lightly tap the axle with a rubber mallet to seat the axle into the transaxle. Did the axle seat easily? _____ YES _____ NO

If not, what caused the difficulty? _____

_____

13. Place the outer CV joint back into the hub. Push the hub over the CV joint as far as it will go. A light spray of penetrating oil on the hub or axle can ease the reinstallation.

14. Align the lower ball joint stud and lower control arm, and install.

15. If the axle is bolted to the transaxle, begin tightening the bolts in the order specified by the service information. If no sequence is given, tighten in a crisscrossing pattern. Note the axle-to-transaxle tightening sequence.

_____

16. Tighten all fasteners to specifications. Replace cotter pins as necessary.

17. Install a new axle nut cotter pin. If the axle nut does not use a cotter pin, use a hammer and a punch to stake the edge of the nut that covers the key groove of the CV axle shaft.

18. Reinstall the tire and test-drive the vehicle.

Problems Encountered

_____

_____

_____

Instructor's Comments

_____

_____

_____

☐ JOB SHEET 40–3

Inspecting U-Joints and the Drive Shaft

Name _____ Station _____ Date _____

Objective

Upon completion of this job sheet, you will have demonstrated the ability to inspect the components of a RWD driveline, check U-joint angles and drive shaft runout, and balance a drive shaft. This task applies to ASE Education Foundation Task AST/MAST 3.D.2 (P-2).

Tools and Materials

Brass drift   Torque wrench

Chalk, crayon, or paint stick   Transmission jack

Hose clamps   Inclinometer

Dial indicator

Description of Vehicle

Year _____ Make _____ Model _____

VIN _____ Engine Type and Size _____

Mileage _____

PROCEDURE (VISUAL INSPECTION)

1. Place the transmission in neutral, and raise the vehicle on a drive-on hoist. Inspect for leaks at the slip joint, U-joints, final drive pinion seal, and pinion companion flange. Record your results.

   _____

2. Shake and twist the drive shaft to locate worn or loose parts. Pry with a screwdriver around the U-joints. Record your results.

   _____

3. Check for dirt, undercoating, dents, or missing balancing weights on the drive shaft. Inspect the center-bearing rubber bushing and support bracket, if equipped. Record your results.

   _____

   CAUTION: *Before attempting to check a center bearing, be sure the driving wheels and drive shaft are free to rotate.*

4. Check the center bearing, if equipped. Record your results.

   _____

   WARNING: *Extreme care should be taken when working around a rotating drive shaft. Severe injury can result from touching a moving shaft.*

PROCEDURE (CHECK U-JOINT ANGLES)

1. Locate the specifications for U-joint angles in the service manual. Clean the surfaces where the inclinometer will be mounted.

   Angle Specs _____

   CAUTION: *Do not force the inclinometer when setting it into position, or a false reading will be recorded (**Figure 40-2**).*

2. Check the front U-joint angle and record the reading. _____

3. Check the rear U-joint angle and record the reading. _____

4. If necessary, how could you correct the U-joint angles? _____

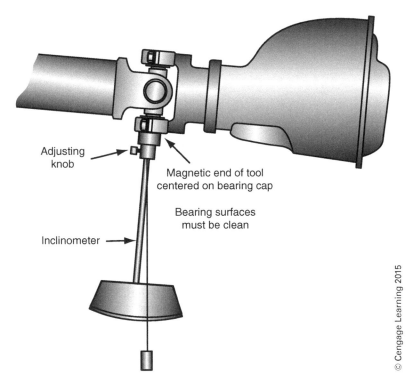

Adjusting knob

Magnetic end of tool
centered on bearing cap

Bearing surfaces
must be clean

Inclinometer

© Cengage Learning 2015

Figure 40-2   Checking drive shaft angles with an inclinometer.

CAUTION: *Do not use too many shims. Measure at the center of each shim. It should be no thicker than 1/4 inch (6.35 mm). If the rear U-joint angle is not correct, other problems may exist in the suspension. These problems include broken springs or an improperly placed spring seat.*

PROCEDURE (CHECK DRIVE SHAFT RUNOUT)

1. Locate the specification for drive shaft runout in the service manual. Clean the areas on the drive shaft where the dial indicator plunger will ride.

   Runout Spec _____

2. Mount the dial indicator. Take runout readings at each end and at the center of the drive shaft (**Figure 40-3**).

Runout readings _____

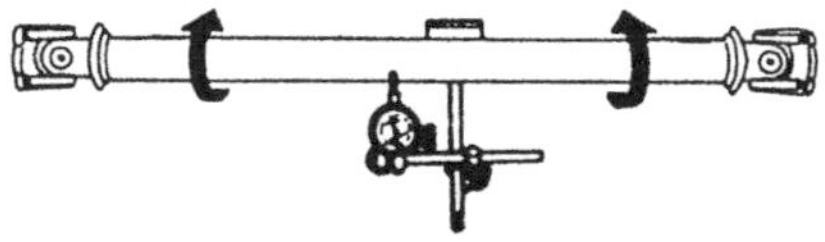

Figure 40-3 Taking runout readings on a drive shaft.

3. If necessary, disconnect the drive shaft, rotate it 180 degrees on the differential companion flange, and reinstall it. Recheck the runout readings. Is the runout within specs? _____

4. What may be required to correct the runout? _____

Problems Encountered

_____

_____

_____

Instructor's Comments

_____

_____

_____

☐ JOB SHEET 40–4

Measuring and Adjusting Pinion Depth, Bearing Preload, and Backlash

Name _____ Station _____ Date _____

Objective

Upon completion of this job sheet, you will have demonstrated the ability to measure and adjust drive pinion depth, bearing preload, and ring and pinion gear backlash. This task applies to ASE Education Foundation Tasks MAST 3.E.8 (P-3), 3.E.9 (P-3), and 3.E.10 (P-3).

Tools and Materials

Dial indicator

Pinion depth gauge

Torque wrench

Pinion gear flange holding tool

Micrometer

Description of Vehicle

Year _____ Make _____ Model _____

VIN _____ Engine Type and Size _____

Mileage _____

PROCEDURE (DRIVE PINION DEPTH)

1. Inspect the condition of the pinion bearings; replace them if necessary.

2. Inspect the pinion gear for any markings indicating additional adjustments. Record the markings.

   _____

3. Set up the pinion depth gauge according to the procedure outlined in the service manual.

4. Set up the dial indicator on the carrier housing.

5. Make the necessary readings with the indicator, and record the results.

   _____

   How much needs to be added or subtracted to achieve proper pinion depth?

   _____

6. Refer to the service manual to determine the correct size of pinion shim that should be used.

   Recommended Shim Size _____

7. Install the shim and bearing on the pinion gear shaft.

8. Install the pinion gear into the carrier housing.

PROCEDURE (PINION BEARING PRELOAD)

1. Install the pinion gear, crush sleeve, and bearing into the carrier housing.

2. Install the pinion seal into the housing.

3. Install the pinion flange, washer, and nut on the pinion.

4. Using the flange holding tool, tighten the pinion nut. Pinion nut torque spec is _____
   _____

5. Using a torque wrench, measure the torque required to turn the pinion gear. Required torque is
   _____.

6. Refer to the service manual for the proper torque required to turn the pinion. Specified torque is
   _____.

7. Tighten the pinion nut until the proper torque reading is reached.

PROCEDURE (RING AND PINION GEAR BACKLASH)

1. Check the ring gear for runout by setting the dial indicator on the back side of the ring gear.

2. Rotate the ring gear one complete revolution, and note the movement on the dial indicator. Describe
   what you observed.
   _____
   _____

   What was the highest reading? _____

   What was the lowest reading? _____

   Subtract the lowest from the highest; this indicates the total runout of the ring gear. What was it?
   _____

3. If the runout was not within specifications, check the runout of the carrier before replacing the ring
   gear. Note the results.
   _____

4. Now, install the differential case and ring gear into the carrier housing.

5. Mount the dial indicator onto the carrier housing.

6. Set the dial indicator on a ring gear tooth.

7. Look up the specifications for backlash and record them here.
   _____

8. Rock the ring gear back and forth against the teeth of the pinion gear. Observe the total movement of
   the indicator; this is the total backlash. Your readings were _____.

9. Measure backlash at four different spots on the ring gear.

10. Describe what needs to happen to correct the backlash.

_____

_____

_____

11. Using knock-in shims or adjusting nuts (depending on axle design), move the ring gear in reference to the pinion gear to achieve proper backlash.

12. When proper backlash is reached, torque the retaining caps to specifications.

13. Verify the backlash to make sure it is still within specifications.

Problems Encountered

_____

_____

_____

Instructor's Comments

_____

_____

_____

 **REVIEW QUESTIONS**

Review Chapter 40 of the textbook to answer these questions:

1. What type of final drive gearset requires timing?

   _____

   _____

2. How is rear axle shaft end play adjusted?

   _____

   _____

   _____

3. What is an inclinometer used for?

   _____

   _____

   _____

4. A clicking noise is heard from the front of a FWD vehicle when turning corners. What is the most likely cause?

   a. Worn inner CV joint

   b. Worn outer CV joint

   c. Improper ring and pinion adjustment

   d. Excessive ring gear preload

5. A clunking noise is heard each time the gears are changes while driving a FWD vehicle. Technician A says worn U-joints could be the cause. Technician B say worn CV joints could be the cause. Who is correct?

   a. Technician A

   b. Technician B

   c. Both A and B

   d. Neither A nor B

 ASE PREP TEST

1. Technician A says the axle hub nut on a FWD vehicle is used to set the amount of bearing play at the wheel. Technician B says most FWD hub and bearing assemblies are replaced as a unit. Who is correct?

   a. Technician A
   b. Technician B
   c. Both A and B
   d. Neither A nor B

2. While discussing the possible causes for a clunking noise heard during acceleration, Technician A says a CV joint could be the cause. Technician B says insufficient ring and pinion gear backlash could be the cause. Who is correct?

   a. Technician A
   b. Technician B
   c. Both A and B
   d. Neither A nor B

3. While reviewing the procedure for replacing an axle boot, Technician A says the position of the old boot should be marked on the shaft prior to removing it. Technician B says that if the boot is dimpled or collapsed after installing it, a dulled screwdriver should be used to allow air to enter the boot. Who is correct?

   a. Technician A
   b. Technician B
   c. Both A and B
   d. Neither A nor B

4. A vehicle makes a loud clunking sound during a left turn; during a right turn, the noise is still apparent but is less pronounced. Technician A says that the left inner CV joint may be the source of the noise. Technician B says the right outer CV joint may be causing the problem. Who is correct?

   a. Technician A
   b. Technician B
   c. Both A and B
   d. Neither A nor B

5. When reviewing the procedure for installing a clutch disc, Technician A says the clutch disc should always be installed with its hub and springs toward the engine. Technician B says the splines of the disc should be clean and never greased or oiled. Who is correct?

   a. Technician A
   b. Technician B
   c. Both A and B
   d. Neither A nor B

6. A bearing-type noise is heard immediately when the clutch pedal is depressed; as the pedal is pushed to the floor the noise remains about the same. When the clutch pedal is released, the noise disappears. Technician A says the clutch pilot bearing could be at fault. Technician B says the clutch release bearing could be at fault. Who is correct?

   a. Technician A
   b. Technician B
   c. Both A and B
   d. Neither A nor B

7. While discussing dual clutch transmission, Technician A says most DCTs require no special maintenance other than fluid changes. Technician B says some DCTs use two different fluids, one for the transmission bearings and gears, and another for the clutches. Who is correct?

   a. Technician A
   b. Technician B
   c. Both A and B
   d. Neither A nor B

8. A customer says that occasionally when he is at a stoplight with the transmission in first gear and the clutch pedal depressed, the vehicle will begin to move by itself. Technician A says the problem could be caused by insufficient clutch pedal free play. Technician B says the problem could be caused by an internal leak in the clutch master cylinder. Who is correct?

   a. Technician A
   b. Technician B
   c. Both A and B
   d. Neither A nor B

9. Technician A says transmission noise that is only heard when the vehicle is driven in first gear is probably caused by a bad input shaft bearing. Technician B says noise in first gear only could be caused by bad output shaft bearings. Who is correct?

   a. Technician A
   b. Technician B
   c. Both A and B
   d. Neither A nor B

10. Which of the following would not cause hard shifting?

   a. Worn shift forks
   b. Improperly adjusted clutch
   c. Weak pressure plate
   d. Worn synchronizers

11. Technician A says the transmission may jump out of gear if the transmission is not mounted squarely and securely to the engine. Technician B says excessive input shaft endplay may cause a transmission to jump out of gear. Who is correct?

   a. Technician A
   b. Technician B
   c. Both A and B
   d. Neither A nor B

12. A RWD vehicle exhibits a vibration that is most evident at 32–34 mph (51–54 km/h). Technician A says that this may be caused by incorrect vehicle height. Technician B says that an improper drive pinion angle may be the cause of this problem. Who is correct?

   a. Technician A
   b. Technician B
   c. Both A and B
   d. Neither A nor B

13. Technician A says the driveline balance should be inspected by a specialty shop. Technician B says if the driveline angles are wrong, the rear axle has moved. Who is correct?

   a. Technician A
   b. Technician B
   c. Both A and B
   d. Neither A nor B

14. While diagnosing a vibration that becomes more noticeable with vehicle speed, Technician A says the drive shaft may be out of balance. Technician B says the tires may be out of balance. Who is correct?

   a. Technician A
   b. Technician B
   c. Both A and B
   d. Neither A nor B

15. Which of the following should be checked before disassembling the differential assembly?

   a. Ring and pinion gears
   b. Side play
   c. Ring gear runout
   d. All of the above

16. A transmission is stuck in gear. Technician A says worn or misadjusted clutch linkage may be the cause. Technician B says the fluid level should be checked. Who is correct?

   a. Technician A
   b. Technician B
   c. Both A and B
   d. Neither A nor B

17. The interpretation of a ring and pinion gear contact pattern test is being discussed. Technician A says that excessive toe contact will require that the drive pinion be moved closer to the ring gear. Technician B says that the ideal contact pattern will be centered on the coast side of the ring gear. Who is correct?

   a. Technician A
   b. Technician B
   c. Both A and B
   d. Neither A nor B

18. Which of the following is not a likely cause of a fluid leak at the axle shaft seals in a RWD axle housing?

   a. Incorrect installation of the seal
   b. Damaged companion flange journal
   c. Damaged axle tube
   d. Damaged axle shaft journal

19. Technician A says that clutch drag is caused by a warped disc. Technician B says that clutch drag is caused by liquid contaminate on the friction surfaces. Who is correct?

   a. Technician A
   b. Technician B
   c. Both A and B
   d. Neither A nor B

20. Which of the following would not typically cause a rear axle shaft to break?

   a. Incorrect wheel bearing adjustments
   b. Misaligned axle housing
   c. Overloaded vehicle
   d. Slipping clutch

CHAPTER
41

© Cengage Learning 2015

AUTOMATIC TRANSMISSIONS AND TRANSAXLES

OVERVIEW

Chapters 41 covers theory and the operation of automatic transmissions and transaxles. Modern automatic transmissions and transaxles rely on hydraulics, electronics, and basic gear operation to transfer the power from the engine to the driving wheels. Before electronic control of on-board systems, automatic transmissions relied solely on hydraulic and mechanical components to control shifting. Modern transmissions and transaxles utilize the on-board computer network of sensors and computer-controlled solenoids within the transmission to control shifting and adjust for wear. A typical automatic transmission/transaxle consists of a torque converter, planetary gearsets, and a valve body.

Unlike a manual transmission, which uses a clutch to engage power flow to the transmission, an automatic transmission uses a torque converter. Sandwiched between the rear of the engine and the transmission, the torque converter is a fluid coupling. In a manual transmission, the clutch provides a mechanical connection for the engine and transmission. In a torque converter, fluid and the rotary motion of the fluid within the torque converter make the connection. In a manual transmission, a properly operating clutch transmits all of the power delivered by the engine without loss. However, because a fluid is used to apply the force within the torque converter, only about 90% efficiency is achieved. Some energy is wasted as heat and turbulence from the fluid's movement. Most modern torque converters are equipped with a lockup clutch. When the clutch is applied, the convert achieves 100% efficiency.

Concept Activity 41-1
Torque Converters

To understand how fluid power is used in a torque converter, face two similar size fans toward each other. Turn one fan on, and watch as the airflow from the first fan turns the blades of the second fan.

1. Why does the air cause the blades of the second fan to turn?

   _____

2. Do the blades of the second fan turn as fast as those of the first fan? _____ YES _____ NO

3. Why might the blades be turning at different speeds?

   _____

4. Are the fan blades turning in the same direction?

_____

5. Move the driving fan away from the driven fan. At what distance does the driven fan stop rotating?

_____

6. What could be done to get the highest possible speed from the driven fan?

_____

_____

■

A torque converter operates under the same principle as the fans do in the activity above. Instead of using air to transmit motion, automatic transmission fluid is used. Inside of a torque converter, there is a fluid pump, called the impeller, a driven member called the turbine, and a stator **(Figure 41-1)**.

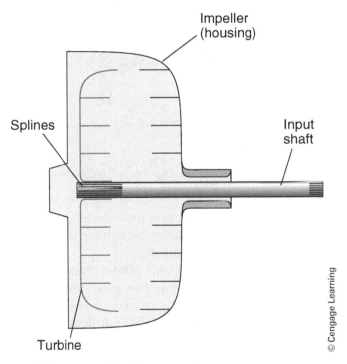

Impeller
(housing)

Splines

Input
shaft

Turbine

© Cengage Learning

Figure 41-1 Parts of a torque converter.

Attached to the inside of the torque converter shell are fins to move the transmission fluid. These fins form the impeller. As soon as the torque converter starts turning, the impeller moves fluid to the turbine. The turbine connects to the transmission input shaft. When the impeller moves enough fluid against the turbine, the input shaft begins to turn in the transmission. At idle speeds, there is very little fluid flow against the turbine and very little forward motion of the vehicle. As engine speed increases, so does the flow against the turbine, causing more torque output by the transmission to the drivetrain. The stator redirects the fluid returning to the impeller. The stator has wing-shaped fins, which change the direction of the fluid. Otherwise, the fluid returning from the turbine would work against the flow from the impeller **(Figure 41-2)**.

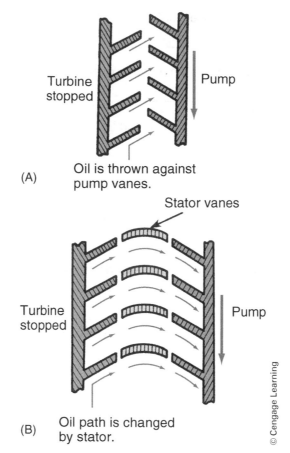

(A) Oil is thrown against pump vanes.

(B) Oil path is changed by stator.

© Cengage Learning

Figure 41-2 (A) Without a stator, fluid leaving the turbine works against the direction in which the impeller or pump is rotating. (B) With a stator in its locked (noncoupling) mode, fluid is directed to help push the impeller in its rotating direction.

Once the input shaft is rotating, planetary gearsets provide torque multiplication, also called gear reduction, or overdrive gearing. Gear reduction allows the vehicle to accelerate quickly. Once the vehicle is cruising, overdrive is used to increase fuel economy.

A planetary gearset is composed of a central sun gear, planetary gears that rotate around the sun gear, and an internal ring gear around the outside. The planetary gears are located between the ring gear and the sun gear **(Figure 41-3)**. Each part of the planetary gearset can be the driving gear or the driven gear, or can be held in place. Using this combination of possible configurations makes gear reduction, overdrive, direct drive, and reverse possible. Modern transmissions/transaxles use two or three planetary gearsets for the forward speeds and reverse.

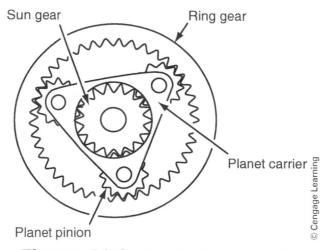

© Cengage Learning

Figure 41-3 A simple planetary gearset.

Concept Activity 41-2
Planetary Gearsets

Label the following figures as gear reduction, overdrive, direct drive, or reverse.

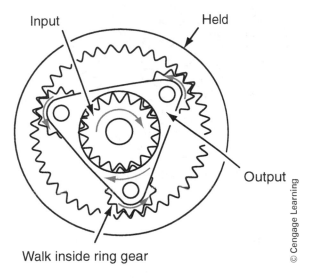

Input

Held

Output

Walk inside ring gear

© Cengage Learning

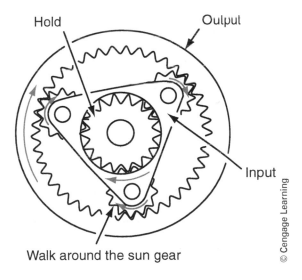

Hold

Output

Input

Walk around the sun gear

© Cengage Learning

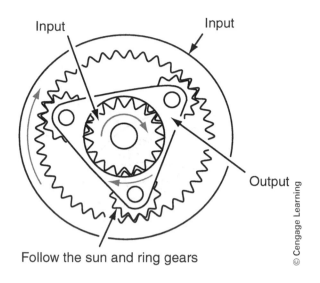

Input   Input

Output

© Cengage Learning

Follow the sun and ring gears

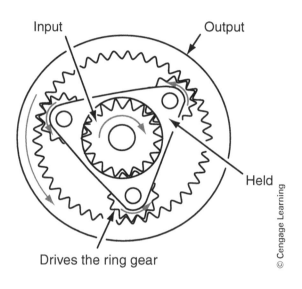

Input   Output

Held

© Cengage Learning

Drives the ring gear

1. What would be the result if none of the parts of the planetary are held? _____

2. How many forward speeds are available from a simple planetary gearset? _____

3. What are the three types of planetary gearsets? _____

■

The transmission valve body controls shift timing and feel. Numerous hydraulic circuits are contained in the valve body. In modern transmissions and transaxles, fluid flow is controlled by solenoids that are controlled by a control module. The powertrain control module (PCM) or transmission control module (TCM) controls shift points by monitoring inputs from sensors such as coolant temperature, throttle position, and the vehicle speed sensor. On older cars and trucks, the transmission may have required an occasional adjustment to maintain proper shift feel and to reduce slipping. On computer-controlled transmissions, the PCM can detect wear in the internal components and adapt the shift solenoid application to compensate for wear. Hydraulics is discussed in Chapter 3.

Concept Activity 41-3
Transmission Control

Examine two vehicles' transmissions/transaxles as provided by your instructor.

Year _____ Make _____ Model _____ Transmission _____

Year _____ Make _____ Model _____ Transmission _____

1. Does either vehicle use a vacuum modulator? _____

2. Describe the location of the modulator. _____

3. Inspect the vacuum hose to the modulator. Note your findings. _____
_____

4. Remove the vacuum hose from the modulator and inspect for fluid leaking from the vacuum port. Note your findings. _____
_____

5. Does this modulator allow for adjustment? _____ YES _____ NO

6. Why and how would the modulator need to be adjusted? _____
_____

7. Does either transmission use throttle valve (TV) cable? _____ YES _____ NO

8. Describe the function of the TV. _____
_____

9. What would be some symptoms of an improperly adjusted TV cable? _____
_____

10. Describe how to check and adjust the TV cable. _____
_____

11. Does either transmission use solenoids to controls shifting? _____

12. How many solenoids are used? _____

13. Describe how the output of solenoids controls shifting. _____
_____

14. Which gear range is the limp-in mode gear? _____

15. What is the output state of the solenoids in limp-in mode? _____

16. Are these transmissions/transaxles equipped with lockup torque converter clutches (TCCs)? _____ YES _____ NO

17. What are the criteria for the PCM to command the TCC solenoid on? _____
_____

18. What are the criteria for the PCM to command the TCC solenoid off? _____
_____

19. Is the TCC commanded from off to completely on, or is it applied over several seconds? _____

20. What problems can result from a TCC sticking in the applied position? _____
_____
_____

■

The following job sheets will prepare you for transmission/transaxle service and repair.

☐ JOB SHEET 41–1

Visually Inspecting an Automatic Transmission

Name _____ Station _____ Date _____

Objective

Upon completion of this job sheet, you will have demonstrated the ability to conduct a preliminary inspection of an automatic transmission or transaxle. This task applies to ASE Education Foundation Tasks MLR 2.B.2 (P-2), AST/MAST 2.A.2 – 2.A.5 (P-1), 2.B.1 (P-1), and 2.B.2 (P-2).

Description of Vehicle

Year _____ Make _____ Model _____

VIN _____ Engine Type and Size _____

Mileage _____

Model and type of transmission _____

PROCEDURE

1. Check the transmission housing for damage, cracks, and signs of leaks. Record your findings and recommendations.

   _____

   _____

2. Inspect the slip joint area in the transmission extension housing for leaks. If the transmission is a transaxle, check the area where the inner CV joint is attached to the housing. Record your findings and recommendations.

   _____

   _____

3. Check for leaks wherever something is attached to the housing. Record your findings and recommendations.

   _____

   _____

4. Check the shift and throttle linkages for looseness, wear, and/or damage. Record your findings and recommendations.

   _____

   _____

5. Inspect any cables for binding, wear, and/or damage. Record your findings and recommendations.

   _____

   _____

6. Check the condition of all cooler lines and hoses. Record your findings and recommendations.

_____

_____

7. Check the condition of the fluid; pay attention to its level, color, smell, and feel. Record your findings and recommendations.

_____

Problems Encountered

_____

_____

_____

Instructor's Comments

_____

_____

_____

☐ JOB SHEET 41–2

Determining Transmission/Transaxle Fluid Requirements

Name _____ Station _____ Date _____

Objective

Upon completion of this job sheet, you will have demonstrated the ability to determine the correct fluid type, amount, and fluid level. This task applies to ASE Education Foundation Tasks MLR 2.A.1 (P-1), AST/MAST 2.A.2 (P-1), and MAST 2.A.4 (P-1).

Description of Vehicle

Year _____ Make _____ Model _____

VIN _____ Engine Type and Size _____

Mileage _____

Transmission/transaxle model _____

_____

PROCEDURE

1. Using the service information, determine the correct type of fluid as specified by the manufacturer. Note it here.

   _____

2. Locate the fluid service refill amount and note it here.

   _____

3. What is the correct procedure for checking the fluid level?

   _____
   _____

4. Visually inspect the transmission/transaxle for signs of leaks. Note your findings.

   _____

5. Prepare the vehicle as necessary to accurately check the fluid level. Record the steps taken.

   _____
   _____
   _____

6. Is the fluid level within specifications?

   _____

7. Examine the fluid closely. Note the color, smell, and texture.

_____

_____

_____

8. Does the fluid present any indication of possible transmission/transaxle problems?

_____

_____

_____

Problems Encountered

_____

_____

_____

Instructor's Comments

_____

_____

_____

REVIEW QUESTIONS

Review Chapter 41 of the textbook to answer these questions:

1. Which of the following statements about transmission solenoids is not true?

    a. They can be tested with an ohmmeter.
    b. They can be checked with a lab scope.
    c. They can be air tested.
    d. They can be checked by listening and feeling for movement.

2. Explain the differences between the three major types of planetary gearsets used in modern automatic transmissions/transaxles.

    _____

    _____

    _____

3. What job does the TV cable perform?

    _____

    _____

    _____

4. What effect does a 20°F increase in operating temperature have on the automatic transmission fluid?

    _____

    _____

    _____

5. Explain the purpose and operation of the TCC.

    _____

    _____

    _____

ELECTRONIC AUTOMATIC TRANSMISSIONS

OVERVIEW

Before the adoption of electronic controls for the various powertrain systems, automatic transmissions/ transaxles performed shifting based solely on mechanical and hydraulic indicators. Since the addition of electronic controls, the PCM or transmission control module (TCM) controls shifting via hydraulic shift solenoids. Newer performance vehicles have electronic clutchless automatic transmissions capable of being shifted manually or by letting the on-board computer have control.

Concept Activity 42-1
Electronic Transmissions

Using a vehicle assigned to you by your instructor, determine the following.

Year _____ Make _____ Model _____

VIN _____ Engine Type and Size _____

Mileage _____

Transmission/transaxle model _____

Connect a scan tool to the DLC, and begin communication with the on-board network.

1. List the available data parameters for the automatic transmission. _____

_____

_____

_____

2. List the inputs used for the transmission/transaxle. _____

_____

3. List the outputs for the transmission/transaxle. _____

_____

4. List any active commands available for the transmission/transaxle. _____

_____

5. How many shift solenoids does this transmission/transaxle use?

_____

6. What is the default or limp-in gear for this transmission?

_____

7. Explain how the manual gear-position sensor operates on this vehicle. _____

_____

_____

8. Does this transmission/transaxle use a torque converter clutch (TCC)? _____

_____

9. What are the parameters for the PCM/TCM to apply the TCC? _____

_____

_____

_____

■

 Concept Activity 42-2
Constantly Variable Transmissions

Many newer vehicles have constantly variable transmissions, called CVTs. The CVT is an automatic transmission that uses a set of variable pulleys instead of fixed gearsets. **Figure 42-1** shows the basic CVT arrangement. By varying the diameters of both the driving and driven pulleys, the transmission can adapt the gear ratio to the driving conditions.

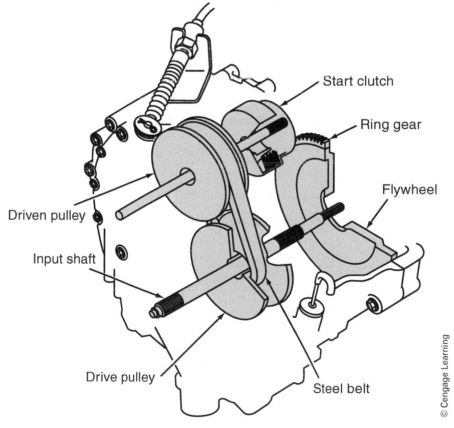

Figure 42-1   An example of a CVT.

1. List three to five late-model vehicles that may have a CVT.

_____

_____

2. What are the advantages of CVTs? _____

_____

3. How does the operation of a vehicle with a CVT differ than that of a manual or traditional automatic transmission?

_____

_____

◾

Full hybrids, such as the Toyota Prius, use an internal combustion engine (ICE) and electric motors to drive the vehicle. The ICE and electric motors are coupled with a power splitting device. This allows for ICE, the electric motor, or a combination of both to power the drive wheels. **Figure 42-2** shows an example of a basic hybrid drive system.

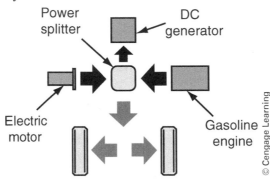

Figure 42-2   The layout of the main components for Toyota's hybrid vehicles.

 **Concept Activity 42-3**
Hybrid Drive Systems

1. List the components that make up a nontraditional hybrid drive arrangement.

_____

_____

_____

2. Explain how the power splitter can be used to drive the wheels and also allow for recharging of the high-voltage battery pack.

_____

_____

3. Under what conditions are both the ICE and electric motors used to propel the vehicle?

_____

_____

4. Under what conditions does the electric motor provide the propulsion for the vehicle?

_____

_____

_____

◾

The following job sheets will prepare you for electronic transmission/transaxle service and repair.

☐ JOB SHEET 42–1

Testing Transmission/Transaxle Electronic Controls

Name _____ Station _____ Date _____

Objective

Upon completion of this job sheet, you will have demonstrated the ability to monitor and test electronic transmission/transaxle controls. This task applies to ASE Education Foundation Task AST/MAST 2.A.11 (P-1).

Tools and Materials
Scan tool

Description of Vehicle
Year _____ Make _____ Model _____

VIN _____ Engine Type and Size _____

Mileage _____

Transmission/transaxle model _____

PROCEDURE

1. Connect a scan tool to the DLC and enter communications.

   Scan Tool Used _____

2. Check for any transmission/transaxle related DTCs.

   _____

3. With the key on engine running (KOER) in park, determine and note the status of the manual valve position sensor, shift solenoids, and TCC solenoid.

   _____

4. Place the gear selector into reverse and note manual valve position sensor, shift solenoids, and TCC solenoid.

   _____

5. Place the gear selector into drive and note manual valve position sensor, shift solenoids, and TCC solenoid.

6. List any other parameter ID (PID) data that is available for the transmission/transaxle.

   _____

   _____

   _____

7. List any output command functions available for the transmission/transaxle.

_____

_____

_____

If possible, and with permission from your instructor, perform the next steps.

Instructor Check _____

8. Test drive the vehicle while noting the operation of the solenoids.

First gear: Solenoid # _____ ON _____ OFF _____

Second gear: Solenoid # _____ ON _____ OFF _____

Third gear: Solenoid # _____ ON _____ OFF _____

Fourth gear: Solenoid # _____ ON _____ OFF _____

Fifth gear: Solenoid # _____ ON _____ OFF _____

Sixth gear: Solenoid # _____ ON _____ OFF _____

TCC command on at what speed? _____

TCC command on in which gear? _____

TCC command off in which gear? _____

TCC command on 0%–100% or incrementally? _____

9. Why are some TCC solenoids commanded on in steps instead of all at once?

_____

10. Did all solenoids and gear ranges operate correctly?

_____

Problems Encountered

_____

_____

_____

Instructor's Comments

_____

_____

_____

☐ JOB SHEET 42–2

Testing Electronic Transmission/Transaxles Solenoids

Name _____ Station _____ Date _____

Objective

Upon completion of this job sheet, you will have demonstrated the ability to test automatic transmission/transaxle solenoids. This task applies to ASE Education Foundation Tasks AST/MAST 2.A.11 (P-1), and 2.B.3 (P-1).

Tools and Materials

Digital multimeter (DMM)

Digital storage oscilloscope (DSO) / graphing multimeter (GMM)

Description of Vehicle

Year _____ Make _____ Model _____

VIN _____ Engine Type and Size _____

Mileage _____

Transmission/transaxle model _____

PROCEDURE

When testing solenoids of the automatic transmission/transaxle, perform basic system tests first.

1. Check battery voltage and charging system voltage. Record your results.

   _____

   _____

2. Verify proper fluid level, condition, and type. Record your findings.

   _____

   _____

3. Verify the power and ground supply to the solenoid. Record your findings.

   _____

   _____

4. Connect a DSO/GMM to the solenoid wiring. Activate the solenoid and check the waveform. Some solenoids are controlled by a pulse-width modulated signal. Draw the waveform you captured.

5. Does the waveform exhibit noise or glitches?

_____

6. Solenoids can be checked for proper internal resistance with a DMM. Locate the resistance specification, and note it here. _____

    Measure the solenoid resistance, and record it here. _____

    Is the solenoid resistance within specifications? _____

7. What problems could result from the solenoid resistance being out of specification?

_____

_____

_____

Problems Encountered

_____

_____

_____

Instructor's Comments

_____

_____

_____

REVIEW QUESTIONS

Review Chapter 42 of the textbook to answer these questions:

1. Define transmission adaptive control.

_____

_____

_____

2. Explain the importance of proper gearshift lever position sensor adjustment.

_____

_____

_____

3. Explain the operation of a CVT.

_____

_____

_____

4. Explain how hybrid CVTs operate in conjunction with the traction motors.

_____

_____

_____

5. Define limp-in mode.

_____

_____

_____

CHAPTER 43

AUTOMATIC TRANSMISSION AND TRANSAXLE SERVICE

OVERVIEW

Many automatic transmission/transaxle concerns are not really problems with the transmission/transaxle at all. Problems such as misfires, worn powertrain mounts, and suspension system problems can be misinterpreted as transmission/transaxle faults. This chapter discusses the proper identification and basic tests of the transmission/transaxle assemblies.

 Concept Activity 43-1
Transmission/Transaxle Identification

To accurately diagnose and service a transmission/transaxle, you must know exactly what model transmission/transaxle you are working on.

1. Locate the transmission/transaxle identification tag and record the information. The tag is usually located on the bell housing or other visible area **(Figure 43-1)**.

_____

_____

_____

_____

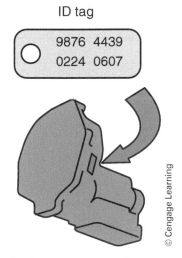

ID tag

9876 4439
0224 0607

© Cengage Learning

Figure 43-1   An example of a transmission ID tag.

A common transmission/transaxle service is checking and adjusting the fluid level. All manufacturers have specific fluids that they recommend for particular transmissions/transaxles. Failure to follow these recommendations can lead to shortened transmission/transaxle life and complete transmission/transaxle failure.

Concept Activity 43-2
Transmission Fluids

Match the following common transmission fluids with the manufacturer.

Dexron	Honda
Mercon	General Motors
ATF +4	Toyota
ATF −Z1	Chrysler
T-III & T-IV	Ford

Always refer to the manufacturer's service information regarding the proper automatic transmission/transaxle fluids.

■

The torque converter acts as the link between the engine and the transmission/transaxle. A faulty torque converter can mimic several more serious problems, so proper diagnosis is important. Torque converters are not serviceable and must be replaced if faulty.

Concept Activity 43-3
Torque Converter Testing

Using a vehicle assigned to you by your instructor, determine the proper procedures to perform a torque converter stall test. Always refer to the service information to determine whether the manufacturer recommends stall testing for a particular vehicle.

1. Manufacturer's warnings and precautions:

_____

_____

2. Vehicle preparation:

_____

_____

_____

■

The following job sheets will prepare you for automatic transmission/transaxle service and repair.

☐ JOB SHEET 43–1

Determining Transmission/Transaxle Fluid Requirements

Name _____ Station _____ Date _____

Objective

Upon completion of this job sheet, you will have demonstrated the ability to determine the correct fluid type, amount, and fluid level. This task applies to ASE Education Foundation Tasks MLR 2.A.1 – 2.A.3 (P-1), AST/MAST 2.A.2 (P-1), 2.A.4 (P-1), and 2.A.5 (P-1).

Description of Vehicle

Year _____ Make _____ Model _____

VIN _____ Engine Type and Size _____

Mileage _____

Transmission/transaxle model _____

PROCEDURE

1. Using the service information, determine the correct type of fluid as specified by the manufacturer. Note it here. _____ _____

2. Locate the fluid service refill amount and note it here. _____

3. Determine the correct procedure for checking the fluid level, and record it here. _____
   _____

4. Visually inspect the transmission/transaxle for signs of leaks. Note your findings.
   _____
   _____
   _____

5. Prepare the vehicle as necessary to accurately check the fluid level. Record the steps taken.
   _____

6. Is the fluid level within specifications?
   _____

7. Examine the fluid closely. Note the color, smell, and texture.
   _____
   _____
   _____

8. Does the fluid present any indication of possible transmission/transaxle problems? If so, note them here.

_____

_____

_____

Problems Encountered

_____

_____

_____

Instructor's Comments

_____

_____

_____

☐ JOB SHEET 43–2

Conducting a Stall Test

Name _____ Station _____ Date _____

Objective

Upon completion of this job sheet, you will have demonstrated the ability to safely conduct a stall test. This task applies to ASE Education Foundation Tasks AST 2.A.6 (P-2), and MAST 2.A.8 (P-2).

> **WARNING:** *Stall testing is not recommended for many newer vehicles. Always refer to the service manual before conducting a stall test.*

Tools and Materials

Tachometer

Stethoscope

Grease pencil or paint stick

Flashlight

Description of Vehicle

Year _____ Make _____ Model _____

VIN _____ Engine Type and Size _____

Mileage _____

Transmission/transaxle model _____

PROCEDURE

1. Park the vehicle on a level surface. Run the engine until it reaches normal operating temperature.

2. Remove the dipstick and observe the condition and level of the transmission fluid. Record your findings.

   _____

   _____

   What is indicated by the fluid's condition?

   _____

   _____

3. Connect a tachometer or scan tool to the engine. Place the tachometer inside the vehicle or at a place where it is easy to read from the driver's seat. If the vehicle has one in the instrument cluster, there is no need to connect another one.

4. Note the stall speed specification.

5. Block the front wheels and set the parking brake.

6. Put the transmission into park, start the engine, and allow the engine and transmission to reach normal operating temperature.

7. Put the gear selector into the gear that is recommended for this test. What is the recommended gear?

   _____

8. Press the throttle pedal to the floor with your right foot while firmly pressing the brake pedal with your left. Make sure you keep the brake pedal down and that no one and nothing is in front of the vehicle.

9. Note the highest speed attained, and immediately release the throttle. What was the highest rpm during the test? _____   Specification _____

10. Put the gear selector into neutral and allow the engine to run at about 1,000 rpm for at least one minute. What is the purpose of doing this?

    _____

    _____

11. If a noise was heard during the stall test, raise the vehicle on a hoist.

12. Use a stethoscope to determine whether the noise is from the transmission or torque converter.

13. What are your conclusions from this test?

    _____

    _____

Problems Encountered

_____

_____

_____

Instructor's Comments

_____

_____

_____

☐ JOB SHEET 43–3

Pressure Testing an Automatic Transmission

Name _____ Station _____ Date _____

Objective

Upon completion of this job sheet, you will have demonstrated the ability to conduct a pressure test on a transmission. This task applies to ASE Education Foundation Task MAST 2.A.6 (P-1).

Tools and Materials

Tachometer   Vacuum gauge

Pressure gauges with necessary adapters   T-fitting and miscellaneous hoses

Description of Vehicle

Year _____ Make _____ Model _____

VIN _____ Engine Type and Size _____

Mileage _____

Transmission/transaxle model _____

PROCEDURE

1. List the pressure specifications with the required operating conditions.

   _____

   _____

2. Start the engine, and allow it to reach normal operating temperature. Then turn off the engine.

3. Connect a tachometer or scan tool if the vehicle is not equipped with one.

4. Have a fellow student sit in the vehicle to operate the throttle, brakes, and transmission during the test.

5. Raise the vehicle on the hoist to a comfortable working height.

6. Describe the location of the pressure taps on the transmission.

   _____

   _____

7. Connect the pressure gauges to the appropriate service ports.

8. If the transmission has a vacuum modulator, use the T-fitting and vacuum hoses to connect the vacuum gauge into the modulator circuit.

9. Start the engine. Have your helper press firmly on the brake pedal and apply the parking brake. Then the helper should move the gear selector into the first test position.

   Gear Selector Position _____

10. Run the engine at the specified speed. Move the gear selector as required by the service manual. Observe and record the pressure and vacuum readings at the various test conditions.

_____

_____

_____

11. Turn off the engine, and move the pressure gauges to the appropriate test ports for the next transmission range to be tested. Describe this location.

_____

_____

_____

12. Restart the engine, and have your helper move the gear selector into the range to be tested and increase the engine's speed to the required test speed. Observe and record the pressure and vacuum readings in the various test conditions.

_____

_____

_____

13. Allow the engine to return to idle, and turn it off.

14. Repeat this sequence until all transmission ranges have been tested.

15. Summarize the results of this test, and compare them to specifications.

_____

_____

_____

_____

_____

_____

Problems Encountered

_____

_____

_____

Instructor's Comments

_____

_____

_____

REVIEW QUESTIONS

Review Chapter 43 of the textbook to answer these questions:

1. Describe the purpose of performing transmission fluid pressures tests.

   _____

   _____

   _____

2. List and explain the components of a modern torque converter.

   _____

   _____

   _____

3. Define two types of transmission holding devices.

   _____

   _____

   _____

4. Air-pressure testing is used to verify the operation of what transmission components?

   _____

   _____

   _____

5. Explain the operation of a one-way clutch.

   _____

   _____

   _____

CHAPTER
44

FOUR- AND ALL-WHEEL DRIVE

OVERVIEW

Many of the best-selling vehicles in North America are sport utility vehicles and pickup trucks. A majority of these vehicles are equipped with four-wheel-drive systems. A few passenger cars also have systems that provide torque to all four wheels. Although these systems are based on either FWD or RWD designs that are familiar to most technicians, they have some unique components that must be diagnosed and serviced.

To accomplish four-wheel drive and all-wheel drive, additional gearing is needed over and above the transmission/transaxle gearing. This gearing is usually contained in a transfer case. Heavy-duty systems based on rear-wheel-drive setups usually use a separate transfer case remote from the transmission. Vehicles not intended for heavy-duty off-road exposure usually locate this additional gearing in a smaller transfer case mounted on the back or bottom of the transaxle case.

Gearing principles are similar to those used in manual transmissions. Many systems also use an interaxle or third differential to help equalize power transmission to both the front and rear drivelines during all types of operating conditions.

Work safely! Many transfer cases are extremely heavy, and they contain transmission fluid that can become very hot. All drivelines must be properly supported when disconnected. Rebuilding transfer cases is similar to transmission/transaxle work and is often undertaken as a specialized field. Service performed by general service technicians may include inspection and road testing, unit removal and replacement, servicing of oil seals and bushings, and replacement of worn hub locking mechanisms.

There are many different mechanisms used to engage and disengage an axle to engage or disengage four-wheel drive. Some of these systems are directly controlled by the driver, whereas others happen automatically in response to axle speed. Before diagnosing a complaint of poor engagement or disengagement, use a service manual to determine how the unit is controlled.

The following job sheets will prepare you to diagnose and service 4WD and AWD systems.

☐ JOB SHEET 44–1

Inspecting the Fluid Level in a Transfer Case

Name _____ Station _____ Date _____

Objective

Upon completion of this job sheet, you will have demonstrated the ability to inspect the fluid in a transfer case properly. This task applies to ASE Education Foundation Tasks AST/MAST 3.F.3 (P-3).

Description of Vehicle

Year _____ Make _____ Model _____

VIN _____ Engine Type and Size _____

Mileage _____

Describe the type of system and model of the transfer case.

_____

PROCEDURE

1. Raise and support the vehicle.

2. Locate the fill plug on the transfer case. (Refer to the service manual for the location of the plug.)

   Plug Location _____

3. Remove the filler plug.

4. Using your little finger, feel in the hole to determine the fluid level. Describe the fluid level.

   _____

5. If you cannot touch the fluid, refer to the service manual for fluid type. Fill the transfer case. The recommended fluid is _____.

6. If the transfer case is low on fluid, visually inspect it to locate the leaks. Note your findings.

   _____

7. If you can reach the fluid with your finger, note the smell, color, and texture of the fluid. Note your findings.

   _____

8. If the fluid is contaminated, determine what is contaminating it. Then correct that problem, drain the fluid from the transfer case, and refill it with clean fluid.

Problems Encountered

_____

_____

_____

Instructor's Comments

_____

_____

_____

☐ JOB SHEET 44–2

Road Testing a Transfer Case

Name _____ Station _____ Date _____

Objective

Upon completion of this job sheet, you will have demonstrated the ability to road test a vehicle and determine the operating condition of a transfer case. This task applies to ASE Education Foundation Task AST/MAST 2.A.1 (P-1).

Description of Vehicle

Year _____ Make _____ Model _____

VIN _____ Engine Type and Size _____

Mileage _____

Describe the transfer case controls and list the type of transfer case found on the vehicle.

_____

PROCEDURE

Obtain permission from your instructor before attempting this job sheet.

Instructor Check _____

Road test the vehicle and attempt to operate it in all of the modes of the transfer case. Preview the following criteria. Record your findings on return to the shop.

1. Does the transfer case make noise in

   a. low drive rear-wheel drive? _____ YES _____ NO _____ N/A

   b. high drive in rear-wheel drive? _____ YES _____ NO _____ N/A

   c. low drive in four-wheel drive? _____ YES _____ NO _____ N/A

   d. high drive in four-wheel drive? _____ YES _____ NO _____ N/A

2. Does it make noise as you turn corners with it in four-wheel drive? _____ YES _____ NO

3. Does it make noise as you drive down the road in high gear? _____ YES _____ NO

4. Does it make a noise when the transfer case is in neutral? _____ YES _____ NO

5. What are your conclusions and recommendations regarding the road test?

_____

_____

_____

Problems Encountered

_____

_____

_____

Instructor's Comments

_____

_____

_____

☐ JOB SHEET 44–3

Replacing a Transfer Case Output Shaft Bushing and Seal

Name _____ Station _____ Date _____

Objective

Upon completion of this job sheet, you will have demonstrated the ability to safely and properly replace the bearing oil seal on the rear driveline shaft of a typical four-wheel-drive transfer case (**Figure 44-1**). This task applies to ASE Education Foundation Task AST/MAST 3.F.3 (P-3).

Tools and Materials

Chalk or marker   Seal installation tool
Extension housing bushing installation tool   Slide hammer and oil seal removal tool
Extension housing bushing removal tool   Wire
Multipurpose lubricant or transmission fluid

Description of Vehicle

Year _____ Make _____ Model _____

VIN _____ Engine Type and Size _____

Mileage _____

Describe the type of transfer case.

_____

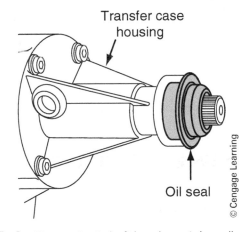

Figure 44-1 Rear output shaft bearing retainer oil seal and housing.

PROCEDURE (REMOVE OIL SEAL AND BEARING)

1. Raise the vehicle on a hoist.

   WARNING: *Lifting a four-wheel-drive truck on some hoists requires the use of adapters. Ensure proper pad contact by shaking the vehicle when it is a few inches off the floor. If the vehicle appears to be unstable, lower it and reset the pads.*

2. Using chalk or a marker, index the driveline shaft at the rear axle flange, and remove the driveshaft. If required, remove the flange from the transfer case.

3. Remove the oil seal from the bearing retainer by prying it with a screwdriver or using a slide hammer and seal removal tool **(Figure 44-2)**.

4. Remove the bushing from the retainer, using an extension housing bushing removal tool **(Figure 44-3)**. Discard both the bushing and the old oil seal.

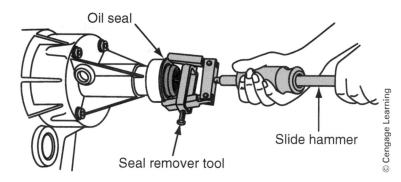

Figure 44-2   Removing the oil seal from the bearing retainer.

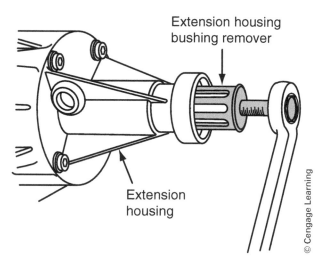

Figure 44-3   Removing the bushing from the retainer, using an extension housing bushing removal tool.

PROCEDURE (INSTALL NEW BUSHING AND OIL SEAL)

1. Drive the bushing into place in the retainer using an extension housing bushing installation tool **(Figure 44-4)**.

2. Position the seal in the retainer so the notch on the seal faces upward and the drain hole in the rubber dust boot faces downward. Drive the seal in the retainer using a seal installation tool **(Figure 44-5)**. Replace the flange and torque the nut to specifications.

Torque Specification _____

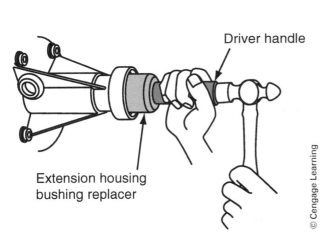

Figure 44-4 Driving the bushing into place using a bushing installation tool.

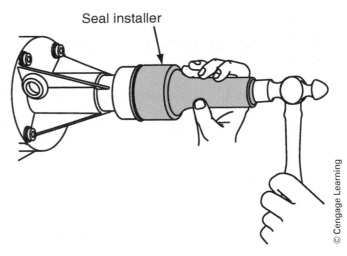

Figure 44-5 Driving the seal into place with a seal installation tool.

3. Position the driveline shaft so it aligns with the index marks on the transfer case and rear axles. Install the driveline shaft in the slip splines in the transfer case retainer. Install the driveshaft to the rear axle flange. Tighten the fasteners to the torque specified in the service manual.

Torque Specification _____

Problems Encountered

_____

_____

_____

Instructor's Comments

_____

_____

_____

 **REVIEW QUESTIONS**

Review Chapter 44 of the textbook to answer these questions:

1. What is the advantage of using limited-slip differentials?_____

   _____

   _____

2. What is the purpose of a viscous clutch in some transfer case designs? _____

   _____

   _____

3. Explain the importance of having the proper tire sizes on 4WD and AWD vehicles. _____

   _____

   _____

4. Explain two methods of engaging the 4WD system. _____

   _____

   _____

5. Technician A says all transfer cases are 1:1 ratio cases. Technician B says some transfer cases allow for gear reduction. Who is correct?

   a. Technician A   c. Both A and B
   b. Technician B   d. Neither A nor B

ASE PREP TEST

1. Technician A says the color of ATF on a dipstick will be dark brown if the transmission has seen normal use. Technician B says a milky color indicates that engine coolant has been leaking into the transmission's cooler in the transmission. Who is correct?

   a. Technician A
   b. Technician B
   c. Both A and B
   d. Neither A nor B

2. Technician A says that some transmissions use accumulators to increase the pressure to apply devices. Technician B says accumulators are used to dampen or cushion the application of servos. Who is correct?

   a. Technician A
   b. Technician B
   c. Both A and B
   d. Neither A nor B

3. The fuse that protects a 4WD vacuum solenoid fails after a short period of driving. Technician A says the solenoid may be faulty. Technician B says the connector that plugs into the solenoid may have high resistance. Who is correct?

   a. Technician A
   b. Technician B
   c. Both A and B
   d. Neither A nor B

4. Technician A says worn planetary gearset members can cause delayed shifting. Technician B says leaking hydraulic circuits or sticking spool valves in the valve body may be caused by delayed shifts or slippage. Who is correct?

   a. Technician A
   b. Technician B
   c. Both A and B
   d. Neither A nor B

5. A stall speed test shows the rpm is less than specifications. This could be caused by any of the following *except* which?

   a. Low ATF level
   b. Slipping stator clutch
   c. Slipping clutch or bands
   d. Restricted exhaust system

6. Technician A says a seized one-way stator clutch will cause the vehicle to have good low-speed operation but poor high-speed operation. Technician B says a freewheeling or nonlocking one-way stator clutch will cause the vehicle to have poor acceleration. Who is correct?

   a. Technician A
   b. Technician B
   c. Both A and B
   d. Neither A nor B

7. Which of the following can be caused by a faulty or misadjusted transmission range switch?

   a. Improper upshifts
   b. Delayed gear engagement
   c. Not staying in geard
   d. All of the above

8. Which of the following would *not* cause the 4WD not to engage?

   a. Loose-fitting front and rear drive shaft slip yokes
   b. Faulty actuator motor
   c. Vacuum leak to the vacuum solenoid
   d. Faulty control module

9. While diagnosing a ratcheting noise from the front hubs of a 4WD vehicle, Technician A says a likely cause is bad U-joints. Technician B says a likely cause is worn or damaged suspension parts. Who is correct?

    a. Technician A
    b. Technician B
    c. Both A and B
    d. Neither A nor B

10. Technician A says a shudder after converter clutch lockup indicates a problem with the TCC solenoid. Technician B says driveline or converter shudder can be isolated by disconnecting the converter's clutch solenoid. Who is correct?

    a. Technician A
    b. Technician B
    c. Both A and B
    d. Neither A nor B

11. Which of the following can cause a buzzing noise from a transmission?

    a. Worn bands or clutches
    b. Improper fluid level or condition
    c. Damaged planetary gearset
    d. Defective oil pump

12. Technician A says doing an endplay check on a transmission before disassembling it will give an indication of the condition of the bearings and seals inside the transmission. Technician B says excessive endplay will allow the clutch drums to move back and forth too much and will cause transmission case wear. Who is correct?

    a. Technician A
    b. Technician B
    c. Both A and B
    d. Neither A nor B

13. A vehicle with an automatic transmission starts when the gear selector is put in the drive position. Technician A says the shift linkage may be improperly adjusted. Technician B says the neutral-start (safety) switch may be improperly adjusted. Who is correct?

    a. Technician A
    b. Technician B
    c. Both A and B
    d. Neither A nor B

14. While discussing servicing a transmission, Technician A says a little bit of dark fluid or contamination in the fluid is normal. Technician B says a small amount of metal particles in the fluid is normal. Who is correct?

    a. Technician A
    b. Technician B
    c. Both A and B
    d. Neither A nor B

15. While discussing testing a noise from the transmission, Technician A says a noise only present in park or neutral is likely caused by a faulty oil pump. Technician B says noise that changes with engine speed in all gears is likely a faulty oil pump. Who is correct?

    a. Technician A
    b. Technician B
    c. Both A and B
    d. Neither A nor B

16. While discussing tires on 4WD and AWD vehicles, Technician A recommends replacing both tires on an axle if one tire needs to be replaced. Technician B says damage to 4WD or AWD components can result if there are excessive tire diameter differences on the vehicle. Who is correct?

    a. Technician A
    b. Technician B
    c. Both A and B
    d. Neither A nor B

17. Which of the following input sensors are used to determine shift timing?

    a. Throttle position sensor
    b. Engine coolant temperature sensor
    c. Vehicle speed sensor
    d. All of the above

18. Which of the following would not be a typical cause for incorrect shift points?

    a. Sticking valves or a dirty valve body
    b. Defective vehicle speed or MAP sensor
    c. Internal fluid leaks in the transmission
    d. Band or clutch not working properly

19. Technician A says early shifts in all gears can be caused by faulty input sensors. Technician B says if a valve body is dirty, all shifts may occur early. Who is correct?

    a. Technician A
    b. Technician B
    c. Both A and B
    d. Neither A nor B

20. Which of the following would not cause the transmission to slip in all forward gears?

    a. Improper fluid level
    b. Locked-up torque converter one-way clutch
    c. Malfunctioning electronic controls
    d. Sticking valves or dirty valve body

CHAPTER 45

TIRES AND WHEELS

© Cengage Learning 2015

OVERVIEW

The tires are the vehicle's only direct contact with the road. Although the primary purpose of tires is to provide traction, they also help absorb road shocks and transfer braking and driving torque to the road. Periodic inspection and maintenance is important to prevent uneven tire wear, prolong tread life, and ensure safe operation. Wheel bearings connect the wheel to the suspension and allow for smooth rotation of the wheel and tire. Wheel bearings can be either serviceable, that is, requiring periodic service and adjustment, or sealed units. FWD vehicles typically have sealed bearings, whereas some RWD vehicles still use serviceable bearings on the nondrive wheels.

Tires have one of the hardest jobs on the automobile; they must provide traction and help the suspension absorb road shocks. In addition, they are also designed to carry the weight of the vehicle, withstand side thrust over varying speeds and conditions, and transfer braking and driving torque to the road.

A tire provides traction against the road surface because of the materials that make up the tire. Tires are constructed in layers of different materials. Various rubber, synthetic compounds, and ply materials such as steel and rayon are molded together under heat and pressure to create the final product. The tire size, construction type, and materials play an important role in overall traction. Because tires are flexible, they are able to deform slightly when impacting bumps and holes in the road surface. This acts to soften the vehicle's ride and reduce road shock.

Concept Activity 45-1
Tire Construction

1. Identify the components of the tire shown in the following figure.

   a. _____

   b. _____

   c. _____

   d. _____

   e. _____

   f. _____

   g. _____

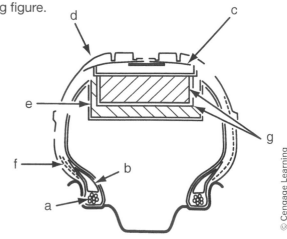

2. Examine several tires in your lab, and note their traction ratings. Compare the traction ratings and the tire construction from the sidewall information.

Tire 1: Traction _____ Construction Materials _____

Tire 2: Traction _____ Construction Materials _____

Tire 3: Traction _____ Construction Materials _____

3. How do you think tire construction affects its traction rating? _____

_____

■

_____

Tires must also carry the weight and load of the vehicle. Most passenger car tires are inflated to between 26 and 35 psi, though some are inflated to 44 psi or more.

Concept Activity 45-2
Tires and Air Pressure

1. How can the tires, with only 30 psi of air pressure, support the weight of a 4,000-pound vehicle?

_____

2. How does tire air pressure affect vehicle ride quality? _____

3. How does the friction from the tire contact with the road affect the tire's air pressure?

_____

4. Why does tire air pressure decrease in cold weather? _____

5. How does tire pressure affect tread wear? _____

Some owners have their tires inflated with nitrogen instead of compressed air.

6. How can you tell if a tire is inflated with nitrogen? _____

7. What are some of the benefits of nitrogen inflation? _____

_____

■

_____

Concept Activity 45-3
Tire Pull

Tires have to be able to roll in a straight path perpendicular to the road surface. If a tire is slightly misshaped or if the belts within the tire are not parallel, tire pull can occur. To illustrate this problem, take a plastic or Styrofoam cup and place it on its side on a workbench. Attempt to roll the cup in a straight line.

1. Describe how the cup rolls. _____

_____

2. Why does the cup roll as is does? _____

_____

3. If a vehicle has a pull to one side, and tire pressure and alignment are correct, how would you diagnose the concern? _____

_____

■

Vibration complaints are common customer concerns. When a vehicle has a vibration at highway speeds, out-of-balance wheels and tires are the common cause. Anytime there is a rotating mass, such as a wheel and tire assembly, it must be balanced so that the weight distribution around the assembly is equal. An out-of-balance component can cause rapid wear and destruction of parts.

Concept Activity 45-4
Balance

Obtain a plastic lid from a coffee can and, in the exact center, drill a hole just large enough for a pencil or similar straight rod to pass through. The hole needs to allow the lid to rotate easily on the pencil.

1. Spin the lid and note any wobble from side to side. _____

Tape a small weight, such as a quarter, to the lid near the outside edge.

2. Does the lid rotate down on its own? _____

3. Why does the lid rotate with the weight down? _____

4. Spin the lid, and note how well it rotates. _____

5. What effect will the added weight have on the rotation of the lid? _____

6. If this were a wheel and tire assembly, how would this heavy spot affect the rotation of the assembly?

7. What symptom(s) could this imbalance cause? _____

8. How could this problem be corrected? _____

_____

■

Wheels connect the tire to the vehicle. Modern wheels can be of stamped steel, aluminum, or alloy construction. All wheels share the same basic features and service needs.

Concept Activity 45-5
Wheel Construction

1. Identify the sections of the wheel in the following figure.

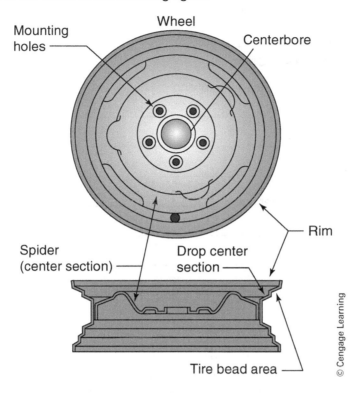

a. Hub

b. Bead

c. Drop center

d. Diameter

e. Width

f. Valve stem

g. TPMS sensor

2. Describe wheel offset. _____

_____

3. What other components or systems can be affected by changing wheel offset? _____

_____

Wheels have specific bolt patterns, the spacing between the holes for the wheel fasteners. Even though a set of wheels may appear similar, their hub openings and lug patterns may be different. Installing an incorrect wheel assembly could lead to serious injury or death.

Locate a selection of wheels as supplied by your instructor.

4. Draw and label the measurement of the lug pattern and hub opening diameter for three different rims in the space below.

Wheel 1   **Wheel 2**   **Wheel 3**

_____   _____   _____

Wheels and tires are often victims of the hazards associated with driving. Potholes, curbs, and debris on the road can all damage the wheels and tires. Small nicks and surface marking usually does not cause a problem. Dents and bending of the wheel and cuts or punctures of the tire can lead to problems that are more serious.

Concept Activity 45-6
Wheel and Tire Inspection

Inspect several wheels and tires and note the following:

1. Visible dents or bending of the wheel. _____

2. Cuts in the tire tread or sidewall. _____

3. Damage to the valve stem. _____

4. Condition of hub and wheel fastener holes. _____

5. Tread wear pattern. _____

6. Bulges or swelling of the tire. _____

Wheels and/or tires that exhibit signs of serious damage or wear should be replaced.

■

Wheel bearings attach the wheel and tire to the vehicles suspension system. Bearings are designed to support loads and reduce friction. Wheel bearings can be either the serviceable or sealed type. Serviceable bearings require periodic cleaning, repacking with new grease, and adjustment. Sealed bearings are lubricated for life and are replaced as an assembly.

Concept Activity 45-7
Wheel Bearing Inspection

Obtain a sealed wheel bearing and a serviceable tapered roller bearing from your instructor.

1. Rotate each bearing by hand, and compare how easily each moves.

   _____

2. Examine each bearing, and describe how each is able to handle the loads placed upon it.

   _____

3. Inspect each bearing for wear and signs of damage. _____

4. What components or actions can cause premature bearing wear or failure?

   _____

   _____

   _____

■

The following job sheets will prepare you to inspect and service wheels and tires.

☐ JOB SHEET 45–1

Inspecting Tires for Proper Inflation

Name _____ Station _____ Date _____

Objective

Upon completion of this job sheet, you will have demonstrated the ability to inspect and set tire air pressure. This task applies to ASE Education Foundation Tasks MLR 4.D.1 (P-1), and AST/MAST 4.F.1 (P-1).

Description of Vehicle

Year _____ Make _____ Model _____

VIN _____ Engine Type and Size _____

Mileage _____

PROCEDURE

1. Locate the tire information placard. Location _____

2. Determine the recommended inflation pressure for the tires installed on this vehicle.

   Recommended pressure _____

   Recommended spare pressure _____

3. Check the air pressure of each tire and record your readings.

   RF _____ LF _____ RR _____ LR _____ Spare _____

4. Does the pressure of any tire vary greatly from the others? _____

5. If so, what action should be taken? _____

   _____

6. Adjust the air pressure to the recommended setting.

Problems Encountered

_____

_____

_____

Instructor's Comments

_____

_____

_____

☐ JOB SHEET 45–2

Diagnosing Wheel/Tire Vibration, Shimmy, and Noise

Name _____ Station _____ Date _____

Objective

Upon completion of this job sheet, you will have demonstrated the ability to inspect a wheel and tire assembly for causes of vibration, shimmy, and noise. This task applies to ASE Education Foundation Tasks AST/MAST 4.F.1 (P-1), and 4.F.2 (P-2).

Tools and Materials

Tire pressure gauge

Tread depth gauge

Description of Vehicle

Year _____ Make _____ Model _____

VIN _____ Engine Type and Size _____

Mileage _____

PROCEDURE

1. Raise and support the vehicle so that all four tires are off the ground.

2. Check and adjust the tire pressure as necessary. Tire pressure specification _____

   RF _____ LF _____ RR _____ LR _____ Spare _____

3. Visually check each tire for wear patterns. Note your findings.

   RF _____ LF _____ RR _____ LR _____ Spare _____

4. Are all four tires of the same size, type, and tread design? _____ YES _____ NO

5. How could different tread designs affect vibration and noise concerns?

   _____

6. Check the tread depth of each tire. Minimum tread depth specification _____

   RF _____ LF _____ RR _____ LR _____ Spare _____

7. Rotate each wheel, and note the ease with which each wheel rotates. Describe your findings.

   _____

8. Rotate each wheel, and look for signs of bent wheels; watch for side-to-side wobble of the wheel and tire assembly. Note your findings. _____

9. Grasp each wheel at the three and nine o'clock positions and feel for looseness. Note your findings.

   _____

10. If any looseness is felt in step 9, what could be the cause(s)? _____

11. Could looseness cause a vibration or noise concern? _____ YES _____ NO

Problems Encountered

_____

_____

_____

Instructor's Comments

_____

_____

_____

    If no problems are noted in this inspection, but a vibration concern is confirmed, check the wheel balance. Continue to Job Sheet 45-3.

☐ JOB SHEET 45–3

Balancing a Wheel and Tire Assembly

Name _____ Station _____ Date _____

Objective

Upon completion of this job sheet, you will have demonstrated the ability to balance a wheel and tire assembly. This task applies to ASE Education Foundation Tasks MLR 4.D.3 (P-1), and AST/MAST 4.F.6 (P-1).

Description of Vehicle

Year _____ _____ Make _____ Model _____

VIN _____ Engine Type and Size _____

Mileage _____

PROCEDURE

When removing all four wheels from the vehicle, mark the location of each wheel before removing it so that it can be returned to the correct location.

1. Mount the wheel and tire on the wheel balancer.

2. Determine the wheel/tire dimensions as needed, and program the balancer.

   Wheel Width _____ Wheel Diameter _____ Wheel Offset _____

3. Engage the balancer and note the required weights.

   Left _____ Right _____

4. Remove the old weights if you have not done so already and spin the wheel again.

   Note the required weights.   Left _____ Right _____

5. Place the necessary weights at the locations indicated by the balancer.

6. Spin the wheel again. Is the wheel/tire in balance? _____ YES _____ NO

7. If correction is required, note the additional weight(s) needed and their location to those already installed.   Left/Location _____ Right/Location _____

8. Once the wheel/tire is in balance, remove it from the wheel balancer, and reinstall it on the vehicle.

9. Tighten lug nuts or studs, using the proper sequence and torque as specified in the service information.

   Tightening pattern _____

   Torque specification _____

Problems Encountered

_____

_____

_____

Instructor's Comments

_____

_____

_____

 REVIEW QUESTIONS

Review Chapter 45 of the textbook to answer these questions:

1. What are the primary purposes of tires?

   _____

   _____

   _____

2. What is the difference between a wheel bearing and an axle bearing?

   _____

   _____

   _____

3. Explain how to check wheel/tire runout.

   _____

   _____

   _____

4. Explain the types of wheel balance.

   _____

   _____

   _____

5. How can improper wheel torque affect the hub and bearing assemblies?

   _____

   _____

   _____

SUSPENSION SYSTEMS

OVERVIEW

The suspension system is responsible for carrying the weight of the vehicle and providing a smooth, comfortable, and safe ride. Suspensions can be dependent or independent, and vary greatly in design and in the use of components, but all try to accomplish the same tasks.

The primary purpose of the suspension is to safely and comfortably carry the weight of the vehicle. To handle this task, springs are used. A spring is a flexible object that is able to store mechanical energy. When a vehicle is assembled, a load is placed on the springs. As the vehicle moves over the road surface, the springs react to vehicle movement by absorbing and releasing tension. This is called spring oscillation. An undampened spring will continue to oscillate until all of the energy has been released. The amount of load a spring can handle and still return to its original condition is called spring rate.

 Concept Activity 46-1
Oscillations

Clamp a wooden or plastic ruler to a workbench so that approximately 10 inches (25.4 cm) of the ruler is extended off the bench top. Push the end of the ruler down about 1 inch (2.54 cm) and let go.

1. Try to count the number of ruler oscillations before the movement stops. If the oscillations are too quick, measure the amount of time it takes to stop moving instead.

   Oscillations _____ Length of time _____

2. Now move the ruler so that only five inches (12.7 cm) is extended from the table, and retest.

   Oscillations _____ Length of time _____

   Compare the results of the two experiments.

3. Which test produced more oscillations? _____

4. Describe why you think this occurred. _____

5. Why does changing the length of the ruler affect its oscillation rate? _____

   Move the ruler back so that about 10 inches (25.4 cm) is extended from the table. Place a weight of known quantity on the end of the ruler and note the amount of deflection.

6. Weight _____ Deflection _____

7. Double the weight, and remeasure. Weight _____ Deflection _____

8. Is the amount of deflection twice as much as the first amount? _____

9. Whether your answer is yes or no, explain why you think this is so. _____

_____

10. Quickly remove the weight and again note the oscillations or length of oscillation time.

 Oscillations _____ Length of time _____

11. Place the original weight back on the edge of the ruler. Quickly remove the weight, and again note the oscillations or length of oscillation time.

 Oscillations _____ Length of time _____

12. Was the result of the second test one-half of the first test? _____ YES _____ NO

13. Explain your results. _____

14. How does the load placed on the ruler affect the oscillation and deflection rates? _____

_____

_____

■

Springs can operate under tension/extension, by compression, or by twisting under load. Springs that twist are called torsion springs. Coil springs, leaf springs, air springs, and seat springs all operate under compression. When a load is placed on these springs, they compress. When the load is removed, they extend back to their original position. Torsion springs absorb energy by twisting. Once the load is removed, they untwist back to their original position.

Concept Activity 46-2
Spring Types

Locate as many of the following springs on a vehicle assigned to you by your instructor. Note the type of each spring below.

 Year _____ Make _____ Model _____

1. Front suspension springs _____

2. Rear suspension springs _____

3. Hood springs _____

4. Trunk/hatch springs _____

5. Hood/trunk/fuel door release _____

6. What factors would need to be considered when choosing a spring type for a particular application on a vehicle? _____

_____

_____

■

To control spring oscillations in the suspension, shock absorbers are used. A shock is a hydraulic device that dampens spring oscillations. Inside a shock absorber, oil moves between chambers via valves and pistons. The sizes of the valves and pistons determine how much and how quickly the oil can flow. The control of the oil flow allows the shock to help control the movement of the spring. As the spring tries to oscillate, the shock resists the movement, dampening out the spring.

Concept Activity 46-3
Shock Absorbers

Examine several shock absorbers provided by your instructor.

1. Note the physical differences in the body and piston rod sizes. _____
   _____

2. Why are some shocks larger in diameter than others? _____

   Manually compress and extend each shock.

3. Do the shocks compress and extend smoothly? _____ YES _____ NO

4. Do any compress or extend more easily or harder than the others? _____ YES _____ NO

5. Do any require a different amount of force to either compress or extend the shock? _____ YES _____ NO

6. If you answered yes to question 5, why do you think a shock would react differently between compression and extension? _____
   _____

7. How is the energy absorbed by the shock dissipated? _____
   _____

8. Attempt to cycle the shock rapidly between extension and compression. Did the shock get warm, and if so, why? _____
   _____

9. How can replacing the shocks on a vehicle affect the vehicle's ride quality?
   _____

Some vehicles are equipped with electronically controlled suspension systems. These systems allow changing the ride quality based on driver preference. Some systems control the vehicle ride height by adjusting the amount of air pressure in air springs, whereas other systems electronically vary the action of the shock absorber.

Although many suspension systems are similar in appearance, and all provide for the same basic functions, manufacturers design the suspension system on each vehicle in order to achieve a desired level of ride quality and handling performance.

Concept Activity 46-4
Suspension System Layout

Examine the suspension systems, front and rear, on two vehicles as directed by your instructor, and record the following information.

Vehicle 1

Year _____ Make _____ Model _____

1. Front suspension type _____

2. Rear suspension type _____

3. Spring used front/rear _____

4. Placement of springs relative to suspension components _____

5. Shock absorber location relative to spring location _____

6. Diameter of front/rear sway bars _____

Vehicle 2

Year _____ Make _____ Model _____

1. Front suspension type _____

2. Rear suspension type _____

3. Spring used front/rear _____

4. Placement of springs relative to suspension components _____

_____

5. Shock absorber location relative to spring location _____

_____

6. Diameter of front/rear sway bars: _____

_____

■

Concept Activity 46-5
Electronically Controlled Suspensions

Connect a scan tool to a vehicle with an electronically controlled suspension system.

Year _____ Make _____ Model _____

Scan tool _____

Enter communication with the on-board computer system, and navigate to the suspension inputs and outputs.

1. List the inputs for the suspension system. _____

   _____

2. List the outputs controlled by the computer system. _____

   _____

3. List any active commands the scan tool can control. _____

   _____

   Obtain your instructor's permission to attempt to perform any active commands of the suspension system.

   Instructor's Check _____

4. If equipped, command the suspension air compressor on, and note any change in ride height. _____

   _____

5. If equipped, command the air ride system to VENT, and note any change in ride height. _____

   _____

   _____

   ■

The remaining components of the suspension system support the operation of the springs and shocks, as well as maintain the correct tire positioning. The following job sheets will prepare you for suspension system service and repair.

☐ JOB SHEET 46-1

Identifying Types of Suspension Systems

Name _____ Station _____ Date _____

Objective

Upon completion of this job sheet, you will have demonstrated the ability to identify the different types of suspension systems. This task applies to ASE Education Foundation Task MLR/AST/MAST 4.A.1 (P-1).

Several types of suspensions are used in modern vehicles. These include the short-long arm (SLA), I-beam, MacPherson strut, modified strut, multilink, live axle, and dead axle.

PROCEDURE

1. Locate and record the information for as many of the different styles of suspensions as possible.

Vehicle 1

Year _____ Make _____ Model _____

Suspension Type _____ Spring Type _____

_____ Independent _____ Non-independent

Load carrying ball joint location _____

Rear Suspension Type _____

Vehicle 2

Year _____ Make _____ Model _____

Suspension Type _____ Spring Type _____

_____ Independent _____ Non-independent

Load carrying ball joint location _____

Rear Suspension Type _____

Vehicle 3

Year _____ Make _____ Model _____

Suspension Type _____ Spring Type _____

_____ Independent _____ Non-independent

Load carrying ball joint location _____

Rear Suspension Type _____

Vehicle 4

Year _____ Make _____ Model _____

Suspension Type _____ Spring Type _____

_____ Independent _____ Non-independent

Load carrying ball joint location _____

Rear Suspension Type _____

Problems Encountered

_____

_____

_____

Instructor's Comments

_____

_____

_____

☐ JOB SHEET 46–2

Measuring Vehicle Ride Height

Name _____ Station _____ Date _____

Objective

Upon completion of this job sheet, you will have demonstrated the ability to measure vehicle ride height. This task applies to ASE Education Foundation Task AST/MAST 4.C.1 (P-1), and 4.C.2 (P-1).

Tools and Materials

Tape measure

Description of Vehicle

Year _____ Make _____ Model _____

VIN _____ Engine Type and Size _____

Mileage _____

PROCEDURE

Before measuring the vehicle ride height, verify correct tire inflation and that there is not excessive weight in the vehicle's trunk or cargo area.

1. Locate the ride height specifications. _____ Front _____ Rear

2. Describe the location where ride height is to be measured.

   _____

3. Measure and record the ride height at each corner of the vehicle.

   RF _____ LF _____ RR _____ LR _____

4. Is the ride height within specifications? _____ YES _____ NO

5. If the ride height is not within specifications, what components should be inspected?

   _____

   _____

6. How does the ride height affect vehicle performance? _____

Problems Encountered

_____

_____

_____

Instructor's Comments

_____

_____

_____

☐ JOB SHEET 46–3

Inspecting Shock Absorbers

Name _____ Station _____ Date _____

Objective

Upon completion of this job sheet, you will have demonstrated the ability to inspect and test shock absorbers. This task applies to ASE Education Foundation Tasks MLR 4.B.20, AST/MAST 4.C.1 (P-1), 4.C.2 (P-1), and 4.D.1 (P-1).

Description of Vehicle

Year _____ Make _____ Model _____

VIN _____ Engine Type and Size _____

Mileage _____

PROCEDURE

1. Bounce each corner of the vehicle several times, and note the number of times the vehicle rebounds up and down.

   RF _____   LF _____

   RR _____   LR _____

2. Does the vehicle bounce excessively? _____ YES _____ NO

3. Were any noises noticed during the bounce test? _____ YES _____ NO

4. If noises were noticed, what components could be responsible? _____

5. Raise and support the vehicle. _____

6. Inspect each shock absorber for signs of oil leakage. Note your findings. _____

   _____

   Note: *A very slight amount of oil leakage is considered normal, especially in cold weather. An oil covered shock or a dripping shock needs replaced.*

7. Inspect each shock mounting and bushings. Note your findings. _____

   _____

8. Weak shocks can cause what type of tire wear? _____

9. Inspect the tires for signs of wear from weak shock absorbers. Note your findings. _____

   _____

Problems Encountered

_____

_____

_____

Instructor's Comments

_____

_____

_____

☐ JOB SHEET 46–4

Identifying Lifting and Jacking Precautions

Name _____ Station _____ Date _____

Objective

Upon completion of this job sheet, you will be able to demonstrate the ability to identify lifting and jacking precautions on vehicles with air springs and other electronically controlled suspensions. This task applies to ASE Education Foundation Task MLR/AST/MAST 4.A.1 (P-1).

Description of Vehicle

Year _____ Make _____ Model _____

VIN _____ Engine Type and Size _____

Mileage _____

PROCEDURE

1. Using a vehicle assigned to you by your instructor, determine the type of front and rear suspension systems used. _____

2. Locate any suspension, lifting, or jacking decals located on the vehicle.

   Location _____ Information _____

   Location _____ Information _____

   Location _____ Information _____

3. Describe what is required to safely jack or lift the vehicle. Locate any switches for the suspension system that may need turned off to raise the vehicle. _____

   _____

   Complete the steps necessary to jack or lift the vehicle. Obtain your instructor's permission before continuing.

   Instructor's Check _____

4. Jack and support or lift the vehicle. Note where the jacking equipment or lift contacts were placed. _____

5. Are any special procedures required to safely lower the vehicle? _____

6. Lower the vehicle to the ground. Perform the required steps to enable the ride control system. Describe how this is done. _____

   _____

Problems Encountered

_____

_____

_____

Instructor's Comments

_____

_____

_____

REVIEW QUESTIONS

Review Chapter 46 of the textbook to answer these questions:

1. Explain the purpose of the stabilizer bar/bushings.

_____

_____

_____

2. Describe the construction and operation of a shock absorber.

_____

_____

_____

3. What is the most likely cause for lower than specified ride height?

_____

_____

_____

4. What suspension components are typically eliminated by the MacPherson strut suspension?

_____

_____

_____

5. What is meant by a "wet" strut?

_____

_____

_____

OVERVIEW

The steering system allows the driver to control the direction of the vehicle. The steering system includes three major subsystems: the steering linkage, steering gear, and steering column and wheel. As the steering wheel is turned, the steering gear transfers this motion to the steering linkage. The steering linkage turns the wheels to control the vehicle's direction. Some newer vehicles have electronic steering systems that allow for four-wheel steering. This reduces the load on the engine from the power-steering hydraulic system and improves vehicle stability and safety.

The steering linkage connects the steering gear to the wheels. On some systems, such as the parallelogram linkage, there are many different pieces all linked together. Vehicles that use rack-and-pinion steering have fewer linkage components.

Concept Activity 47-1
Steering Linkage

1. Identify the components of the parallelogram linkage system in the following figure.

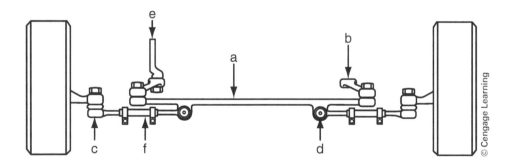

© Cengage Learning

a. _____   b. _____

c. _____   d. _____

e. _____   f. _____

2. Identify the components of the rack-and-pinion linkage system in the following figure.

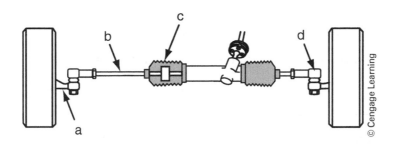

a. _____   b. _____

c. _____   d. _____

_____

■

There are two major types of steering gearboxes: the recirculating ball gearbox and the rack and pinion. Both steering gears convert the rotary motion of the steering wheel into a linear motion to turn the wheels. The steering gear also provides mechanical advantage to decrease the effort required by the driver to turn the vehicle. This is accomplished by the gear ratio of the gearbox. A box with a high gear ratio will provide for easy steering but will require more turns of the steering wheel to make turns. A gearbox with a low gear ratio will require fewer turns of the wheel, but more effort will be needed to make turns. Most vehicles are designed with a gearbox ratio that is a compromise between ease and performance.

Concept Activity 47-2
Gearbox Ratio

Steering ratio is calculated by dividing the degrees of steering wheel rotation by the degrees of movement at the wheels. For example, if 20 degrees of rotation of the steering wheel provided 1 degree of movement at the wheels, the ratio would be 20:1.

1. Determine the steering ratio of the gearboxes in the following figure.

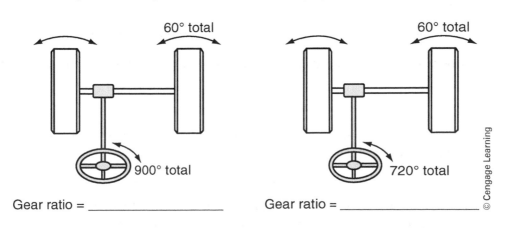

Gear ratio = _____   Gear ratio = _____

2. What type of vehicles would you expect to have a high gear ratio, and why?

_____

_____

3. What type of vehicles would you expect to have a low gear ratio, and why?

_____

_____

4. What are the advantages and disadvantages of a high gear ratio?

_____

_____

5. What are the advantages and disadvantages of a low gear ratio?

_____

_____

_____

Even though the main function of the steering column is to provide a means to turn the wheels, in modern vehicles the column contains electrical switches for the turn signals, horn, and often other components. The driver-side air bag is located in the steering wheel, and the column itself is designed to collapse in the event of an accident, in order to reduce driver injury. The steering wheel plays a large part in how much effort is required to turn the vehicle.

Concept Activity 47-3
Manual and Power-Steering

Measure the diameter of the steering wheels in several lab vehicles. Note whether each vehicle has manual or power-steering.

Vehicle 1: Wheel diameter _____   Manual/power-steering

Vehicle 1: Wheel diameter _____   Manual/power-steering

Vehicle 1: Wheel diameter _____   Manual/power-steering

1. Does there appear to be any correlation between the size of the steering wheels and the size of the vehicles?

_____

2. Calculate the difference in torque between the smallest and largest diameter wheel, based on 20 pounds of force being used to turn the wheel.

_____

Nearly all cars and trucks sold today are equipped with power-steering. Power-steering is really an added-on feature to traditional steering systems. The purpose of power-steering is to reduce steering effort. Most systems use a hydraulic pump to supply fluid to the steering gearbox and increase the force supplied by the driver. Some newer vehicles and hybrid vehicles are using electrically operated power-steering systems.

3. If you examine two identical vehicles, and one has manual steering and one has power-steering, would they have the same gearbox steering ratio?

_____

4. Explain your answer to question 3.

_____

5. What are the advantages to electrically operated power-steering?

_____

_____

■

_____

The following job sheets will prepare you for testing and servicing steering systems.

☐ JOB SHEET 47–1

Inspecting Steering Linkage for Wear

Name _____ Station _____ Date _____

Objective

Upon completion of this job sheet, you will have demonstrated the ability to inspect the steering linkage for wear. This task applies to ASE Education Foundation Tasks MLR 4.B.7 (P-1), 4.B.8 (P-1), AST/MAST 4.B.4 (P-2), 4.B.5 (P-2), 4.B.7 (P-2), 4.B.8 (P-2), 4.B.16 (P-2), and 4.B.17 (P-1).

Tools and Materials

Dial indicator

Description of Vehicle

Year _____ Make _____ Model _____

VIN _____ Engine Type and Size _____

Mileage _____

PROCEDURE

1. Locate any specifications for allowable movement of the steering linkage components, and record them here.

   _____

   A dry park check can often be used to locate loose steering linkage components. Have a fellow student sit in the driver's seat of the vehicle and raise the vehicle on a drive-on style lift.

2. Have your fellow student rock the steering wheel back and forth from the straight-ahead position with the engine off. Watch the linkage ball sockets for movement, and listen for any popping or other noises. Note your findings.

   _____

3. On parallelogram systems, firmly grasp each component near the ball socket, and attempt to move the component to compress and extend the ball socket. Do any of the sockets show any movement?

   _____

4. If the idler arm shows movement, connect the dial indicator so that the movement can be measured. Dial indicator reading _____

5. How does this movement compare to specifications?

   _____

6. On rack-and-pinion systems, raise the front wheels off the ramps. Grasp each tire at the three and nine o'clock position, and move back and forth. Watch for movement of the inner tie rods in the rack bellows. Note your findings.

   _____

Problems Encountered

_____

_____

_____

Instructor's Comments

_____

_____

_____

☐ JOB SHEET 47–2

Inspecting the Steering Gearbox for Leaks

Name _____ Station _____ Date _____

Objective

Upon completion of this job sheet, you will have demonstrated the ability to inspect a power-steering gearbox for fluid loss. This task applies to ASE Education Foundation Tasks MLR 4.B.4 (P-1), and AST/MAST 4.B.11 (P-1).

Description of Vehicle

Year _____ Make _____ Model _____

VIN _____ Engine Type and Size _____

Mileage _____

PROCEDURE

1. Locate the power-steering fluid reservoir, and determine the fluid level. Note your findings.

   _____

2. What is the recommended power-steering fluid?

   _____

3. Top off the fluid level with the correct fluid, if necessary.

4. Raise and support the vehicle.

   For recirculating gearboxes:

5. Inspect the pitman shaft and worm shaft seals for leakage. Note your findings.

   _____

   _____

6. Inspect the area where the power-steering lines connect to the gearbox. Note your findings.

   For rack-and-pinion gearboxes:

7. Inspect the pinion shaft seal for signs of leakage. Note your findings.

   _____

   _____

8. Inspect the area around the bellows for signs of leakage. If possible, pull the bellows back at either end to inspect, or remove the crossover tube and inspect. Note your findings.

   _____

   _____

9. Inspect the area where the power-steering hoses and lines connect to the rack. Note your findings.

_____

10. If the power-steering fluid level was low, but there is no sign of leakage from the gearbox, what are other possible causes of the fluid loss?

_____

_____

Problems Encountered

_____

_____

_____

Instructor's Comments

_____

_____

_____

☐ JOB SHEET 47–3

Inspecting Power-Steering Fluid Level and Condition

Name _____ Station _____ Date _____

Objective

Upon completion of this job sheet, you will have demonstrated the ability to inspect the power-steering fluid. This task applies to ASE Education Foundation Tasks MLR 4.B.2 (P-1), and AST/MAST 4.B.9 (P-1).

Description of Vehicle

Year _____ Make _____ Model _____

VIN _____ Engine Type and Size _____

Mileage _____

PROCEDURE

1. Using the proper service information or owner's manual, determine the recommended power-steering fluid, and note it here.

   _____

2. Locate the power-steering fluid reservoir, and determine the fluid level. Note your findings.

   _____

   If the power-steering cap has a built-in dipstick, wipe the fluid on a clean white shop towel. Otherwise, use a straw or similar item to remove a sample of fluid.

3. Inspect the fluid for signs of metal content. Note your findings.

   _____

4. If metal is present in the fluid, what does this indicate?

   _____

5. Inspect the fluid in the reservoir for signs of foam or bubbles. Note your findings.

   _____

   _____

6. If foam or bubbles are present, what does this indicate?

   _____

Problems Encountered

_____

_____

_____

Instructor's Comments

_____

_____

_____

☐ JOB SHEET 47–4

Inspecting Steering Column for Noises, Looseness, and Binding

Name _____ Station _____ Date _____

Objective

Upon completion of this job sheet, you will have demonstrated the ability to inspect a steering column for noise, looseness, and binding. This task applies to ASE Education Foundation Task AST/MAST 4.B.3 (P-2).

Description of Vehicle

Year _____ Make _____ Model _____

VIN _____ Engine Type and Size _____

Mileage _____

PROCEDURE

1. Turn the ignition key to the ON position with the engine off. Turn the steering wheel in both directions, and listen for any noises. Note your findings.

   _____

2. Have a fellow student turn the wheel while you listen for noises along the steering shaft and flexible couplings. Note your findings.

   _____

3. If noises are present, what could be the cause(s)?

   _____

4. If the steering column is equipped with the tilt-wheel function, move the wheel through its entire range of up and down movement. Feel for evidence of looseness or binding in the tilt mechanism. Note your findings.

   _____

5. If the column is loose or binding, what actions will be necessary to correct these concerns?

   _____

6. Start the engine, and note the effort necessary to turn the steering wheel from lock to lock. Is the effort consistent?

   _____

7. While turning the wheels, is there evidence of binding or tightness?

   _____

8. If the steering column and steering shaft are acceptable, what other components could cause the steering to bind?

   _____

Problems Encountered

_____

_____

_____

Instructor's Comments

_____

_____

_____

☐ JOB SHEET 47–5

Disabling and Enabling the Supplemental Restraint System

Name _____ Station _____ Date _____

Objective

Upon completion of this job sheet, you will have demonstrated the ability to temporarily disable and enable the supplemental restraint system. This task applies to ASE Education Foundation Tasks MLR 4.A.2 (P-1), and AST/MAST 4.B.1 (P-1).

Description of Vehicle

Year _____ Make _____ Model _____

VIN _____ Engine Type and Size _____

Mileage _____

PROCEDURE

1. Locate the manufacturer's procedure to disable and enable the supplemental restraint (air bag) system. Record the procedure here.

   _____

   _____

   _____

2. If a specific fuse is listed to be removed, remove the fuse and note its location.

   _____

   _____

3. Does the procedure specify disconnecting the air bag harness connector? If yes, where is the connector located? _____

4. Disconnect the air bag harness connector (if applicable).

5. How long is the wait time before service can be started on the system? _____

6. Why does the procedure specify that a certain amount of time pass between disconnecting the circuit and servicing the system? _____

   _____

   Instructor's Check _____

7. Perform the system enabling procedure as provided in the service information.

   Instructor's Check _____

8. Start the vehicle, and note the supplemental restraint system (SRS) warning light. Does the light go off after several seconds? _____

9. What would be indicated if the warning light stays illuminated?

   _____

Problems Encountered

_____

_____

_____

Instructor's Comments

_____

_____

_____

REVIEW QUESTIONS

Review Chapter 47 of the textbook to answer these questions:

1. What are the primary functions of the steering gearbox?

   _____

   _____

   _____

2. What precautions must be taken when servicing a steering column?

   _____

   _____

   _____

3. Explain how and why to perform a dry-park check.

   _____

   _____

   _____

4. Which of the following is the least likely to cause hard steering?

   _____

   a. Binding steering column U-joints
   b. Overinflated tires
   c. Loose power-steering pump belt
   d. Damaged steering column bearings

5. List three methods of breaking a ball-and-socket joint taper.

   _____

   _____

   _____

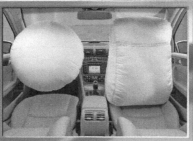

Courtesy of Chrysler LLC

CHAPTER 48

RESTRAINT SYSTEMS: THEORY, DIAGNOSIS, AND SERVICE

OVERVIEW

There are many automotive safety features now available as standard equipment or as options. Some of these include side impact barriers, crumple zone body construction, antilock brakes, traction and stability control, and air bags. Many safety items, such as tempered glass and padded dash panels, have been standard for many years.

Common restraint systems items—seat belts and air bags—are covered in this chapter. It is important to understand how these systems operate and how to diagnose and service them.

All vehicles sold in the United States must have a passenger restraint system. Seat belts became standard equipment in the 1960s, and dual front air bags have been mandatory since 1998. Today's automobiles use a combination of active and passive safety systems. An active safety system is used by or requires the occupant to make an effort to use, such as a seat belt. A passive system does not require any action by the occupant. An air bag system is an example of a passive system. Other passive safety items include reinforced side impact beams, crumple zones, and collapsible steering columns.

Concept Activity 48-1
Identify Restraints

Using a vehicle owner's manual, list the active and passive safety features for a vehicle.

Year _____ Make _____ Model _____

1. Passive restraints _____

_____

_____

2. Active restraints _____

_____

_____

3. Does the vehicle have a passenger side air bag on–off switch? _____

4. Does the vehicle have adaptive (smart) restraints? _____

5. If equipped with adaptive restraints, what sensors are used to determine the number of occupants and their positions? _____

_____

_____

■

Concept Activity 48-2
Restraint Inspection

Proper operation of the restraint system is vital for the safety of the occupants of the vehicle. An inspection of the entire restraint system, seat belts, and air bags, should be performed as part of routine maintenance and if the vehicle has been in any sort of collision.

1. Inspect the condition of the seat belts, anchors, and latches. Record your findings.

   _____

   _____

2. Turn the key to the RUN position, and record the action of the restraint system warning light.

   _____

   _____

3. With your instructor's permission, start the vehicle, and note the restraint warning light. Does the warning light indicate there is a fault with the restraint system?

   _____

   If the restraint light remains on or blinking, a fault is present in the system. Connect a scan tool to the vehicle and enter into the restraint system for diagnostic trouble codes (DTCs). If a scan tool is not required for code retrieval, proceed to step 5.

4. Scan tool used _____

5. Restraint system DTCs _____

6. Describe the restraint system deactivation process for the vehicle being inspected.

   _____

   _____

   _____

7. What color is the wiring identifying air bag system components?

   _____

■

Concept Activity 48-3
Air Bag Service

Here are some very important guidelines to follow when working with and around air bag systems. These are listed with some key words left out. Read through these and fill in the blanks with the correct words.

1. Wear _____ _____ when servicing an air bag system and when handling an air bag module.

2. Wait at least _____ minutes after disconnecting the battery and before beginning any service. The reserve _____ module is capable of storing enough power to deploy the air bag after battery voltage is lost.

3. Always handle all _____ and other components with extreme care. Never strike or jar a sensor, especially when the battery is connected. Doing so can cause deployment of the air bag.

4. Never carry an air bag module by its _____ or _____, and when carrying it, always face the trim and air bag _____ from your body. When placing a module on a bench, always face the trim and air bag _____.

5. Deployed air bags may have a powdery residue on them. _____ _____ is produced by the deployment reaction and is converted to _____ _____ when it comes into contact with the moisture in the atmosphere. Although it is unlikely that harmful chemicals will still be on the bag, it is wise to wear _____ _____ and _____ when handling a deployed air bag.

   Immediately wash your hands after handling a deployed air bag.

6. A live air bag must be _____ before it is disposed. A deployed air bag should be disposed of in a manner consistent with the _____ and manufacturer's procedures.

7. Never use a battery- or AC-powered _____, _____, or any other type of test equipment in the system unless the manufacturer specifically says to. Never probe with a _____ _____ for voltage.

■

The following job sheets will prepare you for restraint system diagnosis and service.

☐ JOB SHEET 48–1

Inspecting Seat Belts

Name _____ Station _____ Date _____

Objective

Upon completion of this job sheet, you will have demonstrated the ability to carefully and thoroughly inspect seat belts. This task applies to ASE Education Foundation Task AST/MAST 6.G.4 (P-1).

Description of Vehicle

Year _____ Make _____ Model _____

VIN _____ Engine Type and Size _____

Mileage _____

List all of the restraint systems found on this vehicle.

_____

_____

_____

PROCEDURE (FRONT SEAT WEBBING INSPECTION)

1. Inspect for twisted webbing due to improper alignment when connecting the buckle. Note problems found.

   _____

2. Fully extend the webbing from the retractor. Inspect the webbing and replace it with a new assembly if the following conditions are noted (**Figure 48-1**):

   Cut or damaged webbing

   Broken or pulled threads

   Cut loops at belt edge

   Color fading as a result of exposure to sun or chemical agents

   Bowed webbing

   Record the results on the accompanying "Report Sheet for Seat-Belt Inspection."

3. If the webbing cannot be pulled out of the retractor or will not retract to its stowed position, check for the following conditions, and clean or correct it as necessary:

   Dirty webbing that is coated with gum, syrup, grease, or other foreign material

   Twisted webbing

   Retractor or loop on B-pillar out of position

   Action Required _____

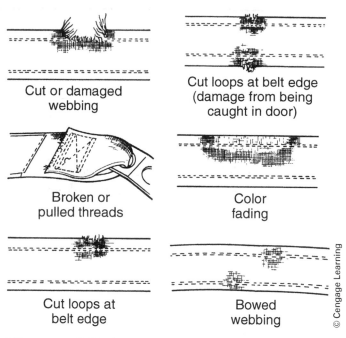

Figure 48-1 Examples of seat-belt webbing defects.

PROCEDURE (BUCKLE INSPECTION)

1. Insert the tongue of the seat belt into the buckle until a click is heard. Pull back on the webbing quickly to assure that the buckle is latched properly. Follow the service manual procedures to replace the seat-belt assembly if the buckle will not latch. Record the results on the accompanying "Report Sheet for Seat-Belt Inspection."

2. Depress the button on the buckle to release the belt. The belt should release with a pressure of approximately 2 pounds (0.9 kg). Follow the service manual procedures to replace the seat-belt assembly if the buckle cover is cracked, the push button is loose, or the pressure required to release the buckle is too high.

PROCEDURE (RETRACTOR INSPECTION)

1. Grasp the seat belt and, while pulling it from the retractor, give the belt a fast jerk. The belt should lock up.

   Belt Reaction _____

2. Drive the vehicle in an open area away from other vehicles at a speed of approximately 5–15 mph (8–24 km/h) and quickly apply the foot brake. The belt should lock up.

   Belt Reaction _____

3. If the retractor does not lock up under these conditions, remove it and replace the seat-belt assembly, following the service manual procedures.

PROCEDURE (ANCHORAGE INSPECTION)

1. Inspect the seat-belt anchorage for signs of movement or deformation. Replace the anchorage, if necessary. Position the replacement anchorage exactly as in the original installation.

Action Required _____

Problems Encountered

_____

_____

_____

Instructor's Comments

_____

_____

_____

Name _____ Station _____ Date _____

REPORT SHEET FOR SEAT-BELT INSPECTION		
	Yes	**No**
1. Webbing inspection		
Broken threads		
Cut loops		
Color fade		
Bowed webbing		
2. Buckle inspection		
Lock properly		
Release properly		
3. Retractor inspection		
Lock by hand		
15-mph (24 km/h) lock test		
4. Anchorage inspection		
Looseness		
Movement		
Conclusions and Recommendations _____		

© Cengage Learning 2015

☐ JOB SHEET 48–2

Disable and Enable an Air Bag System

Name _____ Station _____ Date _____

Objective

Upon completion of this job sheet, you will have demonstrated the ability to disable and enable an airbag system. This task applies to ASE Education Foundation Tasks MLR 4.A.2 (P-1), and AST/MAST 4.B.1 (P-1).

Description of Vehicle

Year _____ Make _____ Model _____

VIN _____ Engine Type and Size _____

Mileage _____

PROCEDURE

1. Turn the ignition key to the RUN position, and note the restraint system warning light.

   _____

2. Start the engine, and note the restraint system light. _____

3. If the light remains on after the vehicle is started, what does this indicate? _____

4. Turn the ignition off.

5. Locate the air bag disabling procedure in the service information. Record the procedure below.

   _____

   _____

   _____

   Instructor's Check _____

6. List the location and circuit protected by any fuses that need to be removed.

   _____

   _____

7. List the location and circuit of any electrical connections that need to be disconnected.

   _____

8. How long does the manufacturer state to allow the air bag system to discharge before attempting service?

   _____

9. List the steps to rearm the air bag system.

_____

_____

Instructor's Check _____

10. Complete the steps to enable the air bag system.

11. Start the engine and note the restraint system warning light. _____

Instructor's Check _____

Problems Encountered

_____

_____

_____

Instructor's Comments

_____

_____

_____

REVIEW QUESTIONS

Review Chapter 48 of the textbook to answer these questions:

1. Explain the differences between passive and active restraint systems.

_____

_____

_____

2. Why must you wait for a period of time after disabling the air bag system before service?

_____

_____

_____

3. How can the wiring and connectors of SIR systems be identified?

_____

_____

_____

4. What service precautions must be taken when working with seat-belt pretensioners?

_____

_____

_____

5. Explain how to safely handle and dispose of an inflated air bag module.

_____

_____

_____

WHEEL ALIGNMENT

OVERVIEW

With the improvements in vehicle performance, the steering and suspension systems have had to evolve to accommodate increased vehicle speeds and driving demands. Wheel alignment has also changed to meet the needs of modern vehicles. The alignment must allow the wheels to roll without scuffing, dragging, or slipping. Proper front and rear alignment ensures greater safety, easier steering, longer tire life, reduced fuel consumption, and less strain on the steering and suspension components.

Wheel angles are based on several different baselines, the centerline, and vertical reference lines. Before discussing wheel alignment, it is important to understand what the references for the alignment angles are and how they are determined.

Concept Activity 49-1
Reference Lines

An imaginary line running through the center of the vehicle is called the geometric centerline, often referenced as **(Figure 49-1)**. The centerline, also called the thrust line, should run straight through the center of the vehicle.

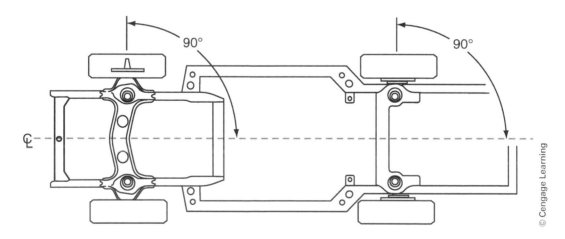

Figure 49-1 Vehicle centerline.

1. How would the vehicle respond if the thrust line deviated from the center of the vehicle? _____

_____

2. List at least three problems that can cause the thrust line to be off center. _____

_____

3. What might the customer complaint be for a vehicle with this problem? _____

_____

_____

■

 Concept Activity 49-2
Camber

For maximum tread life and tire traction, the tires should be perpendicular to the ground. This means that viewed from the front or the rear, the tires would be vertical, or 90 degrees from the ground. The tilt away from vertical is called camber. Camber is often adjusted on both the front and rear wheels.

1. Identify the positive and negative camber in the figure below **(Figure 49-2)**.

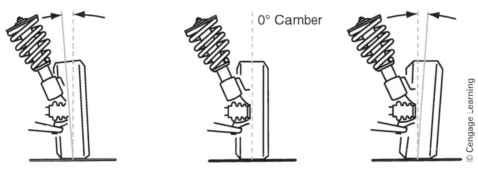

© Cengage Learning

Figure 49-2   Label camber.

2. What problems will excessive negative camber cause? _____

_____

3. What problems will excessive positive camber cause? _____

_____

4. Explain how excessive camber can cause a vehicle to pull. _____

_____

5. What components will affect camber? _____

_____

6. A tire that is vertical has how much camber? _____

_____

_____

■

Concept Activity 49-3
Caster

Nearly every person has experience with two very common examples of the caster angle: a bicycle and the wheels on a shopping cart **(Figure 49-3** and **Figure 49-4)**. Caster is only adjusted on the front wheels.

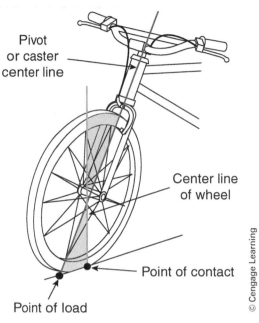

Pivot or caster center line

Center line of wheel

Point of contact

Point of load

© Cengage Learning

Negative ← 0° → Positive

30 20 10 5 10

True vertical

© Cengage Learning

Figure 49-3 An example of caster on the wheels of a bicycle.

Figure 49-4 An example of caster on the vehicle.

1. Explain how the front forks of a bicycle use caster. _____

_____

Note: *The following questions are not intended to cause anyone to operate a bicycle in any unsafe manner. Do not ride a bicycle without the proper safety apparel, and follow all of the manufacturer's warnings and guidelines.*

2. If you have ever ridden a bicycle and removed your hands from the handlebars, how did the bike respond? _____

3. Why did the bike behave as you described? _____

_____

4. How do shopping cart wheels, called casters, respond when pushing a cart?

_____

_____

5. Why do some casters flop around wildly when the cart is being pushed? _____

_____

Caster is measured as positive, negative, or at zero degrees. Positive caster allows for better tracking and stability, whereas negative caster allows for easier steering.

6. The bicycle is an example of _____ caster.

7. The shopping cart caster is an example of _____ caster.

8. Which angle do you think is used by most vehicles, and why? _____

_____

9. What could be the customer complaint on a vehicle with the caster angles out of specification?

_____

10. What components affect caster? _____

11. What affect will caster have on tire wear? _____

_____

_____

■

 ### Concept Activity 49-4
Toe

The angle that is most responsible for tire wear is the toe angle. Toe can be either in or out, or measured in degrees positive and negative **(Figure 49-5)**. Toe is often adjusted on both the front and rear wheels.

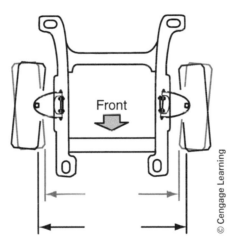

© Cengage Learning

Figure 49-5 Front wheel toe.

1. Explain why toe is the most tire-wearing angle. _____

_____

2. If the front edges of the tires are closer together than the rear edges, this would indicate toe _____.

3. If the rear edges of the tires are closer together than the front edges, this would indicate toe _____.

4. What is the best toe setting as the vehicle is driving on the road? _____

5. Why do some vehicles have a toe angle specification for toe-in and some for toe-out? _____

_____

6. What components affect toe? _____

_____

■

There are also several other alignment angles used for diagnostic purposes that are not generally adjustable.

Concept Activity 49-5
Diagnostic Angles

Steering axis inclination (SAI), included angle, scrub radius, and turning radius (sometimes called toe-out-on-turns), is not adjustable, but can be used for diagnosing alignment and tire wear concerns.

SAI is the line through the steering axis, either the upper and lower ball joints or upper strut mount and lower ball joint, and true vertical. SAI plus or minus camber equals the included angle. SAI can be used to diagnose bent suspension components. Included angle is determined from the SAI readings. SAI is also used for determining scrub radius. Scrub radius is the distance from the center of the tire to where the SAI angle intersects the ground **(Figure 49-6** and **Figure 49-7)**.

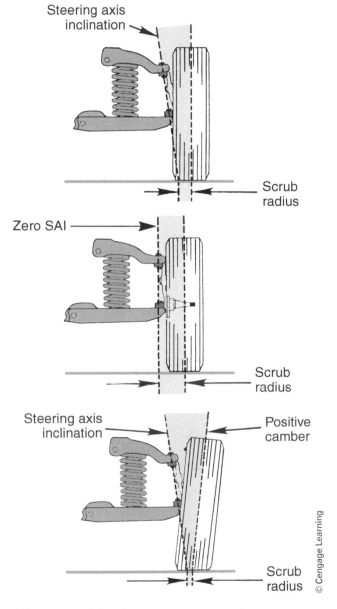

Figure 49-6 SAI and scrub radius.

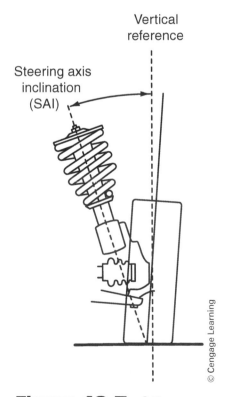

Figure 49-7 SAI.

Turning radius is built into the vehicle so that during turns the inside wheel turns at a slightly sharper angle. This is necessary because the inside tire has to turn in a smaller radius than the outside tire. This angle prevents the tire from scrubbing across the road when turning **(Figure 49-8)**.

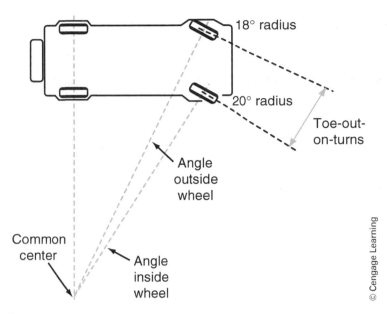

Figure 49-8 Turning radius is the amount one wheel turns more sharply than the other.

1. List three suspension components that will affect SAI. _____

_____

2. Explain how changing wheels can affect scrub radius. _____

_____

_____

3. Explain how the total toe setting could be correct but the turning radius is incorrect.

_____

■

_____

 Concept Activity 49-6
Prealignment Inspection

When the wheel alignment is being checked and adjusted, it is important that certain procedures be performed before checking the alignment.

1. Tire pressure has to set correctly. How can incorrect tire pressure affect the alignment? _____

_____

2. Wheel bearings must be in good condition and not have excessive play. How can a loose or improperly adjusted wheel bearing affect the alignment? _____

_____

3. List the suspension components that must be checked and in good condition before an alignment can be performed. _____

_____

4. List the steering components that must be checked and in good condition before an alignment can be performed. _____

_____

5. If a vehicle has worn-out springs, what symptom may be evident? _____

_____

6. Explain why the wheel alignment cannot be performed if the springs are too weak. _____

_____

■

_____

The following job sheets will prepare you to diagnose wheel alignment concerns and perform wheel alignment service.

☐ JOB SHEET 49-1

Prealignment Inspection

Name _____ Station _____ Date _____

Objective

Upon completion of this job sheet, you will have demonstrated the ability to perform a prealignment inspection. This task applies to ASE Education Foundation Tasks MLR 4.C.1 (P-1), and AST/MAST 4.E.2 (P-1).

Tools and Materials

Tire pressure gauge

Description of Vehicle

Year _____ Make _____ Model _____

VIN _____ Engine Type and Size _____

Mileage _____

PROCEDURE

Complete a prealignment inspection on a vehicle supplied by your instructor. Complete the attached checklist **(Figure 49-9)**.

PREALIGNMENT INSPECTION CHECKLIST

Owner _____ Phone _____ Date _____

Address _____ VIN _____

Make _____ Model _____ Year _____ Lic. number _____ Mileage _____

1. Road test results	Yes	No	Right	Left
Above 30 MPH				
Below 30 MPH				
Bump steer				
When braking				
Steering wheel movement				
Stopping from 2–3 MPH (Front)				
Vehicle steers hard				
Strg wheel returnability normal				
Strg wheel position				

Vibration	Yes	No	Frnt	Rear

2. Tire pressure	Specs	Frnt ___	Rear ___

Record pressure found

RF ____ LF ____ RR ____ LR ____

3. Chassis height	Specs	Frnt ___	Rear ___

Record height found

RF ____ LF ____ RR ____ LR ____

	Yes	No
Springs sagged		
Torsion bars adjusted		

4. Rubber bushings	OK
Upper control arm	
Lower control arm	
Sway bar/stabilizer link	
Strut rod	
Rear bushing	

5. Shock absorbers/struts	Frnt	Rear

6. Steering linkage	Frnt OK	Rear OK
Tie-rod ends		
Idler arm		
Center link		
Sector shaft		
Pitman arm		
Gearbox/rack adjustment		
Gearbox/rack mounting		

7. Ball joints	OK
Load bearings	

Specs	Readings
Right ____ Left ____	Right ____ Left ____

Follower	
Upper strut bearing mount	
Rear	

8. Power steering	OK
Belt tension	
Fluid level	
Leaks/hose fittings	
Spool valve centered	

9. Tires/wheels	OK
Wheel runout	
Condition	
Equal tread depth	
Wheel bearing	

10. Brakes operating properly	

11. Alignment	Spec		Initial reading		Adjusted reading	
	R	L	R	L	R	L
Camber						
Caster						
Toe						

Bump steer	Toe change right wheel		Toe change left wheel	
	Amount	Direction	Amount	Direction
Chassis down 3"				
Chassis up 3"				

	Spec		Initial reading		Adjusted reading	
	R	L	R	L	R	L
Toe-out on turns						
SAI						
Rear camber						
Rear total toe						
Rear indiv. toe						
Wheel balance						
Radial tire pull						

© Cengage Learning

Figure 49-9 Prealignment inspection checklist.

Problems Encountered

_____

_____

_____

Instructor's Comments

_____

_____

_____

☐ JOB SHEET 49–2

Measuring Front- and Rear-Wheel Alignment Angles

Name _____ Station _____ Date _____

Objective

Upon completion of this job sheet, you will have demonstrated the ability to check the alignment angles on the front and rear wheels of a vehicle. This task applies to ASE Education Foundation Tasks AST/MAST 4.E.3 (P-1), 4.E.4 (P-2), 4.E.5 (P-2), and 4.E.6 (P-1).

Description of Vehicle

Year _____ Make _____ Model _____

VIN _____ Engine Type and Size _____

Mileage _____

Describe the type and model of alignment machine being used.

_____

PROCEDURE

1. Lock the front and rear turntables, and drive the vehicle onto the alignment rack.

2. Center the front wheels on the turntables.

3. Make sure the rear wheels are properly positioned on the slip plates.

4. Install the rim clamps and wheel sensors.

5. Perform wheel sensor leveling and wheel runout compensation procedures.

6. Select the specifications for the vehicle on the computer.

7. A prealignment inspection should already have been performed. Note any concerns that needed correction.

_____

_____

8. Measure ride height and compare to specifications.

Left front: Specified _____ Measured _____

Left rear: Specified _____ Measured _____

Right front: Specified _____ Measured _____

Right rear: Specified _____ Measured _____

9. Describe what can be done to correct ride height.

_____

_____

_____

10. Measure front and rear suspension alignment angles, following the prompts on the alignment machine's screen. Record the specifications and your measurements.

Specified front camber _____

    Measured left front camber _____

    Measured right front camber _____

Specified cross camber _____

    Measured cross front camber _____

Specified front caster _____

    Measured left front caster _____

    Measured right front caster _____

Specified cross caster _____

    Measured cross front caster _____

Specified front SAI _____

    Measured left front SAI _____

    Measured right front SAI _____

    Measured included angle _____

Specified thrust angle _____

    Measured thrust angle _____

Specified front toe _____

    Measured left front toe _____

    Measured right front toe _____

    Measured total front toe _____

Specified rear camber _____

    Measured left rear camber _____

    Measured right rear camber _____

Specified rear toe _____

    Measured left rear toe _____

    Measured right rear toe _____

    Measured total rear toe _____

11. Measure the turning radius and compare to specifications. The specified turning radius for this vehicle is _____

    For a left turn, the turning radius of the right front wheel was _____

    For a left turn, the turning radius of the left front wheel was _____

    For a right turn, the turning radius of the right front wheel was _____

    For a right turn, the turning radius of the left front wheel was _____

12. What are your conclusions from this test? _____

    _____

    _____

    _____

13. State the necessary adjustments required to correct front and rear suspension alignment angles.

    _____

    _____

    _____

Problems Encountered

_____

_____

_____

Instructor's Comments

_____

_____

_____

REVIEW QUESTIONS

Review Chapter 49 of the textbook to answer these questions:

1. What is the correct order to perform alignment adjustments?

_____

_____

_____

2. What items should be inspected on a vehicle with incorrect caster and camber readings but for which no provision for adjustment is provided?

_____

_____

_____

3. Which alignment angle is typically used to compensate for road crown and why?

_____

_____

_____

4. Technician A says toe is the most critical tire-wearing angle. Technician B says camber is the most critical tire-wearing angle. Who is correct?

_____

   a. Technician A   c. Both A and B
   b. Technician B   d. Neither A nor B

5. Explain how ride height affects wheel alignment.

_____

_____

_____

 ASE PREP TEST

1. While discussing the cause of overheating damage of a front wheel bearing, Technician A says overheating may be caused by insufficient or incorrect bearing lubrication. Technician B says over-tightening the wheel-bearing nut on installation may cause overheating. Who is correct?

   a. Technician A
   b. Technician B
   c. Both A and B
   d. Neither A nor B

2. Technician A says that a worn or damaged idler arm can cause hard steering. Technician B says that a worn or damaged idler arm can cause excessive play in the steering. Who is correct?

   a. Technician A
   b. Technician B
   c. Both A and B
   d. Neither A nor B

3. A front tire has excessive wear in the center of the tire tread. What is the most likely cause of the problem?

   a. Over-inflation
   b. Under-inflation
   c. Improper static balance
   d. Improper dynamic balance

4. While discussing tire wear, Technician A says static imbalance causes feathered tread wear. Technician B says dynamic imbalance causes cupped wear and bald spots on the tire tread. Who is correct?

   a. Technician A
   b. Technician B
   c. Both A and B
   d. Neither A nor B

5. A customer says his strut front suspension is making a chattering noise when the steering wheel is turned hard to the left. He says he also feels a vibration in the steering wheel when the chatter is heard. Which of the following is most likely to cause this complaint?

   a. Front wheel imbalance
   b. Improper lower strut spring seating
   c. Broken stabilizer links
   d. Worn upper bearing and strut mounting

6. While discussing curb riding height, Technician A says worn shock absorbers reduce curb riding height. Technician B says incorrect riding height affects most other front suspension angles. Who is correct?

   a. Technician A
   b. Technician B
   c. Both A and B
   d. Neither A nor B

7. A customer complains that when she drives her car on an irregular road surface, the car tends to dart left or right. Which of the following is the most likely cause of this problem?

   a. Worn stabilizer bar bushings
   b. Incorrect tire pressure
   c. Worn strut shock absorber
   d. Worn tie-rod ends

8. Technician A says a bent strut can cause a car to pull to one side. Technician B says a deteriorated strut mount can cause steering problems. Who is correct?

   a. Technician A
   b. Technician B
   c. Both A and B
   d. Neither A nor B

9. A car with independent rear suspension has excessive rear tire wear. An inspection of the rear tires shows they are worn on the outside edge, and the tread is feathered. Technician A says the problem could be a bent rear control arm. Technician B says the problem could be worn control arm bushings. Who is correct?

   a. Technician A
   b. Technician B
   c. Both A and B
   d. Neither A nor B

10. Which of the following could cause excessive looseness in the steering system?

   a. Worn strut mount
   b. Worn tie-rod sockets
   c. Loose power-steering pump drive belt
   d. Worn ball joints

11. While discussing steering column service on an air bag-equipped vehicle, Technician A says an air bag module should be placed downward on a workbench after it has been removed. Technician B says the backup power supply must be depleted or disconnected before removing the module. Who is correct?

   a. Technician A
   b. Technician B
   c. Both A and B
   d. Neither A nor B

12. Technician A says a bent pitman arm can cause excessive front tire wear. Technician B says a worn idler arm can cause excessive tire wear. Who is correct?

   a. Technician A
   b. Technician B
   c. Both A and B
   d. Neither A nor B

13. There is air in the power-steering fluid. What should you do to correct the problem?

   a. Turn the steering wheel fully to the right and then to the left while the engine is running.
   b. Crack the pressure line at the pump to release the air.
   c. Drain all of the fluid from the system and refill it.
   d. Turn the engine off, then raise the front wheels and remove the return hose from the steering gear. Then turn the steering wheel fully to the right and then to the left. Refill the reservoir and replace the return hose.

14. All of the following may cause hard steering *except*?

   a. Worn strut bearings
   b. Faulty power-steering pump
   c. Worn steering gears or bearings
   d. Excessive fluid in the system

15. A vehicle with electric power-steering (EPS) has a complaint of loss of power steering. All *except* which of the following could cause this problem?

   a. Faulty steering angle sensor
   b. Improperly inflated tires
   c. Overheating of the power steering motor
   d. Blown EPS fuse

16. Which of the following is the most likely cause of wheel shimmy?

   a. Wheels out of balance
   b. Loose or leaky shock absorber
   c. Loose, worn, or damaged steering linkage
   d. Excessive spring tension

17. Technician A says a moaning or whining noise from a power-steering pump may indicate the fluid is low. Technician B says a growling noise from the power-steering pump may indicate a faulty or worn pump. Who is correct?

    a. Technician A
    b. Technician B
    c. Both A and B
    d. Neither A nor B

18. While discussing turning radius measurement, Technician A says a bent steering arm will cause the turning radius to be out of spec. Technician B says if the turning radius is not within specs, tire tread wear will be excessive during cornering. Who is correct?

    a. Technician A
    b. Technician B
    c. Both A and B
    d. Neither A nor B

19. Technician A says a prealignment inspection should include checking the interior and trunk for heavy items. Technician B says tools and other items normally carried in the vehicle should be kept in the vehicle during an alignment. Who is correct?

    a. Technician A
    b. Technician B
    c. Both A and B
    d. Neither A nor B

20. Technician A says excessive toe-in on the left rear wheel moves the thrust line to the left of the geometric centerline. Technician B says when all four wheels are parallel to the geometric center, the rear wheels track directly behind the front wheels. Who is correct?

    a. Technician A
    b. Technician B
    c. Both A and B
    d. Neither A nor B

21. While discussing adaptive suspension, Technician A says a computer places the suspension in the preset operating mode that matches existing conditions. Technician B says adaptive suspensions are less complicated than hydraulic controlled active suspension. Who is correct?

    a. Technician A
    b. Technician B
    c. Both A and B
    d. Neither A nor B

22. Technician A says curb height must be checked and within specs before a wheel alignment is performed. Technician says camber and toe are affected by incorrect vehicle curb height. Who is correct?

    a. Technician A
    b. Technician B
    c. Both A and B
    d. Neither A nor B

23. When compensating for road crown, Technician A says to set the left front toe slightly negative (toe out) to offset the pull to the right. Technician B says to adjust the right front camber slightly positive to offset for road crown. Who is correct?

    a. Technician A
    b. Technician B
    c. Both A and B
    d. Neither A nor B

24. Technician A says front suspensions are either independent or semi-independent. Technician B says semi-independent suspensions are used in many rear suspension applications. Who is correct?

    a. Technician A
    b. Technician B
    c. Both A and B
    d. Neither A nor B

25. Technician A says an electronic leveling control (ELC) system uses a computer to control the ride height when people or cargo are added to or subtracted from the vehicle. Technician B says the ELC compressor run time is a maximum of 3½ minutes. Who is correct?

    a. Technician A
    b. Technician B
    c. Both A and B
    d. Neither A nor B

CHAPTER

50

BRAKE SYSTEMS

OVERVIEW

The brake systems on modern automobiles rely on many different concepts, such as leverage (also known as mechanical advantage), hydraulics, friction, and heat transfer. Complete the activities before performing brake system service, so you can fully understand these principles.

Concept Activity 50-1
Leverage

Anyone who drives can tell you that to slow or stop the vehicle; all you have to do is press the brake pedal. But how many drivers understand what takes place each time that pedal is pressed? The pedal itself is an application of one of the oldest forms of machines, the lever. Known from the ancient world, the lever is not only one of the oldest machines, it is also a very simple machine. A lever is used to apply force to gain mechanical advantage, such as moving a heavy object by increasing or decreasing the distance over which the force is applied. Many playgrounds have a teeter-totter where children can sit on a long board supported in the middle by a fulcrum. If two people of the same weight sit the same distance apart from the middle, the teeter-totter will balance, and both people will be suspended off the ground at equal heights. But what happens if one person weighs more than the other? To offset the imbalance, the heavier person must move forward, closer to the middle, or the lighter person must move further away from the middle, to the very edge of the board **(Figure 50-1)**.

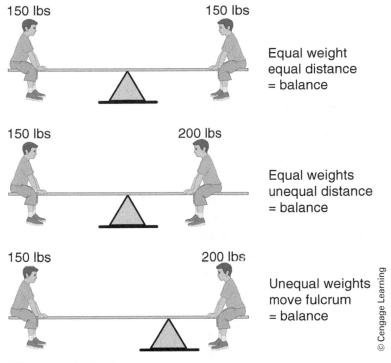

Figure 50-1 Balance on a teeter-totter.

1. If the fulcrum stays in the middle, why must one person move to offset the imbalance?

2. If the two people stay in place, how must the fulcrum move to rebalance the teeter-totter?

A teeter-totter is an example of a first-class lever. In a first-class lever, the fulcrum is located between the load and the effort **(Figure 50-2)**. Using a shovel to dig up a plant is another example of a first-class lever in action.

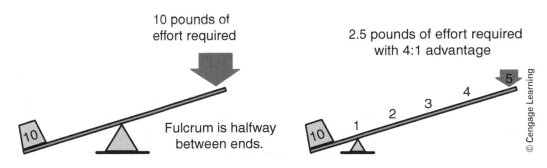

Figure 50-2 Example of a first-class lever.

Construct a teeter-totter balance, using a standard ruler and a C or D battery mounted in the middle as a fulcrum **(Figure 50-3)**. Locate several items from the lab to balance on the ruler.

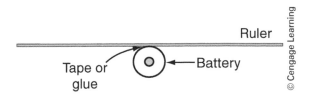

Figure 50-3 Constructing a simple teeter-totter.

3. With the fulcrum in the middle, what were the distances from the center for each item to achieve balance?

_____

_____

4. Place the items to be balanced at the ends of the ruler, and move the fulcrum to achieve balance. How much did the fulcrum move to balance the objects?

_____

_____

The brake pedal in a car operates as a second-class lever. With a second-class lever, the fulcrum is at one end of the lever instead of the middle. The effort is applied to the other end of lever, and the force is applied somewhere in between the effort and the fulcrum **(Figure 50-4)**. Look under the dashboard of several vehicles to see how the brake pedal is mounted. Notice that the top of the pedal is the mounting point, which acts as the fulcrum. Below the fulcrum, the pedal pushrod is mounted and extends forward through the firewall, to the brake booster and/or master cylinder. The footpad at the bottom of the pedal is where the effort is applied.

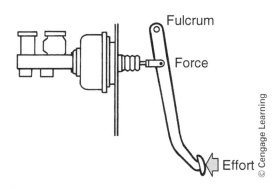

Fulcrum

Force

Effort

© Cengage Learning

Figure 50-4   The brake pedal is a lever.

Measure the length of the pedal between the pushrod and fulcrum and the pushrod and the lowest point of the pedal for several vehicles.

5. Pushrod to fulcrum length _____   Pushrod to pedal length _____

6. Pushrod to fulcrum length _____   Pushrod to pedal length _____

Because the force is between the fulcrum and the effort, the resulting force is going to be determined by the distance below the force divided by the distance above the force **(Figure 50-5)**. If a brake pedal is mounted so that there is 2 inches above the pushrod to the fulcrum and 10 inches from the pushrod down to the end of the pedal, the ratio of force can be determined. Use the following formula: $R = d_a/d_b$, where R is the ratio, d_a is the distance from the pushrod to the end of pedal, and d_b is the distance from the pushrod to the fulcrum. Because distance a is 10 inches and distance b is 2 inches, 10 divided by 2 equals 5. Therefore, the ratio of force applied by this pedal is 5 to 1, or 5:1. This means that the force applied to the brake pedal will increase by a factor of 5. Using the measurements of the brake pedals from questions 5 and 6, determine the ratio of force for each of the two vehicles.

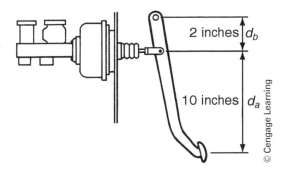

Figure 50-5 Determine the advantage gained by the brake pedal.

7. $d_a/d_b =$ _____   Ratio _____

8. $d_a/d_b =$ _____   Ratio _____

As you can see, the brake pedal and its mounting design reduce the amount of effort needed from the driver. This reduction of effort is beneficial for the driver, as it will reduce fatigue over the time of vehicle operation.

■

Concept Activity 50-2
Hydraulics

Even though the force applied to the brake pedal is increased, it is necessary for the hydraulic system to further increase the force and apply it in a manner that will safely slow and stop the vehicle.

Hydraulics refers to using fluids to perform work. Fluids can be pressurized but cannot practically be compressed. By pressurizing a fluid in a closed system, the fluid can transmit both motion and force. This is accomplished by using pistons in cylinders of different sizes to either increase force or increase movement. Whenever there is an increase in force, there will be a decrease in the amount of movement. Conversely, when there is an increase in movement, there will be a decrease in force. In this manner, a hydraulic system is like the lever we studied before. To increase the force needed to be applied to an object, more distance or movement needs to be applied. Locate two items around your lab that rely on hydraulics to operate.

1. Item 1 _____

   How does this item use hydraulics? _____

2. Item 2 _____

   How does this item use hydraulics? _____

A simple hydraulic system contains two equal-sized containers of a liquid with two equal-sized pistons. The two containers are connected with a hose or tubing **(Figure 50-6)**. If one piston is pushed downward with 100 lb (45 kg) of force and moves down 10 inches, the piston in the second container will move upward 10 inches with the same 100 lb (45 kg) of force. Because the pistons are the same size, any force and movement imparted on one piston will cause the same reaction to the second piston.

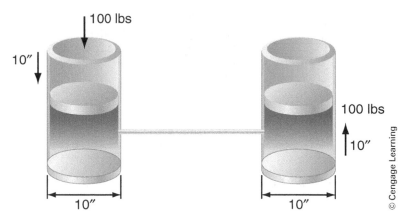

Figure 50-6   Pascal's Law of hydraulics.

Where the use of hydraulics really provides advantage is when the sizes of the pistons are different, and the resulting force and movement can be increased or decreased as needed. The pressure generated by the piston is a factor of the piston size. Input pressure is found by dividing force by piston size, or $P = F/A$. The smaller the input piston surface area, the larger the force will be from that piston. The larger the input-piston surface area, the less the force will be from that piston **(Figure 50-7)**. Conversely, the force of the output piston is proportional to the pressure against the surface area of the piston. An output piston that is larger than the input piston will move with greater force than the input but with less distance moved. To examine this principle, we will look at what is known as Pascal's Principle or Law.

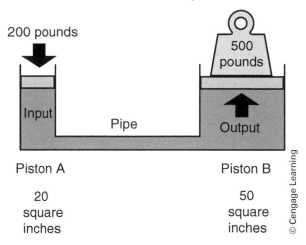

Figure 50-7   Using hydraulics for mechanical advantage.

If the smaller piston on the left is our input piston and has a surface area of 1 square inch (6.45 cm²) and the larger piston is our output piston and has an area of 10 square inches (65.5 cm²), we can calculate the forces and movements quite easily. Suppose a force of 100 lb (45 kg) is exerted on our input piston, moving it downward 1 inch (2.54 cm). Using $P = F/A$, we can calculate P = 100/1, or 100 psi. Our output piston, which has a surface area of 10 square inches, will receive 100 pounds of pressure per square inch. This will result in an output force by the second piston of 1,000 lb. However, because our output force has increased, our output movement will decrease. The larger piston will move 1/10 of the distance of the input piston; because the force was multiplied by 10, the distance will be divided by 10 as well. The 1 inch of movement of piston 1 turns into 1/10 of an inch of movement at piston 2. To calculate the forces and movement in a hydraulic circuit, use the following formulas:

$P_1 = F_1/A_1$ for piston 1 pressure

$F_2 = A_2/A_1 \cdot F_1$ for piston 2 force

$P_2 = F_2/A_2$ for piston 2 pressure

To practice using these principles, complete the following activity.

Calculate the forces and movements of a two-piston hydraulic circuit with an input piston of 2 square inches and an output piston of 10 square inches.

3. Input distance _____ Input force _____

   Output distance _____ Output force _____

Calculate the forces and movements of a two piston hydraulic circuit with an input piston of 2 square inches and an output piston of 1 square inch.

4. Input distance _____ Input force _____

   Output distance _____ Output force _____

The pushrod from the brake pedal pushes on pistons inside the brake master cylinder. The master cylinder provides the pressure in the brake system. The force applied to the brake pedal, the ratio of brake pedal force multiplication, and master cylinder piston size all determine how much pressure will be developed in the brake system. The size of the output pistons at the calipers and wheel cylinders, along with the brake system pressure, determine the final output force for the brakes. Use the diagram below to determine brake system pressure. Remember from Concept Activity 50-1, that d_a is the distance from the pushrod to the end of the pedal, and d_b is the distance from the pushrod to the fulcrum **(Figure 50-8)**.

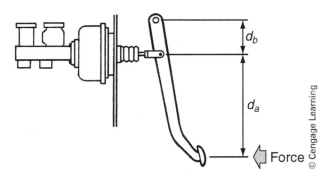

Figure 50-8 Finding the mechanical advantage.

5. Input force to pedal _____ d_a _____ d_b _____

   Output force _____ Hydraulic system pressure _____

6. Input force to pedal _____ d_a _____ d_b _____

   Output force _____ Hydraulic system pressure _____

Obtain a sample of syringes from your instructor. Fill each syringe approximately halfway with water. Connect a hose between two different syringes. Push the plunger of one syringe, and note the reaction of the other. Perform this exercise several times, using different sizes of syringes as the input and output pistons.

7. Syringe 1: Diameter _____ Distance moved _____

   Syringe 2: Diameter _____ Distance moved _____

8. Syringe 1: Diameter _____ Distance moved _____

   Syringe 2: Diameter _____ Distance moved _____

9. Syringe 1: Diameter _____ Distance moved _____

   Syringe 2: Diameter _____ Distance moved _____

10. Syringe 1: Diameter _____   Distance moved _____

    Syringe 2: Diameter _____   Distance moved _____

■

 Concept Activity 50-3
Friction

Once the hydraulic system generates the pressure to actuate the pistons in the calipers and wheel cylinders, it is the job of the lining surfaces to slow the vehicle. The lining surfaces consist of the brake rotor or disc, brake pads, brake drum, and brake shoes. Brake pads are squeezed together against the rotor by the brake caliper. Brake fluid, under pressure from the hydraulic system, pushes on the caliper piston. As the piston moves out, pushing the inner pad against the rotor, the caliper slides inward, pressing the outer pad against the outside of the rotor. On drum brakes, the fluid pushes on two equal-diameter pistons in the wheel cylinder. The pistons move out, pressing the brake shoes against the inside surface of the brake drum.

The pads and shoes create a lot of friction when applied; this in turn creates a lot of heat. Heat is a natural by-product of friction. Rub your hands together briskly, and note how fast your skin warms.

1. Why does rubbing your hands together create heat?

_____

_____

2. Wet your hands with some water, and rub them together briskly again. How did the water change the result?

_____

_____

In automotive applications, there is friction used by the brake system, the friction of moving fluids such as motor oil and transmission fluid; friction between the vehicle and the air in which it is traveling—called drag; and friction between the tires and the ground. Refer to Concept Activity 8-5 about friction.

Friction is created when uneven surfaces in contact with each other start to move relative to each other. Although the two surfaces may appear smooth, there are slight imperfections that resist moving against each other. The amount of force required to move one object along another is called the coefficient of friction (CoF). Simply put, the CoF is equal to the ratio of force to move an object divided by the weight of the object, $CoF = F/M$. Although several factors affect the CoF, such as temperature and speed, we will use a simple example of two bodies moving against each other. Imagine you have a 100 lb (45 kg) block of rubber on the floor of your lab. The force required to slide the block of rubber over concrete would be large. If it takes 100 lb (45 kg) of force to slide the block, the CoF would be 100/100, or 1.0. Imagine the same block of rubber now sitting on the floor of an ice hockey rink. If only takes 25 lb (11 kg) of force to slide the block, what would the CoF be?

3. 25/100 = _____

To experiment with the CoF of various objects in your lab, you can make a small force gauge using an ordinary ballpoint pen, a rubber band, tape, and a paper clip. Assemble the parts as shown in **Figure 50-9**:

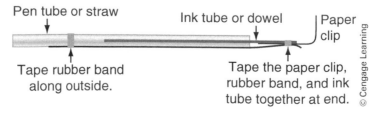

Figure 50-9   Pulling an object with the paper clip will cause the rubber band to lengthen. The amount of lengthening can be measured and recorded for comparison.

Attach the paper clip to an object, and attempt to pull it across a flat surface, using the opposite end of your force meter. Measure the amount of extension of the pen tube from the body of the pen with a ruler. This will give you an idea of how much force is being required to drag each object. Record your results here:

4. Object _____ Surface _____ Length of extension _____

5. Object _____ Surface _____ Length of extension _____

6. Object _____ Surface _____ Length of extension _____

7. Why did some objects require more force than others did?

_____

8. What effect does the surface used to slide across affect the CoF?

_____

9. If a liquid is placed between the two objects, what effect will that have on the CoF?

10. What could happen if the CoF of the brake pads or shoes is too high?

_____

11. What could happen if the CoF of the brake pads or shoes is too low?

_____

12. What factors do you think are involved in determining the correct CoF for a particular vehicle?

_____

The CoF of brake linings have a large impact on how well a vehicle will stop. Using lining materials with too high or low a CoF can cause brake performance issues and rapid wear, and can result in customer dissatisfaction. Similarly, the tires, because they are the contact between the vehicle and the road, also play an important role in brake operation. Even the best performing brakes will not stop a vehicle well if the tires cannot maintain proper traction.

Heat transfer is a very important factor for brake systems. The brake components—pads, shoes, rotors, and drums—can generate an enormous amount of heat. This heat must be dissipated quickly and efficiently, or it can cause damage to the brake components. Heat transferred to the brake fluid can, under the right (or wrong, depending on how you look at it) conditions, cause the fluid to boil. If the brake fluid boils, the gas, being compressible, will allow the brake pedal to sink to the floor. This condition, called brake fade, is very dangerous. Once the brake fluid cools, it will condense back into a liquid state.

Obtain a large rubber band from your instructor. Measure the temperature of the rubber band or hold it against your forehead to feel its temperature. Now, rapidly stretch the rubber band and allow it to retract about a dozen times. Quickly remeasure its temperature.

13. What did the temperature of the rubber band do?

_____

14. Why did the rubber band respond this way?

_____

15. How quickly did the rubber band return to its original temperature?

_____

16. What factors determine how quickly something will dissipate heat?

_____

■

Table 50-1 and Table 50-2 will aid you in brake system diagnosis and repair.

TABLE 50-1 Brake System Diagnosis/Troubleshooting.

Symptom	Possible Cause
Brakes do not apply	Insufficient brake fluid
	Air in system
	Binding or damaged brake pedal linkage
Excessive pedal travel or pedal goes to floor	Air in system
	Loose brake tube end fittings
	Malfunctioning master cylinder
	Malfunctioning ABS hydraulic unit
	Loose wheel bearings—front
	Loose or missing pedal bushings/fasteners
Excessive pedal effort to stop vehicle	Binding or damaged pedal linkage
	Engine vacuum loss
	Booster inoperative
	Worn or contaminated linings
	Brake system
Spongy pedal	Air in system
	Loose or improper brake pedal, pedal support, booster, master cylinder attachment
	Brake system
Brakes drag, slow, or release incompletely	Parking brake cable out of adjustment or binding
	Front wheel bearings worn or damaged
	Blocked master cylinder compensator ports
	On front disc brakes—loose or missing inner shoe clip
	Brake adjustment (rear)
	On front disc brakes—LH or RH shoes misassembled
	Restriction in hydraulic system
	Caliper piston seizure
Noise at wheels when brakes are applied—snap or click	On disc brakes—loose or missing inner antirattle clip
	On front disc brakes—missing pin insulator
	On front disc brakes—missing or loose pins

(Continued)

TABLE 50-1 (*Continued*)

Symptom	Possible Cause
Brakes pull to one side	Unequal air pressure in tires
	Grease or fluid on linings
	Glazed linings
	Loose or missing disc brake caliper retaining pins
	Improper size or type lining on one wheel
	Stuck or seized calipers
	Restricted brake lines or hoses

	Other brake system components:
	    Improper positioning of disc brake shoe and lining in the caliper
	    Damaged or worn wheel bearings

Brakes grab or lock up when applied	Tires worn or incorrect pressure
	Grease or fluid on linings— damaged linings
	Improper size or type of linings

	Other brake system components:
	    Pins for caliper attachment loose or missing
	    Worn or damaged or dry wheel bearings
	    Improperly adjusted parking brake

Brake warning indicator on	Hydraulic system problem
	Shorted indicator circuit
	Parking brake not returned
	Brake warning indicator switch
Intermittent loss of pedal	Loose wheel bearings
Rough engine idle or stall, brakes applied	Vacuum leak in neutral switch
	Vacuum booster
Parking brake control will not latch (manual release).	Kinked or binding release cable
	Control assembly
Parking brake control will not latch (automatic release).	Vacuum leak
	Vacuum switch
	Control assembly
Parking brake will not release or fully return (manual release).	Cable disconnected
	Control assembly binding
	Parking brake linkage binding
	Rear brake problems
Parking brake will not release or fully return (automatic release).	Vacuum line leakage or improper connections
	Neutral switch
	Control assembly

TABLE 50-2 Master Cylinder Diagnosis.

Brake Pedal Feel Conditions	Diagnostic Action
Condition 1—Pedal goes down fast and brake warning indicator comes on.	Fill reservoir with fluid. Pump brake pedal rapidly. If brake pedal height builds up and then sinks down, check for external leak. If brake pedal builds up and holds, check for presence of air.
Condition 2—Pedal eases down slowly and brake warning indicator comes on.	Fill reservoir with fluid. Pump brake pedal rapidly. If the brake pedal height builds up and then sinks, check for external leak. If brake pedal height builds up and holds, check for presence of air.
Condition 3—Pedal is low and brake warning indicator does not come on.	This condition may be caused by air in hydraulic system. Check for presence of air in hydraulic system. If this condition happens occasionally, check front wheel bearings for looseness. (Loose front wheel bearings allow disc rotor to knock caliper piston back, creating excessive lining clearance.)
Condition 4—Pedal feels spongy.	This may be normal because newer brake systems are not designed to produce as hard a pedal as in the past. To verify this, compare pedal feel with that of like vehicle.
Condition 5—Pedal erratic with no brake warning indicator.	This may be caused by incomplete brake release, loose wheel bearings, incorrect parking brake adjustment, or blocked compensator holes in master cylinder. To verify, check wheel rotation, bearing free play, and parking brake tension.

WARNING: *Before test-driving any car, first check the fluid level in the master cylinder; depress the brake pedal to be sure that there is adequate pedal reserve; and make a series of low-speed stops to be sure the brakes are safe for road testing. Always make a preliminary inspection of the brake system in the shop before taking the vehicle on the road.*

The following job sheets will prepare you to diagnose and service the brake system.

☐ JOB SHEET 50–1

Pre-Brake Service Inspection

Name _____ Station _____ Date _____

Objective

Upon completion of this job sheet, you will have demonstrated the ability to perform a brake system inspection. This task applies to ASE Education Foundation task AST/MAST 5.A.1 (P-1).

Description of Vehicle

Year _____ Make _____ Model _____

VIN _____ Engine Type and Size _____

Mileage _____

PROCEDURE

1. A complete brake system inspection should be performed before the customer is given an estimate for repairs.

2. If possible, test-drive the vehicle to confirm the customer's concern. Have the customer test-drive with you if necessary.

3. Inspect the hydraulic system, parking brake system, the power assist, brake system warning lights, stop lamps, and the front and rear service brakes. Partial disassembly of the disc and drum brake is often necessary to thoroughly inspect the pads, shoes, rotors, and drums.

4. Complete the pre-brake service inspection form **(Figure 50-10)**.

Problems Encountered

_____

_____

_____

Conclusions and Recommendations

_____

_____

_____

PRE-BRAKE-JOB INSPECTION CHECKLIST

Owner _____ Phone _____ Date ___|___|___
LAST   FIRST

Address _____ License. No. _____

Make _____ Model _____ Mileage _____ Serial No. _____ Year _____

Special Key for Hubcaps/Wheels   Location _____   Owner Use Parking Brake   Yes ☐   No ☐

4 Drum ☐   4 Disc ☐   Disc/Drum ☐   P/B No ☐ Yes ☐   Vacuum ☐   Hydro ☐   ABS ☐

Owner Comments _____

1. CHECKS BEFORE ROAD TEST	Safe	Unsafe
Stoplight Operation		
Brake Warning Light Operation		
Master Cylinder Checks		
Fluid Level		
Fluid Contamination		
Under Hood Fluid Leaks		
Under Dash Fluid Leaks (No Power)		
Bypassing		

BRAKE PEDAL HEIGHT AND FEEL

Check One		Check One	
Low		Spongy	
Med		Firm	
High			
Power Brake Unit Checks			

VACUUM	Safe	Unsafe	HYDRO	Safe	Unsafe
Vacuum Unit			Hydro Unit		
Engine Vacuum			P/S Fluid		
Vacuum Hose			P/S Belt Tension		
Unit Check Valve			P/S Belt Condition		
Reserve Braking			P/S Fluid Leaks		
			Reserve Braking		

3. In Shop Checks On Hoist	Yes	No	RF	LF	RR	LR
Brake Drag						
Intermittent Brake Drag						
Brake Pedal Linkage Binding						
Wheel Bearing Looseness						
Missing or Broken Wheel Fasteners						
Suspension Looseness						

Tire Pressure Specs	Front	Rear
Record Pressure Found		
RF _____ LF _____ RR _____ LR _____		
Tire Condition		
RF _____ LF _____ RR _____ LR _____		

2. ROAD TEST	Yes	No	RF	LF	RR	LR
Brake Pull						
Brake Clunk						
Brake Scraping						
Brake Squeal						
Brake Grabby						
Brakes Lock Prematurely						
Wheel Bearing Noise						
Vehicle Vibrates						

STEERING WHEEL MOVEMENT WHEN STOPPING
FROM 2–3 MPH   YES / NO / RGT / LFT

	YES	NO
Does ABS Work	YES	NO
Pedal Pulsation when Braking	YES	NO
Steering Wheel Oscillation when Braking	YES	NO
No Stopping Power	YES	NO
Warning Light Comes on when Braking	YES	NO
Difference in Pedal Height after Cornering	YES	NO
Nose Dive	YES	NO

	Front				Rear			
	Right		Left		Right		Left	
	Spec	Safe	Unsafe	Safe	Unsafe	Spec	Safe	Unsafe	Safe	Unsafe
Mark Wheels and Remove										
Caliper/Piston Stuck RF LF RR LR										
Mark Drums and Remove										
Measure Rotor Thickness or Drum Diameter.										
Measure Rotor Thickness Variation.										
Measure Rotor Runout.										
Lining Thickness										
Tubes and Hoses										
Fluid Leaks										
Broken Bleeders										
Leaky Seals										
Self-Adjuster Operation										
									Safe	Unsafe
Parking Brake Cables and Linkage										

Figure 50-10   Pre-brake service inspection checklist.

© Cengage Learning

☐ JOB SHEET 50–2

Measuring Brake Pedal Height, Travel, and Free Play

Name _____ Station _____ Date _____

Objective

Upon completion of this job sheet, you will have demonstrated the ability to measure brake pedal height, travel, and free play. This task applies to ASE Education Foundation Tasks MLR 5.B.1 (P-1), and AST/MAST 5.B.2 (P-1).

Tools and Materials

Rule or tape measure

Brake pedal effort gauge

Description of Vehicle

Year _____ Make _____ Model _____

VIN _____ Engine Type and Size _____

Mileage _____

PROCEDURE

1. Locate the specifications for the following, as shown in **Figure 50-11**:

   Brake pedal height _____

   Brake pedal travel _____

   Brake pedal free play _____

2. With the engine running, press and release the brake pedal several times. Listen for noises. Pedal movement should be smooth and it should return quickly when released. Note your findings.

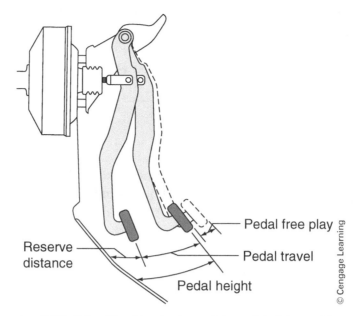

© Cengage Learning

Figure 50-11 Checking brake pedal travel, height, and free play.

3. With the engine off, measure the distance from the floor to the top of the brake pedal pad. Some manufacturers measure from the steering wheel to the brake pedal. Follow the manufacturer's procedure to obtain the correct measurement. Record your measurement.

4. Is the brake pedal height within specifications? _____ YES _____ NO

5. What could cause the released height to be incorrect?

_____

6. Brake pedal travel is also referred to as the pedal reserve distance. Measure the distance from the floor to the top of the brake pedal pad with the brake applied. Some manufacturers measure from the steering wheel to the brake pedal. Follow the manufacturer's procedure to obtain the correct measurement. Record your measurement.

_____

7. Is the brake pedal travel within specifications? _____ YES _____ NO

8. What could cause the brake pedal travel to be incorrect?

_____

9. Brake pedal free play is the amount the pedal moves from the fully released position until the pushrod begins to apply the brakes. This specification is important so the brakes do not drag. Measure the movement of the pedal until the pushrod begins to move. Record your measurement.

_____

10. Is the brake pedal free play within specifications? _____ YES _____ NO

11. If the free play is not within specifications, locate and record the procedure to adjust the free play.

_____

Problems Encountered

_____

_____

_____

Instructor's Comments

_____

_____

_____

☐ JOB SHEET 50–3

Inspecting Brake Fluid

Name _____ Station _____ Date _____

Objective

Upon completion of this job sheet, you will have demonstrated the ability to inspect brake fluid for contamination and condition. This task applies to ASE Education Foundation Tasks MLR 5.B.7 (P-1), and AST/MAST 5.B.13 (P-1).

Tools and Materials
Brake fluid test strips

Brake fluid moisture tester

Description of Vehicle
Year _____ Make _____ Model _____

VIN _____ Engine Type and Size _____

Mileage _____

PROCEDURE

1. Clean the area around the brake fluid reservoir. Inspect the reservoir and cap(s) before removing. Inspect for a tight seal, signs of fluid leaks, and seal/gasket damage. Note your findings.

   _____

2. Remove the reservoir cap(s) and note the condition of the seal/gasket.

   _____

   _____

3. Note the brake fluid level and appearance.

   _____

   _____

4. Brake fluid appearance does not provide a clear indication of the condition of the fluid, unless a petroleum-based liquid has contaminated the fluid. Does the reservoir cap(s) seal/gasket appear swollen or distorted?

   _____

5. If a fluid such as power-steering fluid is introduced into the brake system, the petroleum will react with the rubber parts of the brake system. All rubber seals, cups, and gaskets exposed to the contaminated fluid will begin to swell and break down. This can cause complete brake system failure. Is there evidence of fluid contamination?

   _____

6. Use a fluid test strip to determine the brake fluid condition. Follow the instructions provided by the test strip manufacturer. Note the results of the test.

_____

_____

7. Determine the moisture content of the brake fluid using a moisture tester or test strip. Follow the instructions provided by the tester manufacturer. Note the results of the test.

_____

_____

8. Based on the results of your inspection and testing, what is the condition of the brake fluid?

_____

9. What are your service recommendations?

_____

10. What does the vehicle manufacturer specify as a service interval for the brake fluid?

_____

11. Describe what is included to properly service the brake fluid based on the manufacturers service information.

_____

_____

_____

12. Does the vehicle meet the age and/or mileage requirements for brake fluid service?

_____ YES   _____ NO

13. Has the brake fluid been serviced in the past?   _____ YES   _____ NO

14. Based on the age, mileage, and service history, does the brake fluid require service?

_____ YES   _____ NO

Problems Encountered

_____

_____

_____

Instructor's Comments

_____

_____

_____

☐ JOB SHEET 50–4

Bench Bleeding a Master Cylinder

Name _____ Station _____ Date _____

Objective

Upon completion of this job sheet, you will have demonstrated the ability to bench bleed a tandem master cylinder. This task applies to ASE Education Foundation Task AST/MAST 5.B.4 (P-1).

Tools and Materials

Brake fluid   Wooden dowel or smooth, rounded rod

Vise   Bench bleed bit

Description of Vehicle

Year _____ Make _____ Model _____

VIN _____ Engine Type and Size _____

Mileage _____

PROCEDURE

1. Mount the cylinder in a vise **(Figure 50-12A)**. Do not apply excessive pressure to the casting. Ensure that the bore is horizontal.

   Master cylinder construction material _____

   _____

2. Connect the short lengths of tubing to the outlet ports. Then bend them as shown in **Figure 50-12B**. Fill the reservoirs with fresh brake fluid **(Figure 50-12C)**. What size fittings are used in the ports?

   _____

3. Use a wooden dowel or smooth, rounded rod to slowly pump **(Figure 50-12D)** the master cylinder until bubbles stop appearing at the tube ends. How many times did you have to pump until the air bubbles stopped? _____

4. Using the same tool, push the piston ¼–½ inch (6.30–13 mm) from its fully released position. Allow the piston to return about 1/8 inch (3.17 mm). Then push it back to the initial ¼–½-inch (6.30–13 mm) position. Rapidly move the piston back and forth in this manner until the air bubbles stop coming from the ports located inside the reservoirs **(Figure 50-12E)**.

5. Refill the reservoirs. Secure the master cylinder cover.

6. Install the master cylinder on the vehicle. Remove the tubing and fittings. Attach the lines, but do not torque the tube connections. Master cylinder mounting torque spec

   _____

7. Loosen the fitting, and then slowly depress the brake pedal to force any trapped air out of the connections. Tighten the fitting slightly before releasing the pedal.

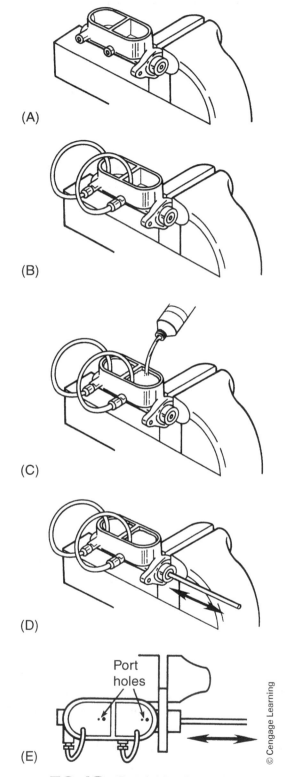

(A)

(B)

(C)

(D)

Port holes

© Cengage Learning

(E)

Figure 50-12 Bench bleeding a master cylinder.

8. Have all air bubbles been removed? If not, repeat the procedure.

9. Tighten the connections to the manufacturer's specifications.

Brake line fitting torque spec _____

10. Make sure the master cylinder reservoirs are adequately filled with brake fluid.

    Recommended brake fluid _____

Problems Encountered

_____

_____

_____

Instructor's Comments

_____

_____

_____

☐ JOB SHEET 50–5

Pressure Bleeding a Hydraulic Brake System

Name _____ Station _____ Date _____

Objective

Upon completion of this job sheet, you will have demonstrated the ability to pressure bleed a front-to-rear split hydraulic system with a pressure bleeder and adapter. This task applies to ASE Education Foundation Tasks MLR 5.B.6 (P-1), and AST/MAST 5.B.12 (P-1).

Tools and Materials

Bleeder hose	Pressure bleeder and adapter
Brake fluid	Pressure bleeder service manual
Glass jar	

Description of Vehicle

Year _____ Make _____ Model _____

VIN _____ Engine Type and Size _____

Mileage _____

PROCEDURE

WARNING: *Always follow the vehicle manufacturer's brake bleeding procedure.*

CAUTION: *Be careful to avoid spills and splashes. Brake fluid can remove paint as well as do serious damage to your eyes.*

1. Place fender covers over the fenders. Study operation of the pressure bleeder in its service manual.

2. Bring the unit to a working pressure of 15–20 psi (103–137 kPa). Be sure enough brake fluid is in the pressure bleeder to complete the bleeding operation.

3. Use a shop towel to clean the master cylinder cover and remove it. Remove the reservoir diaphragm gasket if there is one. Clean the gasket seat and fill the reservoir. Attach the master cylinder bleeder adapter to the reservoir following the manufacturer's instructions.

Instructor's Check _____

4. If the rear wheel cylinders (secondary brake system) are to be bled, use a suitable box wrench on the bleeder fitting at the right rear brake wheel cylinder. Attach the bleeder tube snugly around the bleeder fitting. Open the valve on the bleeder tank to admit pressurized brake fluid into the master cylinder reservoir **(Figure 50-13)**.

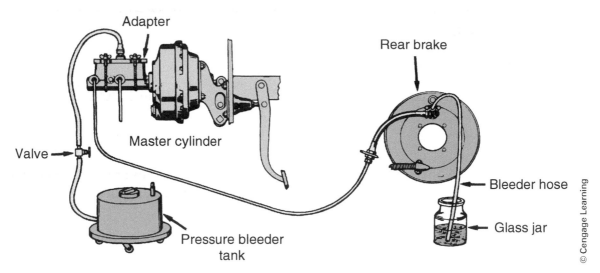

Figure 50-13 Pressure bleeding the brake system.

5. Submerge the free end of the tube into a container partially filled with clean brake fluid, and loosen the bleeder fitting.

6. When air bubbles cease to appear in the fluid at the submerged end of the bleeder tube, close the bleeder fitting. Remove the tube. Replace the rubber dust cap on the bleeder screw.

7. Attach the bleeder tube, and repeat steps 3 and 4 at the left rear wheel cylinder.

8. On the front brakes, repeat steps 3 and 4, starting at the right front disc caliper and ending at the left front disc caliper. The metering valve may require being locked in an open position. Be sure to remove the lock when the task is completed.

9. When the bleeding operation is complete, close the bleeder tank valve and remove the tank hose from the adapter fitting.

10. After disc brake service, ensure the disc brake pistons have returned to their normal positions and the shoe and lining assemblies are properly seated. This is accomplished by depressing the brake pedal several times until normal pedal travel is established.

11. Remove the pressure bleeder adapter tool from the master cylinder. Fill the master cylinder reservoir to the "max" line or ¼ inch (6.30 mm) below the top. Install the master cylinder cover and gasket. Ensure that the diaphragm-type gasket is properly positioned in the master cylinder cover.

CAUTION: *To prevent the air in the pressure tank from getting into the lines, do not shake the tank while air is being added to the tank or after it has been pressurized. Set the tank in the required location, bring the air hose to the tank, and do not move it during the bleeding operation. The tank should be kept at least one-third full.*

Problems Encountered

_____

_____

_____

Instructor's Comments

_____

_____

_____

☐ JOB SHEET 50–6

Performing a Power Vacuum Brake Booster Test

Name _____ Station _____ Date _____

Objective

Upon completion of this job sheet, you will have demonstrated the ability to perform various tests to determine the condition of vacuum power brakes. This task applies to ASE Education Foundation Tasks MLR 5.E.1 (P-2), 5.E.2 (P-1), AST/MAST 5.E.1 (P-2), 5.E.2 (P-1), and 5.E.3 (P-1).

Tools and Materials

Vacuum gauge

Description of Vehicle

Year _____ Make _____ Model _____

VIN _____ Engine Type and Size _____

Mileage _____

PROCEDURE (BASIC OPERATIONAL TEST)

1. Place fender covers over the fenders. With the engine off, pump the brake pedal numerous times to be sure any residual vacuum is exhausted from the booster unit.

2. Hold firm pressure on the brake pedal and start the engine. Describe the brake pedal reaction.

_____

3. Is the booster operating as specified?

_____

PROCEDURE (VACUUM SUPPLY TEST)

1. With the engine idling, attach the vacuum gauge to the intake manifold port **(Figure 50-14)**. Record the vacuum reading _____

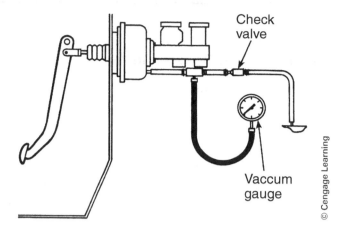

© Cengage Learning

Figure 50-14 A vacuum gauge attached to the intake manifold.

2. Disconnect the vacuum tube or hose that runs from the intake manifold to the booster unit, and place your thumb over it (do this quickly or the engine might stall). If you do not feel a strong vacuum at the end of the tube or hose, shut off the engine, remove the tube or hose, and see whether it is collapsed, crimped, or clogged. Record your findings.

PROCEDURE (VACUUM CHECK VALVE TEST)

1. Shut the engine off, and wait 5 minutes.

2. Apply the brakes. Is there power assist on at least one pedal stroke? _____

3. If no power assist is felt, carefully remove the check valve from the booster unit.

4. Test the check valve by blowing into the intake manifold end of the valve. There should be a complete blockage of airflow. Alternatively, apply vacuum to the booster unit end of the valve. Vacuum should be blocked. Record your findings.

5. If the vacuum is not blocked, replace the check valve.

PROCEDURE (BRAKE DRAG TEST)

1. With the wheels properly raised off the floor, pump the brake pedal to exhaust the vacuum from the booster.

2. Turn the front wheels by hand and note amount of drag that is present.

3. Start the engine and allow it to run for one minute. Then shut it off.

4. Turn the front wheels by hand again. If the drag has increased, the booster control valve is faulty. Record your findings.

PROCEDURE (FLUID LOSS TEST)

1. If the master cylinder reservoir level has fallen noticeably, but there is no sign of an external brake fluid leak, remove the vacuum tube or hose that runs between the intake manifold and booster unit, and inspect it carefully for evidence of brake fluid. If any fluid is found, the master cylinder is leaking. Record your results.

_____

Problems Encountered

_____

_____

_____

Instructor's Comments

_____

_____

_____

REVIEW QUESTIONS

Review Chapter 50 of the textbook to answer these questions:

1. List four safety precautions when working with brake fluid.

   _____

   _____

   _____

2. Explain the purpose of the split-diagonal hydraulic system.

   _____

   _____

   _____

3. Contrast the two types of power brake systems.

   _____

   _____

   _____

4. What is a typical brake fluid change interval specification?

   _____

   _____

   _____

5. What are two methods of testing master cylinder pushrod adjustment?

   _____

   _____

   _____

RUM BRAKES

OVERVIEW

Drum brakes have been used since the earliest days of the automobile. The major advantage of drum brakes is that they are capable of increasing brake application force over that which is supplied by the hydraulic system. The major problem, however, is that they have many heat-related problems. Because they are enclosed in a drum, heat has a hard time leaving the brake shoes. Therefore, the CoF of the linings decreases and the brakes begin to fade. Also, as the drum heats, it expands. This can eventually increase the distance between the shoes and the drum.

There are two types of drum brakes in use today, the duo-servo, sometimes called self-energizing, and non-servo designs. Duo-servo brakes increase the application force of the brake shoes by using one shoe as a lever against the other shoe. Non-servo brakes do not use this technique and therefore rely solely on the hydraulic system pressure to apply the shoes against the drum.

 Concept Activity 51-1
Drum Brakes

1. Identify the components of the duo-servo brake shown in **Figure 51-1**.

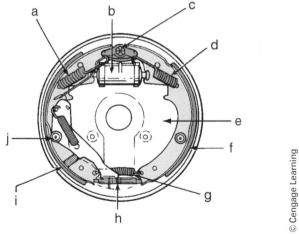

Figure 51-1 Typical duo-servo drum brake.

a. _____   b. _____

c. _____   d. _____

e. _____   f. _____

g. _____   h. _____

i. _____   j. _____

Duo-servo drum brakes operate by using hydraulics, leverage, and friction to slow the rotation of the brake drum. See Concept Activities 50-2, 50-1, and 50-3 to review these principles. In a duo-serve brake system, when the brake pedal is pressed, brake fluid under pressure pushes equally on the two pistons of the wheel cylinder. The pistons in the wheel cylinder push against the brake shoes, forcing them against the inside surface of the brake drum. Once the primary shoe contacts the drum, it attempts to rotate in the same direction as the drum. This reaction force is transferred to the secondary shoe through the self-adjuster. This action increases the braking force created by the hydraulic system alone **(Figure 51-2)**.

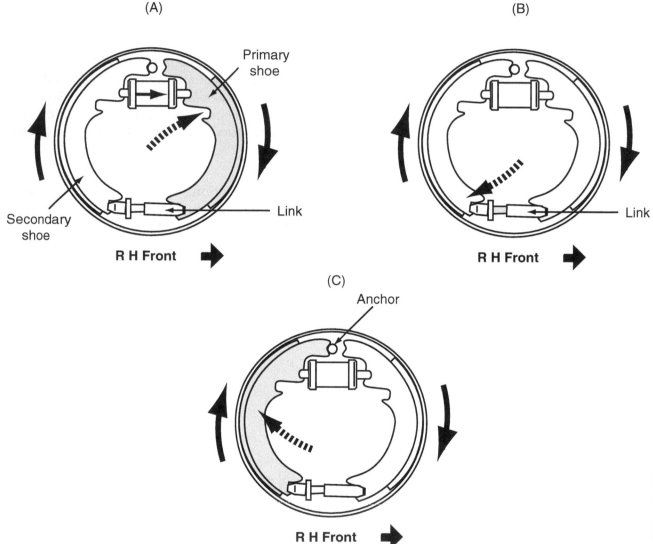

Figure 51-2 Duo-servo action drum brake operation. The primary shoe is forced against the drum, causing the shoe to rotate slightly with the drum (A). The primary shoe pushes against the secondary shoes through the self-adjuster (B). The secondary shoe applies against the drum with increased force (C).

2. What types of vehicles would benefit by using duo-servo brakes?

_____

_____

3. What types of vehicles would not require duo-servo brakes?

_____

_____

_____

■

Many vehicles use non-servo drum brakes. Non-servo brakes do not self-energize the rear shoe, which means there is no increase in the brake force application. The shoes are actuated solely by the hydraulic pressure on the wheel cylinder pistons.

Concept Activity 51-2
Non-Servo Brakes

1. Identify the components of the non-servo drum brake shown is **Figure 51-3**.

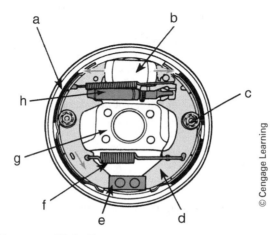

Figure 51-3   Typical non-servo drum brake.

a. _____   b. _____

c. _____   d. _____

e. _____   f. _____

g. _____   h. _____

2. Where would you expect most of the wear to occur on a non-servo drum brake design?

_____

3. Explain three differences in the design and layout of non-servo brakes compared to duo-servo brakes.

_____

_____

_____

Both duo-servo and non-servo drum brakes contain a simple parking brake mechanism. Attached to one of the brake shoes is a lever, which is also attached to the parking brake cables. When the parking brake is applied, the cables pull on the lever, which moves inside of the drum brake assembly. As the lever moves, it pushes on either a parking brake strut or the self-adjuster. This action forces the shoes away from each other and against the drum, keeping the vehicle from rolling **(Figure 51-4)**.

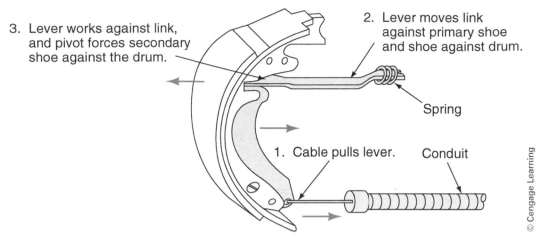

3. Lever works against link, and pivot forces secondary shoe against the drum.

2. Lever moves link against primary shoe and shoe against drum.

Spring

1. Cable pulls lever.

Conduit

© Cengage Learning

Figure 51-4 Parking brake lever and strut operation.

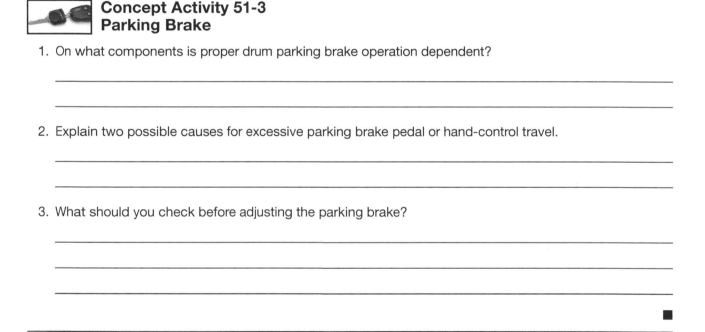

Concept Activity 51-3
Parking Brake

1. On what components is proper drum parking brake operation dependent?

_____

_____

2. Explain two possible causes for excessive parking brake pedal or hand-control travel.

_____

_____

3. What should you check before adjusting the parking brake?

_____

_____

_____

■

Some vehicles with rear disc brakes use a small set of drum brakes as the parking brake. These are known as drum-in-hat parking brakes **(Figure 51-5)**. Although these are very similar to traditional drum brakes, they are used as a parking brake only. Because they do not apply with the service brakes, they seldom need replacement. Anytime the rear brakes are being inspected, the drum-in-hat components should also be thoroughly inspected and adjusted as needed.

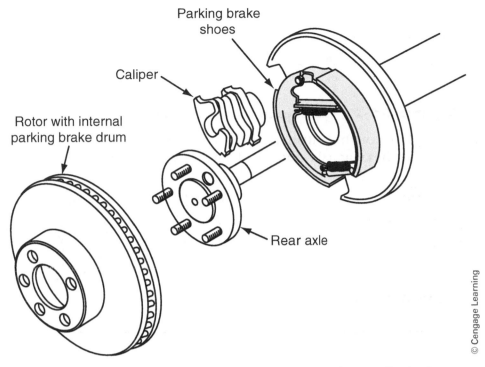

Figure 51-5 An auxiliary parking brake installation for rear disc brakes.

If the customer complains of a brake pulsation or brake noise, the brake drums will need to be measured and possibly machined on a brake lathe. Drums that are worn beyond service specifications must be replaced. A brake drum that is out-of-round can cause pedal pulsation. If the drum is severely out-of-round, the entire vehicle may shake or rock during braking.

Brake drums are typically measured for diameter and out-of-round. Drums usually have a machine-to specification, also called maximum refinish diameter, and a discard diameter. The machine-to diameter is normally between 0.020 and 0.030 inch (0.51 to 0.76 mm) smaller than the discard diameter. Drum nominal or original diameter is the size of the drum from the factory. Drum diameter is checked using a drum micrometer.

 Concept Activity 51-4
Drum Micrometer

To measure a brake drum, you must understand how to set up and read a drum micrometer. **Figure 51-6** shows the parts of a drum micrometer. Both anvils can move over the graduated shaft when the lock screws are loosened. The shaft on an English micrometer is graduated in inches and in 1/8s (0.125) of an inch. The dial reads from 0 to 0.130 inch. **(Figure 51-7)**.

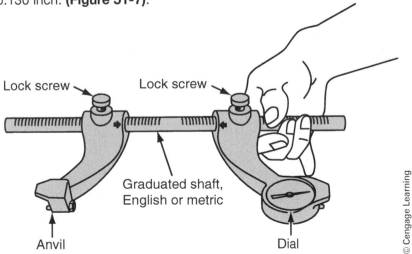

Figure 51-6 Parts of a drum micrometer.

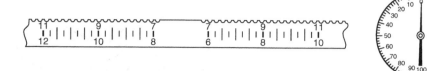

Figure 51-7 The scales on an English micrometer.

If the two anvils are placed on the 9/10 mark on the left and the 9/8 mark on the right, the micrometer will be reading exactly 9 inches when the plunger of the dial is pressed in and the needle is pointing at 0 **(Figure 51-8)**.

Figure 51-8 An example of a drum measurement.

■

Concept Activity 51-5
Drum Micrometer Reading, English

Practice reading an English drum micrometer by completing the activity below **(Figure 51-9)**.

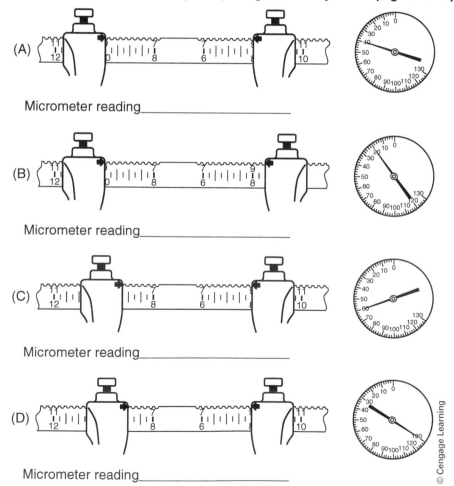

(A)

Micrometer reading_____

(B)

Micrometer reading_____

(C)

Micrometer reading_____

(D)

Micrometer reading_____

Figure 51-9 Practice drum micrometer reading.

To measure a brake drum, first set the micrometer to the drum's nominal size **(Figure 51-10)**. Move the micrometer back and forth until the largest reading is obtained. In this example, the micrometer is set to 11.375 inches **(Figure 51-11)**. When the micrometer is placed into the drum and the plunger on the dial extends, the drum is larger than 11.375 inches **(Figure 51-12)**. Because the plunger extended out until the dial read 0.015 inch, this amount is added to the 11.375 inches for a total of 11.390 inches. To check for out-of-round, turn the micrometer perpendicular to the first measurement and remeasure. The two readings should be identical. If they are not the same, the drum is out-of-round.

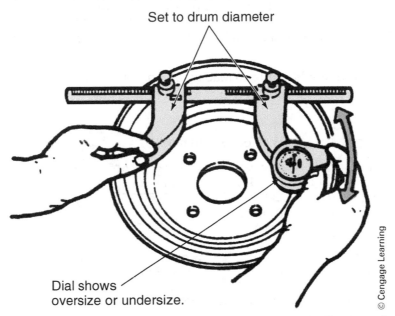

Set to drum diameter

Dial shows oversize or undersize.

© Cengage Learning

Figure 51-10   A drum micrometer is used to measure the inside diameter of a brake drum.

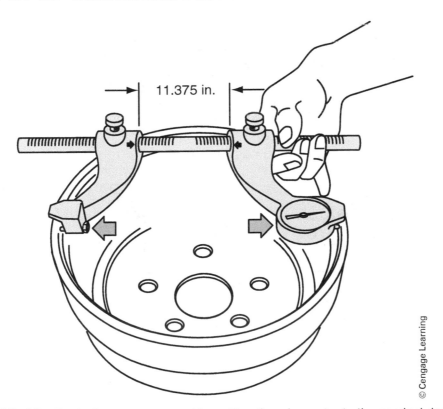

11.375 in.

© Cengage Learning

Figure 51-11   Begin drum measurement by setting the micrometer to the nominal drum diameter.

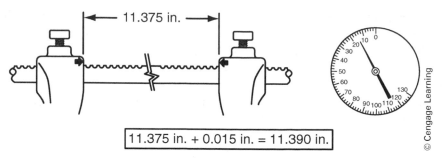

11.375 in. + 0.015 in. = 11.390 in.

Figure 51-12   Add the dial indicator reading to the original micrometer setting to get the drum diameter.

Reading a metric micrometer is similar to reading an English micrometer, except the units are centimeters instead of inches. **Figure 51-13** shows an example of how to read a metric micrometer.

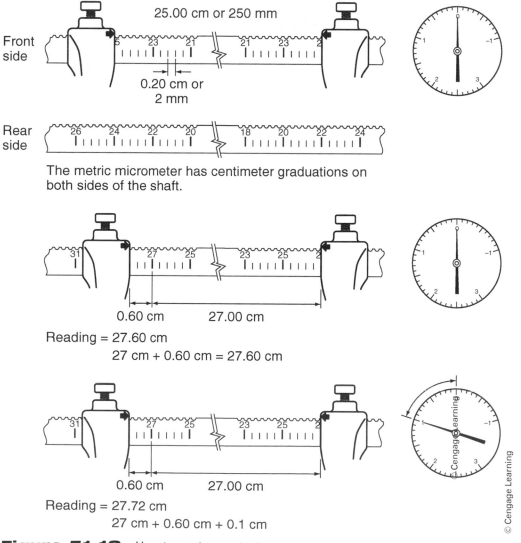

Figure 51-13   How to read a metric drum micrometer.

Concept Activity 51-6
Drum Micrometer Reading, Metric

Practice reading a metric drum micrometer by completing the activity below **(Figure 51-14)**.

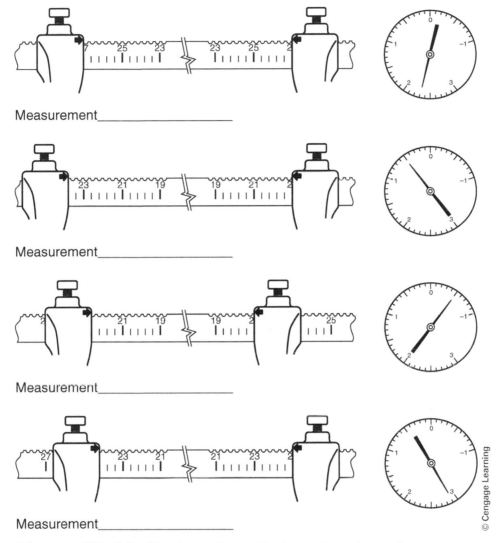

Measurement_____

Measurement_____

Measurement_____

Measurement_____

Figure 51-14   Practice reading metric drum micrometer readings.

The following job sheets will prepare you for drum brake system diagnosis and service.

☐ JOB SHEET 51–1

Inspecting and Servicing Drum Brakes

Name _____ Station _____ Date _____

Objective

Upon completion of this job sheet, you will have demonstrated the ability to inspect and service drum brakes. This task applies to ASE Education Foundation Tasks MLR 5.C.1 (P-1), 5.C.3 (P-1), 5.C.4 (P-1), 5.C.5 (P-1), AST/MAST 5.C.1 (P-1), 5.C.4 (P-1), 5.C.5 (P-2), and 5.C.6 (P-1).

Tools and Materials

Drum brake spring tools

Asbestos containment system

Brake lubricant

Description of Vehicle

Year _____ Make _____ Model _____

VIN _____ Engine Type and Size _____

Mileage _____

PROCEDURE

1. Remove the brake drum and set aside. Using an approved asbestos containment system, remove all brake dust from the drum and drum brake assembly. The types of asbestos containment system used:

    _____

    Instructor's Check _____

2. Once the brake assembly is clean and free of dust, inspect the shoes for wear, cracks, glazing, fluid contamination, or other damage. Note your findings.

    _____

    _____

3. Remove the return and hold-down springs, and the brake shoes. Inspect the springs for deformation, excessive rust, and tension. Many technicians recommend replacing the springs and related hardware each time the shoes are replaced, due to the heat and stresses placed on the springs. Note your findings. _____

    _____

    _____

4. Remove the self-adjuster and related components. Inspect the adjuster's threads, non-threaded end, and star wheel for wear. The self-adjuster should be cleaned and lubricated before reinstallation. Note your findings. _____

    _____

5. Carefully pull back the edge of the wheel cylinder dust boots, and check for evidence of brake fluid. A very slight amount of fluid leakage is considered normal. If the fluid drips out of the dust boot, the cylinder should be rebuilt or replaced. Note your findings.

_____

_____

6. Inspect the backing plate for excessive rust and wear. Check the raised shoe pads for grooves or other defects that can keep the shoes from sliding properly. Ensure the backing plate is not bent or distorted. Note your findings. _____

7. Verify the operation of the parking brake. Ensure the cables apply and release easily. Check the parking brake lever for signs of wear. The parking brake lever is normally reused and is not supplied in a drum brake hardware kit. Note your findings.

_____

_____

8. Apply the recommended brake lubricant to the raised pads on the backing plate, the self-adjuster threads and smooth end, and at contact points on the shoes and parking brake lever where movement occurs. Lubricant used _____

Instructor's Check _____

Before installing the new shoes, some technicians cover the friction surface of the shoe with masking tape. This prevents dirt and lubricant from contaminating the new shoes.

9. Install the new brake shoes, springs, and hardware. Double-check that all the parts are installed correctly. Adjust the shoe-to-drum clearance until there is a slight amount of drag when the drum is rotated.

Instructor's Check _____

10. Check the operation of the parking brake. Locate the specification for the number of clicks the brake should move until fully applied and record it here. _____

11. Apply the parking brake. How many clicks before full application? _____

12. Readjust the shoe-to-drum clearance if necessary. If clearance is correct, adjust the parking brake until the correct number of clicks is achieved. Note your results. _____

Instructor's Check _____

Problems Encountered

_____

_____

_____

Instructor's Comments

_____

_____

☐ JOB SHEET 51–2

Inspecting Drum Brakes

Name _____ Station _____ Date _____

Objective

Upon completion of this job sheet, you will have demonstrated the ability to remove, clean, inspect, and measure a brake drum. This task applies to ASE Education Foundation Task MLR/AST/MAST 5.C.2 (P-1).

Tools and Materials

Drum micrometer

Asbestos containment system

Brake cleaner

Description of Vehicle

Year _____ Make _____ Model _____

VIN _____ Engine Type and Size _____

Mileage _____

PROCEDURE

1. Remove the brake drum. On some models, the rear-wheel bearing will require disassembly. If the drum has threaded holes in the face, insert the correct size bolt and tighten. This will push the drum off the hub. Drums often rust and stick to the hub, cleaning with a wire brush and the application of a penetrant may be required for removal. If the brake shoes are well adjusted or a ridge is worn into the edge of the drum, the brake shoes may need to be adjusted in to aid in drum removal.

   WARNING: *Do not pry on the drum where it surrounds the backing plate.*

   Record the steps you had to take to remove the drum. _____

   _____

2. Once the drum is removed, use the recommended procedures to safely contain the brake dust. Method used _____

3. Clean the brake drum and inspect for grooves, scoring, glazing, hard spots, heat checks, and cracks. Note your findings. _____

4. Locate the brake drum diameter specifications and record them below.

   Machine-to diameter _____

   Maximum diameter _____

   Out-of-round limit _____

5. Use a drum micrometer to measure the brake drum.

   Drum diameter reading _____

6. Once you have made your first measurement, rotate the micrometer at a right angle, and remeasure to determine whether the drum is out-of-round.

Out-of-round reading _____

7. Compare your results to the specifications and determine the necessary action.

_____

_____

Problems Encountered

_____

_____

_____

Instructor's Comments

_____

_____

_____

☐ JOB SHEET 51–3

Adjusting and Replace a Parking Brake Cable

Name _____ Station _____ Date _____

Objective

Upon completion of this job sheet, you will have demonstrated the ability to remove, inspect, and replace a parking brake cable. This task applies to ASE Education Foundation Tasks MLR 5.F.2 (P-1), and AST/MAST 5.F.3 (P-1).

Description of Vehicle

Year _____ Make _____ Model _____

VIN _____ Engine Type and Size _____

Mileage _____

REAR CABLE REMOVAL—DRUM BRAKES

1. Raise the vehicle to a convenient working height on a hoist, and remove the rear wheels.

2. Detach the retaining clip from the brake cable retainer bracket. Remove the brake drum (from the rear axle), brake shoe return springs, and brake shoe retaining springs.

    WARNING: *After removing the brake drum, install the asbestos removal vacuum. Follow the manufacturer's instructions for the removal of asbestos from the brake components.*

3. Remove the brake shoe strut and spring from the brake support plate.

4. Disconnect the brake cable from the operating lever.

5. Compress the retainers on the end of the brake cable housing, and remove the cable from the backing plate.

REAR CABLE INSTALLATION

1. Insert the brake cable and housing into the backing plate, making sure that the housing retainers lock the housing firmly into place.

2. Hold the brake shoes in place on the backing plate. Engage the brake cable into the brake shoe operating lever.

3. Install the parking brake strut and spring, brake shoe retaining springs, brake shoe return springs, and brake drum and wheel.

4. Insert the brake cable and housing into the bracket, and install the retaining clip.

5. Insert the brake cable into the equalizer.

6. Adjust the rear service brakes and parking brake cable.

PARKING BRAKE ADJUSTMENT—REAR DRUM

1. Explain why the rear brakes must be properly adjusted before adjusting the parking brake.

   _____

   _____

2. Raise the vehicle, and support it with jack stands placed under the suspension. Set the transmission shift lever in the neutral position.

3. Release the parking brake operating lever, and loosen the cable adjusting nut to make sure the cable is slack. Clean the cable threads with a wire brush, and lubricate them to prevent rusting.

4. Tighten the cable adjusting nut at the equalizer until a slight drag is felt while rotating the rear wheels.

5. Loosen the cable adjusting nut until both rear wheels can be rotated freely. Back off the cable adjusting nut two full turns. There should be no drag at the rear wheels.

6. Make a final operational check of the parking brake. Then lower the vehicle.

Problems Encountered

_____

_____

_____

Instructor's Comments

_____

_____

_____

REVIEW QUESTIONS

Review Chapter 51 of the textbook to answer these questions:

1. Explain how to check for brake drum out-of-round.

   _____

   _____

2. Explain the difference between machine-to and discard diameters.

   _____

   _____

3. List four items that should be checked when performing a drum brake inspection.

   _____

   _____

   _____

4. Explain the purpose of the spring inside the wheel cylinder.

   _____

   _____

5. Explain problems that can be caused by weak or fatigued brake springs.

   _____

   _____

CHAPTER

52

DISC BRAKES

OVERVIEW

Disc brakes offer many advantages over drum brakes; they are self-adjusting, dissipate heat better, and are less prone to pull. Disc brakes are standard equipment as front brakes and are very common on the rear brakes as well.

Concept Activity 52-1
Disc Brakes

1. Identify the components of the disc brake system shown in **Figure 52-1**.

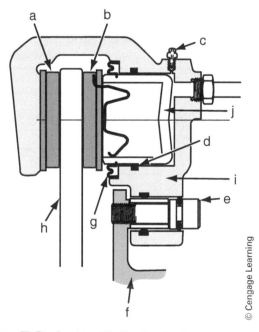

Figure 52-1 Identify the disc brake components.

© Cengage Learning

a. _____   b. _____

c. _____   d. _____

e. _____   f. _____

g. _____   h. _____

i. _____   j. _____

2. Which component allows for the self-adjusting capability of disc brakes? _____

_____

3. Describe the three types of caliper designs. _____

_____

■

_____

The most common disc brake calipers in use today operate by taking advantage of Isaac Newton's Third Law of Motion: For every action there is an equal and opposite reaction. When the brakes are applied, fluid pushes the caliper piston out of its bore slightly. This causes the caliper to respond by sliding backward slightly. This action causes both the inboard and outboard brake pads to apply the same clamping force against the rotor.

 Concept Activity 52-2
Calipers

1. Examine a disc brake assembly, and note how the caliper is attached to the steering knuckle.

   Is this a floating, sliding, or fixed caliper design?

   _____

2. How does the caliper mounting allow for back-and-forth movement?

   _____

   _____

3. Remove the caliper mounting bolts. Describe how the mounting bolts allow for caliper movement.

   _____

4. Examine the mounting bolt sleeves, bushings, O-rings, and other related hardware. Note your findings.

   _____

5. What type of problems could cause the caliper to not move back and forth correctly?

   If this is a fixed caliper, what could cause uneven pad wear?

   _____

6. What type of wear or symptoms could this cause?

   _____

   _____

   _____

■

The brake rotor provides the rotating friction surface for the disc brake system. One of the advantages of disc brakes compared to drum brakes is the ability of disc brakes to dissipate heat more effectively. **Figure 52-2** illustrates this advantage.

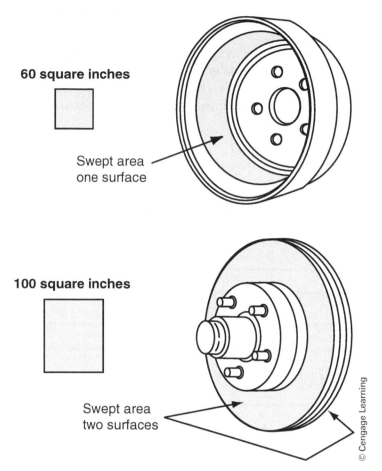

60 square inches

Swept area
one surface

100 square inches

Swept area
two surfaces

© Cengage Learning

Figure 52-2 For any given size wheel, a disc brake always has 35% to 52% more swept area to dissipate heat.

 Concept Activity 52-3
Disc Brake Design

1. In addition to heat dissipation, how does the design of disc brakes reduce brake fade?

_____

2. Explain why disc brakes require greater application force than drum brakes.

_____

_____

3. Explain why some brake rotors are drilled and/or slotted.

_____

_____

_____

■

To perform a thorough disc brake inspection, the rotors must be inspected and measured. Rotors are measured for thickness, thickness variation, also called parallelism or disc taper, and runout.

Brake rotor thickness is critical for safe operation. As the rotor wears, its thickness decreases. This increases the stress on the rotor and reduces its ability to handle heat. Rotors must be checked for thickness whenever disc brake service is being performed. To precisely measure a rotor, a disc brake micrometer is used. **Figure 52-3** shows an example of a typical brake rotor micrometer.

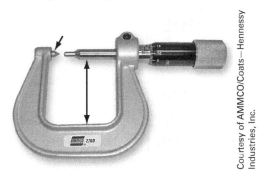

Courtesy of AMMCO/Coats—Hennessy Industries, Inc.

Figure 52-3   Notice the depth of the micrometer and pointed anvil.

 Concept Activity 52-4
Rotor Micrometer Reading

1. Why does a rotor micrometer use a pointed anvil compared to a flat anvil on a standard micrometer?

Brake rotor micrometers are read the same way as standard micrometers. **Figure 52-4** shows how to read a standard English micrometer. Practice your micrometer reading skills by completing the following readings **(Figure 52-5)**.

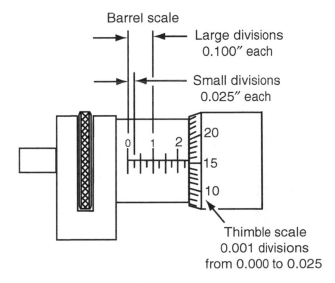

Barrel scale

Large divisions
0.100″ each

Small divisions
0.025″ each

Thimble scale
0.001 divisions
from 0.000 to 0.025

4 steps to read, add together:	Example above:
1. Select frame size	0–1
2. Large barrel divisions	X 0.100 = 0.200″
3. Small barrel divisions	X 0.025 = 0.025″
4. Thimble divisions	X 0.001 = 0.016″
Reading	0.241″

© Cengage Learning

Figure 52-4   A micrometer is read by totaling the number of whole divisions on the barrel scale and adding the thousandths from the thimble scale.

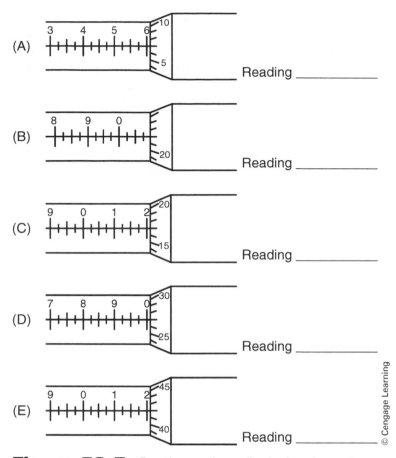

(A) Reading _____

(B) Reading _____

(C) Reading _____

(D) Reading _____

(E) Reading _____

© Cengage Learning

Figure 52-5   Practice reading a disc brake micrometer.

Metric brake micrometers are read as a traditional metric micrometer. **Figure 52-6** shows how to read a metric micrometer. Practice your metric micrometer reading skills by completing the following readings **(Figure 52-7)**.

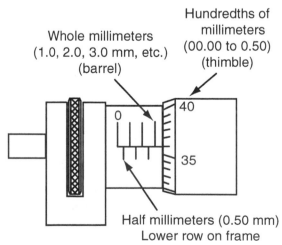

Hundredths of
millimeters
(00.00 to 0.50)
(thimble)

Whole millimeters
(1.0, 2.0, 3.0 mm, etc.)
(barrel)

Half millimeters (0.50 mm)
Lower row on frame

3 steps to read, add together: Example above:

1. Whole mm lines (upper) on barrel   3 = 3.00 mm
2. Half mm line (lower) on barrel   0 = 0.00 mm
3. Lines on thimble   36 = 0.36 mm

Reading   3.36 mm

© Cengage Learning

Figure 52-6   The metric micrometer is read similar to the standard micrometer. The graduations are now read in millimeters rather than inches.

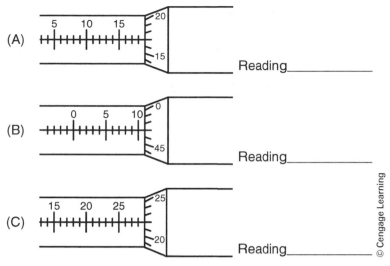

(A) Reading_____

(B) Reading_____

(C) Reading_____

Figure 52-7 Determine the metric micrometer readings.

Brake rotors normally have two thickness specifications, minimum thickness and machine to. The machine to, also called minimum refinish thickness, is typically 0.020 inch to 0.030 inch (0.51–0.76 mm) larger than minimum thickness.

2. Why is there a difference between machine to and the minimum thickness specifications?

_____

Vibration, also called pulsation, from the disc brakes is felt in the brake pedal and steering wheel during brake application. The pulsations can range from very slight to severe depending on the cause and the condition of the brake components. The major cause of disc brake pulsation is brake rotor thickness variation, also called disc brake taper or rotor parallelism.

Rotors must also be checked for thickness variation. **Figure 52-8** illustrates how to check for thickness variation. Subtract the smallest reading from the largest to obtain the actual variation. Many manufacturers specify a very small allowable thickness variation, usually around 0.0005 inch (0.0127 mm). To measure this you will need a micrometer that reads to one ten thousandths of an inch or millimeter. **Figure 52-9** is an example of a micrometer with an additional scale, called a Vernier scale, used to measure to 0.0001 inch (0.00254 mm).

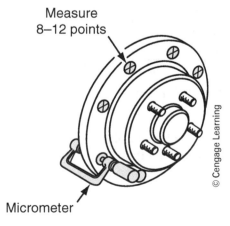

Measure
8–12 points

Micrometer

Figure 52-8 Check for thickness variations by measuring at eight to twelve points around the rotor.

Example

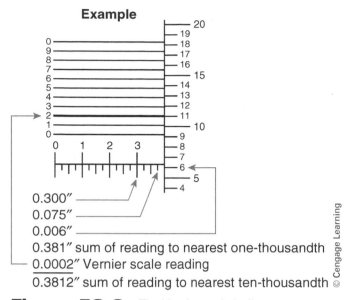

0.300″ ———————
0.075″ ———————
0.006″ ———————
0.381″ sum of reading to nearest one-thousandth
0.0002″ Vernier scale reading
0.3812″ sum of reading to nearest ten-thousandth

Figure 52-9   The Vernier scale indicates 0.0002″.

3. Determine the total reading of this micrometer **(Figure 52-10)**.

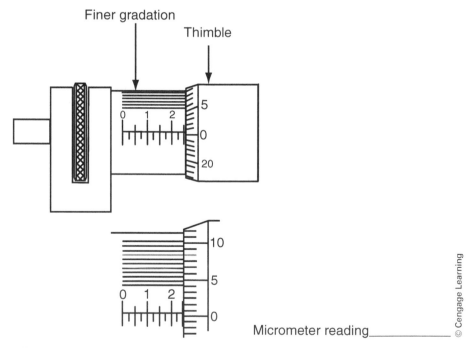

Micrometer reading_____

Figure 52-10   Determining the micrometer reading shown.

The final rotor measurement is to check for lateral runout. Mount a dial indicator as shown in **Figure 52-11**. Secure the rotor with lug nuts and slowly rotate the rotor while watching the dial indicator. Runout specifications vary with manufacturer, but generally, less than 0.003 inch (2.0762 mm) is considered acceptable.

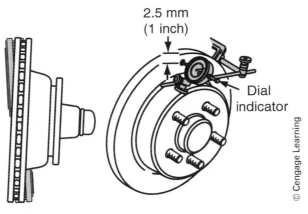

2.5 mm
(1 inch)

Dial
indicator

© Cengage Learning

Figure 52-11 Measuring rotor lateral runout.

4. If a runout reading is obtained that is beyond the specification, what diagnostic steps should be taken next? _____

5. What other components can cause the rotor to show excessive runout?

_____

_____

6. If the rotor has excessive runout, how can it be corrected?

_____

_____

Rotors, like brake drums, are inspected for scoring, cracks, hard spots, and glazing **(Figure 52-12)**. Light scoring is acceptable and may not require refinishing. Inspect the cooling fin area. Excessive rusting of the cooling fins can cause poor heat dissipation and rotor failure.

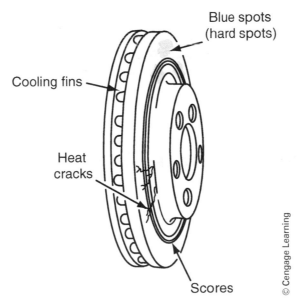

Blue spots
(hard spots)

Cooling fins

Heat
cracks

Scores

© Cengage Learning

Figure 52-12 Typical rotor defects.

7. What can cause scoring of the brake rotor surfaces?

_____

_____

8. What can cause brake glazing?

_____

_____

_____

■

_____

Even though disc brakes have many advantages compared to drum brakes, there are problems common to disc brakes. Noise and vibration are often the most common customer complaints regarding the brake system.

Concept Activity 52-5
Disc Brake Noise Diagnosis

Disc brake noise is common and sometimes difficult to correct. Verify during a test drive that the noise is disc brake related. Many vehicles use audible pad wear indicators, which are designed to emit a loud noise when the pads are worn down to the point of replacement. The vibrations of the brake pads and pad hardware cause brake pad squeal. Many manufacturers use special shims and springs to reduce pad squeal or change the frequency so that it is inaudible to human hearing. **Figure 52-13** and **Figure 52-14** show two examples of brake hardware designed to reduce noise.

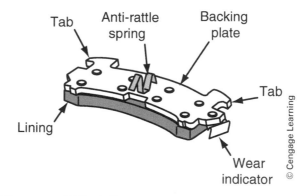

Figure 52-13 Brake pad and antirattle spring.

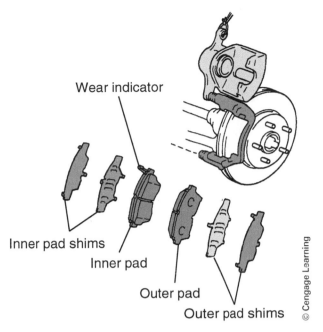

Wear indicator

Inner pad shims

Inner pad

Outer pad

Outer pad shims

© Cengage Learning

Figure 52-14 Brake pad shims.

1. Examine several different disc brake assemblies, and note the use of shims and springs to reduce pad noise. Summarize your findings.

_____

_____

2. As the pads and hardware age, how can the effects of wear and climate affect their operation?

_____

_____

The following job sheet will prepare you for disc brake inspection and service.

☐ JOB SHEET 52–1

Inspecting and Servicing Disc Brakes

Name _____ Station _____ Date _____

Objective

Upon completion of this job sheet, you will have demonstrated the ability to inspect, diagnose, and service the disc brake concerns. This task applies to ASE Education Foundation Tasks MLR/AST/MAST 5.D.1(P-1), 5.D.2 (P-1), 5.D.3 (P-1), 5.D.4 (P-1), and 5.D.5 (P-1).

Tools and Materials

Rotor micrometer

Dial indicator

Description of Vehicle

Year _____ Make _____ Model _____

VIN _____ Engine Type and Size _____

Mileage _____

PROCEDURE

1. If the vehicle has a customer complaint, test-drive to verify the complaint if necessary.

2. Visually inspect the condition of the disc brake components. Record your findings.

   Rotor visual check _____ ____

   Caliper visual check _____

   Brake hoses _____

3. Summarize any problems that you observed from the visual inspection. _____

   _____

4. Loosen the caliper mounting hardware, and remove. Slide the caliper off the rotor, and support it with a strap. Why should the caliper not be allowed to hang by its hose?

   _____

   _____

5. Inspect the caliper mounting hardware. Does the mounting hardware allow the caliper to move easily?

   _____

6. What problems can result from a caliper that does not slide correctly?

   _____

   _____

7. Inspect the caliper piston dust boot(s) for signs of damage or fluid loss past the dust boot. Note your findings. _____

_____

8. If fluid were leaking from around the piston(s), what would this indicate?

_____

_____

9. Closely inspect the brake pads. Check for uneven wear, cracks, glazing, and lining separation. Note your findings. _____

_____

10. Inspect the inside and outside surfaces of the rotor. Inspect for scoring, rust, cracks, glazing, and hard spot. Note your findings. _____

_____

11. Measure the brake rotor for thickness, thickness variation, and runout.

    Thickness specifications _____

    Thickness variation specification _____

    Runout specification _____

    Actual thickness _____

    Actual thickness variation _____

    Actual runout _____

12. Compare the measurements to specifications, and determine the necessary actions.

    _____

    _____

13. Thoroughly clean the caliper mountings, guides, bushings, and related hardware.

14. Lubricate the caliper mountings, guides, bushings, and related hardware as necessary to ensure proper movement.

    Lubricant Used _____

15. Reassemble the pads and related hardware. Lubricate contact points between the pads and mounting points.

    Lubricant Used _____

16. Install the caliper bolts and torque to specifications.

    Torque Specifications _____

17. Pump the brake pedal several times to fully seat the pads. Describe pedal feel after several applications.

    _____

    Instructor's Check _____

Problems Encountered

_____

_____

_____

Instructor's Comments

_____

_____

_____

 **REVIEW QUESTIONS**

Review Chapter 52 of the textbook to answer these questions:

1. List and describe the two major types of caliper design.

    _____

    _____

2. List three advantages of disc brakes.

    _____

    _____

3. Why do disc brake systems require greater application force compared to drum brakes?

    _____

    _____

4. Define rotor parallelism.

    _____

    _____

5. Explain how to test rotor runout.

    _____

    _____

ANTILOCK BRAKE, TRACTION CONTROL, AND STABILITY CONTROL SYSTEMS

CHAPTER 53

© Cengage Learning 2015

OVERVIEW

Antilock brake systems (ABS) and traction and stability control systems are standard equipment on many vehicles and are becoming standard equipment on all new vehicles. Traction control systems utilize many of the components of the ABS to control wheel slipping during acceleration. Stability control systems also use ABS components. These systems apply a wheel brake unit to control understeer and oversteer handling problems.

Antilock brake systems (ABS) can be thought of as electronic/hydraulic pumping of the brakes for stopping under panic conditions. Good drivers have always pumped the brake pedal during panic stops to avoid wheel lockup and the loss of steering control. Antilock brake systems simply get the pumping job done much faster and in a much more precise manner than the fastest human foot. A tire on the verge of slipping produces more friction with respect to the road than one that is locked and skidding. Once a tire loses its grip, friction is reduced, and the vehicle takes longer to stop. By modulating the pressure to the brakes, friction between the tires and the road is maintained, and the vehicle is able to come to a controllable stop.

Many different designs of ABS are found on today's vehicles. These designs vary in their basic layout, operation, and components.

Manufacturers use the components of antilock braking systems to control tire traction and vehicle stability. Automatic traction control (ATC) systems apply the brakes when a drive wheel attempts to spin and lose traction.

Various stability control systems are found on today's vehicles. Like traction control systems, stability controls are based on and linked to the ABS. On some vehicles the stability control system is also linked to the electronic suspension system.

Stability control systems momentarily apply the brakes at any one wheel to correct oversteer or understeer. The control unit receives signals from the typical sensors plus a yaw, lateral acceleration (G-force), and a steering angle sensor. Stability control systems can control the vehicle during acceleration, braking, and coasting. If the brakes are already applied but oversteer or understeer is occurring, the fluid pressure to the appropriate brake is increased **(Figure 53-1)**.

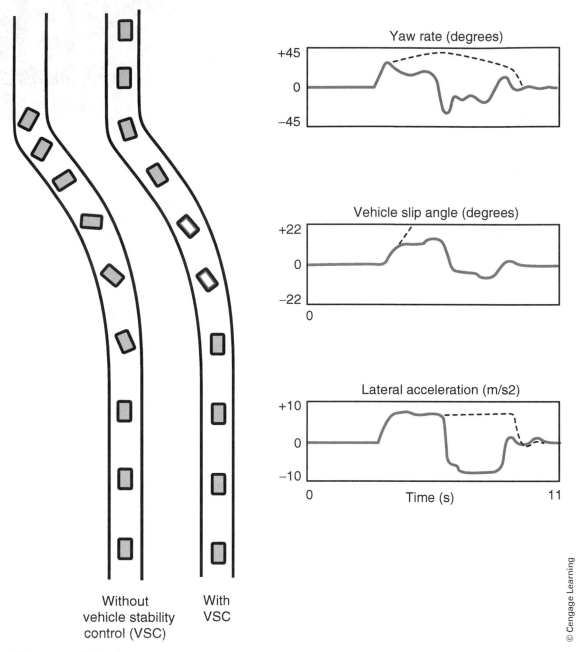

Figure 53-1 Comparison during a turn between a vehicle with a stability control system and a vehicle with no stability control system.

Concept Activity 53-1
Identify ABS/ATC/ESC Function

Match the vehicle condition to the correct control system, antilock brakes (ABS), automatic traction control (ATC), or electronic stability control (ESC).

a. Monitor wheel speeds   _____

b. Control wheel lockup   _____

c. Prevent wheel spin   _____

d. Reduce body roll   _____

e. Monitor steering input   _____

f. Monitor throttle position   _____

SERVICE

Many ABS components are simply remove-and-replace items. Normal brake repairs, such as replacing brake pads, caliper replacement, rotor machining or replacement, brake hose replacement, master cylinder or power booster replacement, or parking brake repair, can all be performed as usual. In other words, brake service on an ABS-equipped vehicle is similar to brake service on a conventional system, with a few exceptions. Before beginning any service, check the service manual. It may be necessary to depressurize the accumulator to prevent personal injury from high-pressure fluid.

Concept Activity 53-2
Identify High-Pressure ABS Accumulators

ABS/ATC systems that use accumulators operate at very high pressures. Use caution whenever servicing these systems.

1. How can a system with high-pressure accumulators be identified? _____

_____

2. How can the pressure in the accumulators be relieved? _____

_____

■

Safety Precautions

■ When replacing brake lines and/or hoses, always use lines and hoses designed for and specifically labeled for use on ABS vehicles.

■ Never use silicone brake fluids in ABS vehicles. Use only the brake fluid type recommended by the manufacturer (normally DOT-3).

■ Never begin to bleed the hydraulic brake system on a vehicle equipped with ABS until you have checked a service manual for the proper procedure.

■ Never open a bleeder screw or loosen a hydraulic brake line or hose while the ABS system is pressurized.

■ Disconnect all vehicle computers, including the ABS control unit, before doing any electrical welding on the vehicle.

■ Never disconnect or reconnect electrical connectors while the ignition switch is on.

■ Never install a telephone or CB antenna close to the ABS control unit or any other control module or computer.

■ Check the air gap between the wheel-speed sensor and the sensor ring after any parts of the wheel-speed circuit have been replaced.

■ Keep the wheel sensor clean. Never cover the sensor with grease unless the manufacturer specifies doing so; in those cases, use only the recommended type.

■ When replacing speed (toothed) rings, never beat them on with a hammer. These rings should be pressed onto their flange. Hammering may result in a loss of polarization or magnetization.

A prediagnosis inspection consists of a visual check of system components. Problems can often be spotted during this inspection, and the need to conduct other more time-consuming procedures can be eliminated. This inspection should include the following:

1. Check the master cylinder fluid level.

2. Inspect all brake hoses, lines, and fittings for signs of damage, deterioration, and leakage. Inspect the hydraulic modulator unit for any leaks or wiring damage.

3. Inspect the brake components at all four wheels. Make sure that no brake drag exists and that all brakes react normally when they are applied.

4. Inspect for worn or damaged wheel bearings that may allow a wheel to wobble.

5. Check the condition of the CV joints.

6. Make sure the tires meet the legal tread depth requirements and that they are the correct size.

7. Inspect ABS electrical connections for signs of corrosion, damage, fraying, and disconnection.

8. Inspect the wheel-speed sensors and their wiring.

Concept Activity 53-3
Identify ABS/ATC/ESC Warning Lights

1. Low brake fluid level can cause which instrument panel (IP) warning light(s) to turn on?

_____

2. If there is a leak in the brake hydraulic system, which IP warning light(s) may illuminate?

_____

3. List the nonbrake system components that can affect the operation of the brake, ABS, ATC, or ESC systems. _____

_____

4. How can excessive rust buildup around the wheel-speed sensor (WSS) cause ABS/ATC malfunctions?

_____

_____

_____

■

The following job sheets will prepare you for diagnosing and servicing the ABS, ATC, and ESC systems.

☐ JOB SHEET 53–1

Performing ABS, ATC, and ESC Warning Light Diagnosis

Name _____ Station _____ Date _____

Objective

Upon completion of this job sheet, you will have demonstrated the ability to determine causes of ABS, ATC, and ESC warning light illumination. This task applies to ASE Education Foundation Task MAST 5.G.4 (P-2).

Tools and Materials
Scan tool

Description of Vehicle

Year _____ Make _____ Model _____

VIN _____ Engine Type and Size _____

Mileage _____

PROCEDURE

To begin diagnosis of the ABS, ATC, and ESC systems, observe the warning indicator lights on the instrument panel.

1. Turn the key to the key on engine off (KOEO) bulb-check position, and note the warning lights. Are the BRAKE, ABS, ATC, and ESC lights illuminated? _____

2. If one of more of the warning lights does not illuminate, refer to the service manual, and locate the diagnostic procedure to test that system.

3. Start the engine, and note the warning lights. _____

4. What conditions can cause the red BRAKE warning light to remain illuminated?

_____

_____

5. If all the warning lights go out, apply the parking brake, and note the red BRAKE light. Does it illuminate? _____

6. Release the parking brake, and note the red BRAKE light.

_____

7. Inspect and note the brake fluid level in the master cylinder reservoir.

_____

_____

8. If the amber ABS light remains on, does either the ATC or ESC remain on also?

_____

9. Why would the ATC and ESC lights illuminate if the ABS light were on?

   _____

   _____

10. If the any of the ABS, ATC, or ESC lights remain on, connect a scan tool to the data link connector (DLC), and enter diagnostics.

    Scan Tool Used _____

11. List any codes stored in memory for the ABS, ATC, and ESC. _____

    _____

    Instructor's Check _____

Problems Encountered

_____

_____

_____

Instructor's Comments

_____

_____

_____

☐ JOB SHEET 53–2

Identifying ABS/ATC System and Components

Name _____ Station _____ Date _____

Objective

Upon completion of this job sheet, you will have demonstrated the ability to identify ABS/ATC systems and components. This task applies to ASE Education Foundation Task MLR 5.G.1 (P-3), and AST/MAST 5.G.1 (P-1).

Description of Vehicle

Year _____ Make _____ Model _____

VIN _____ Engine Type and Size _____

Mileage _____

PROCEDURE

1. Verify the vehicle has ABS and ATC systems. _____ ABS only _____ ABS/ATC

2. Determine whether the ABS system is an integral or nonintegral system.

   _____ Integral   _____ Non-integral

3. If the vehicle has an integral system, explain how to depressurize the ABS system.

   _____

   _____

4. How many channels does this system use?

   _____

5. Describe the location of the ABS electro-hydraulic control unit.

   _____

   _____

6. Describe the location of the WSSs.

   _____

   _____

   _____

7. Connect a scan tool to the DLC, and enter communications with the on-board network.

   Scan Tool Used _____

8. List the sensors used by the ABS/ATC system to monitor vehicle-operating conditions.

   _____

   _____

9. Explain what vehicle functions the ABS/ATC system can control during an ABS/ATC event.

_____

_____

Problems Encountered

_____

_____

_____

Instructor's Comments

_____

_____

_____

☐ JOB SHEET 53–3

Testing a Wheel-Speed Sensor and Adjusting Its Gap

Name _____ Station _____ Date _____

Objective

Upon completion of this job sheet, you will have demonstrated the ability to test a wheel-speed sensor and check and adjust its gap if necessary. This task applies to ASE Education Foundation Task MAST 5.G.7 (P-2).

Tools and Materials

Lab scope

Digital multimeter (DMM)

Feeler gauge

Description of Vehicle

Year _____ Make _____ Model _____

VIN _____ Engine Type and Size _____

Mileage _____

Describe the type of ABS found on the vehicle.

_____

Model lab scope and DMM that will be used _____

PROCEDURE

1. Raise the vehicle on a frame contact hoist. Make sure the wheel that has the sensor that will be tested is free to rotate. Wheel position: LF _____ RF _____ LR _____ RR _____

2. Turn the ignition switch to the run position.

3. Visually inspect the wheel-speed sensor pulsers for chipped or damaged teeth. Record your findings.

_____

_____

_____

4. Inspect the wheel sensor's wiring harness for any damage. Record your findings.

_____

_____

_____

5. Connect the lab scope across the wheel sensor.

6. Spin the wheel by hand, and observe the waveform on the scope. You should notice that as the wheel begins to spin, the waveform of the sensor's output should begin to oscillate above and below zero volts. The oscillations should get taller as speed increases. If the wheel's speed is kept constant, the waveform should also stay constant. Record what you observed.

_____

_____

_____

7. Turn off the ignition.

8. Locate the air gap specifications for the sensor. The specified gap is _____.

9. Use a feeler gauge to measure the air gap between the sensor and its pulser (rotor) all the way around while rotating the drive shaft, wheel, or rear hub bearing unit by hand.

Record the measured gap. _____

Does the gap vary while the wheel is rotated? If so, what does this suggest?

_____

_____

_____

If the gap is not within specifications, what needs to be done?

_____

_____

_____

How would an incorrect air gap at the wheel sensor affect ABS operation?

_____

_____

_____

Problems Encountered

_____

_____

_____

Instructor's Comments

_____

_____

_____

REVIEW QUESTIONS

Review Chapter 53 of the textbook to answer these questions:

1. Explain how the ABS system determines wheel lockup.

_____

_____

2. Describe how to depressurize an ABS system.

_____

_____

3. Explain how to test an AC generating wheel-speed sensor.

_____

_____

4. Describe what actions the powertrain control module (PCM) can take to reduce wheel slip.

_____

_____

5. Explain the purpose of electronic stability control.

_____

_____

ASE PREP TEST

1. Technician A says that when replacing brake lines and/or hoses, always use lines and hoses designed for and specifically labeled for use on ABS vehicles. Technician B says to use only the silicone-type brake fluid in ABS. Who is correct?

   a. Technician A
   b. Technician B
   c. Both A and B
   d. Neither A nor B

2. Technician A says never begin to bleed the hydraulic brake system on a vehicle equipped with ABS until you have checked a service manual for the proper procedure. Technician B says never open a bleeder screw or loosen a hydraulic brake line or hose while the ABS is pressurized. Who is correct?

   a. Technician A
   b. Technician B
   c. Both A and B
   d. Neither A nor B

3. Technician A says wheel sensors produce an analog a/c voltage as the wheels rotate. Technician B says some wheel-speed sensors produce a digital signal. Who is correct?

   a. Technician A
   b. Technician B
   c. Both A and B
   d. Neither A nor B

4. Technician A says many ABS have an accumulator that stores the high pressure for the system. The accumulator must be fully depressurized. To depressurize the system, turn the ignition switch to the off position, and pump the brake pedal between 25 and 50 times. Technician B says the brake pedal should be noticeably softer when the accumulator is discharged. Who is correct?

   a. Technician A
   b. Technician B
   c. Both A and B
   d. Neither A nor B

5. Which of the following is *not* true?

   a. During a test drive of an ABS vehicle, both brake warning lights should remain off.
   b. If the PCM detects a problem with the system, the ABS indicator lamp will either flash or light continuously to alert the driver of the problem.
   c. In some systems, an illuminated amber ABS indicator lamp indicates that the control unit detected a problem but has not suspended ABS operation.
   d. A solid amber ABS indicator lamp indicates that a problem has been detected that affects the operation of the ABS. No antilock braking will be available, but normal, nonantilock brake performance will remain.

6. A power brake system needs bleeding. Technician A says that a pressure bleeder must be used. Technician B says that all the bleeder screws must be loosened, and all wheels bled at the same time. Who is correct?

   a. Technician A
   b. Technician B
   c. Both A and B
   d. Neither A nor B

7. When servicing drum brake assemblies, Technician A washes brake dust off using aerosol brake cleaner and a shop pan. Technician B cleans with an Occupational Safety and Health Administration (OSHA)-approved washer. Who is correct?

   a. Technician A
   b. Technician B
   c. Both A and B
   d. Neither A nor B

8. Technician A says the brakes of an ABS may be bled in the conventional manner. Technician B says bleeding the antilock brakes may require using a scan tool to properly bleed the system. Who is correct?

   a. Technician A
   b. Technician B
   c. Both A and B
   d. Neither A nor B

9. Technician A says many brake linings are riveted or bonded to the backing of the pad or shoe. Technician B says all brake shoe linings are riveted to the shoes. Who is correct?

   a. Technician A
   b. Technician B
   c. Both A and B
   d. Neither A nor B

10. Which of the following statements is *not* true?

    a. Proportioning valves limit pressure to the rear drum brake assemblies.
    b. Metering valves limit pressure to the front disc brakes on combination brake systems.
    c. Proportioning valves reduce pressure to the front disc brakes during hard stops.
    d. A combination valve may include the metering, proportioning, and pressure differential valve in one part.

11. While discussing the possible causes for a lower than normal brake pedal, Technician A says the drum brake self-adjusters may not be working properly. Technician B says worn front brake pads and rotors may be the cause. Who is correct?

    a. Technician A
    b. Technician B
    c. Both A and B
    d. Neither A nor B

12. Technician A says a height-sensing proportional valve provides two different brake balance modes to the rear brakes based on vehicle load. Technician B says when a vehicle is loaded, hydraulic pressure should be reduced to the rear brakes. Who is correct?

    a. Technician A
    b. Technician B
    c. Both A and B
    d. Neither A nor B

13. Technician A says some manufacturers recommend periodic flushing of the brake hydraulic system. Technician B says some manufacturers state that the system must be flushed whenever there is any doubt about the grade of fluid in the system or if the fluid contains the slightest trace of petroleum- or mineral-based fluids. Who is correct?

    a. Technician A
    b. Technician B
    c. Both A and B
    d. Neither A nor B

14. A vehicle is equipped with power disc brakes. The owner says that to stop the car, excessive pedal effort is required. Technician A says that a master cylinder or power brake malfunction could be the reason. Technician B says that air in the hydraulic system is probably the cause. Who is correct?

    a. Technician A
    b. Technician B
    c. Both A and B
    d. Neither A nor B

15. The driver of a car with power brakes says that the brake pedal is very hard to push. Technician A says that this could be caused by a restricted vacuum hose. Technician B says that this could be caused by a damaged booster diaphragm. Who is correct?

    a. Technician A
    b. Technician B
    c. Both A and B
    d. Neither A nor B

16. With the brakes applied, the pedal moves down slightly when the engine is started on a vehicle with power brakes. Technician A says that the cause could be a leaking power brake booster diaphragm. Technician B says that the cause could be a stuck closed check valve on the power brake booster manifold. Who is correct?

    a. Technician A   c. Both A and B

    b. Technician B   d. Neither A nor B

17. On a vehicle with single piston, floating caliper disc brakes, the disc brake pad between the caliper piston and the rotor is badly worn. The other brake pad is only slightly worn. Technician A says that too much rotor runout could be the cause. Technician B says that a binding caliper piston could be the cause. Who is correct?

    a. Technician A   c. Both A and B

    b. Technician B   d. Neither A nor B

18. Which of the following statements concerning the caliper assembly is *incorrect*?

    a. Floating calipers have at least two pistons, one inboard and one outboard.

    b. Caliper action converts hydraulic pressure into mechanical force.

    c. A caliper assembly can contain as many as four cylinder bores and pistons.

    d. The caliper assembly provides a means of forcing the brake pads against the rotor.

19. Which of the following is not likely to cause a pulsating brake pedal?

    a. Loose wheel bearings   c. Excessive lateral runout

    b. Worn brake pad linings   d. Excessive rotor parallelism

20. While discussing how to remove the piston from a brake caliper, Technician A says the dust boot should be removed, and then a large dull screwdriver should be inserted into the piston groove to pry the piston out. Technician B says air pressure should be injected into the bleeder screw's bore to force the piston out of the caliper. Who is correct?

    a. Technician A   c. Both A and B

    b. Technician B   d. Neither A nor B

CHAPTER 54

HEATING AND AIR CONDITIONING

OVERVIEW

Heating, ventilation, and air-conditioning (HVAC) systems are used to provide passenger comfort. Heating and ventilation systems are required as standard equipment on all passenger cars and trucks. Air conditioning is standard on most models and available for nearly all others. As with all other aspects of the automobile, the HVAC system has evolved to meet the needs and expectations of the consumer. Modern systems offer automatic climate control as well as separate controls for the driver and passengers.

A network of ducts and registers provides ventilation through the passenger compartment. Passengers can select outside air or to recirculate air. Air enters via the ram-air effect when the vehicle is moving. A fan circulates the air when the vehicle is stopped. **Figure 54-1** shows an example of how air flows into and through the passenger compartment.

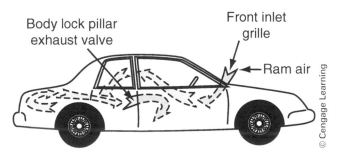

Figure 54-1 A flow-through ventilation system is used on some vehicles. Ram air is forced into the inlet grill and sent throughout the passenger and trunk compartments. The air then flows out of the vehicle through exhaust areas.

The HVAC panel on the dash controls air distribution with a fan and a network of doors. **Figure 54-2** shows an example of the air distribution components. An HVAC system allows for control of the air discharge position, defrost, dash vents, and floor vents. A blend door, which varies the amount of inside and outside air, controls temperature and air either warmed by the heater core or cooled by the air-conditioning evaporator. The recirculation function shuts off outside airflow. Cables, vacuum diaphragms, or electric motors can control the outlet, blend, and recirculation doors.

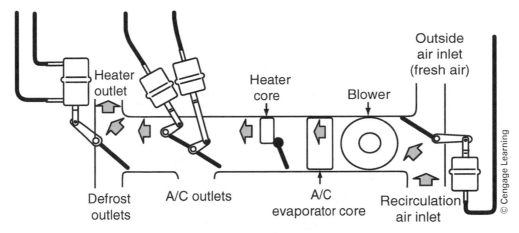

Figure 54-2 A typical vacuum controlled HVAC door system.

 Concept Activity 54-1
Air Circulation

1. Why does the HVAC system allow for outside airflow and recirculated airflow? _____

2. If a vehicle has a complaint of inadequate air circulation, what items should be checked? _____

_____

_____

■

To provide heat for the passenger compartment, hot coolant is circulated from the cooling system into the passenger compartment. Vehicles are equipped with a heater core, which is similar to a small radiator that transfers heat absorbed by the cooling system to the air in the passenger compartment. The heater core is supplied coolant by heater hoses. A heater control valve restricts or stops the flow of coolant into the heater core, depending on the desired inside temperature.

 Concept Activity 54-2
Heater System

Identify the components of the heater system below **(Figure 54-3)**.

  a. _____   b. _____

  c. _____   d. _____

  e. _____   f. _____

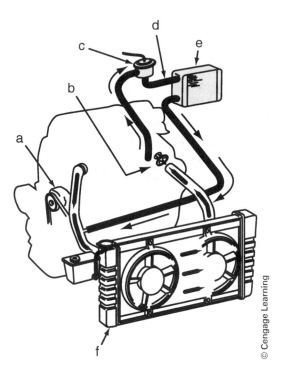

Figure 54-3   Identify the components of the heater system.

Air-conditioning systems apply the principles of heat transfer discussed in Chapter 8 and Concept Activity 8-1.

Heat is transferred in three ways: conduction, convection, and radiation. Conduction takes place when two objects touching each other exchange heat. A pot of water placed on a kitchen stove has heat applied directly to the bottom of the pot. That heat is then transferred to the water inside of the pot. Heat transfer by convection is when heat transfers to the surrounding air. On a hot day, if you stand in front of a fan blowing air, heat is removed from your body by convection. Heat can transfer by radiation, also called thermal or infrared rays. Heat from the filament of an incandescent light bulb can easily be felt. This is an example of radiated heat.

Concept Activity 54-3
Heat Transfer

1. List at least three examples of heat transfer by conduction in the automobile.

_____

_____

2. List at least three examples of heat transfer by convection in the automobile.

_____

_____

_____

In the air-conditioning system, a special refrigerant is used to absorb and release the heat from the passenger compartment. This is accomplished by changing the pressure of the refrigerant. Concept Activity 9-2 discusses the relationship between temperature and pressure.

Though modern air-conditioning systems vary slightly in design and control, all operate similarly. Heat is absorbed by a low-pressure, low-temperature liquid. The addition of the heat to the liquid causes it to boil. The hot vapor is circulated though a condenser, where heat is removed. This cools the vapor to a low-temperature gas. This gas is pumped to a restriction that causes the gas to drop in pressure and temperature. The process repeats as refrigerant is cycled through the passenger compartment and removes the heat of the air inside.

Concept Activity 54-4
Air-Conditioning Components

1. Identify the components of the air-conditioning system in the figure below **(Figure 54-4)**.

a. _____   b. _____

c. _____   d. _____

e. _____   f. _____

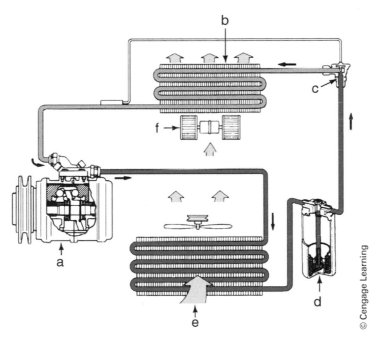

Figure 54-4   Label the components of the A/C system.

The air-conditioning system is divided into two sections: the high side and the low side. The high side includes the components that operate at high pressure and high temperature. The low side includes the components that operate at low pressure and low temperature. Using Figure 54-4 above, identify the components of the high and low sides.

2. High side _____

_____

3. Low side _____

_____

4. Under what condition(s) are the pressures equal in the air-conditioning system? _____

_____

_____

■
_____

The following job sheets will prepare you to inspect and service the HVAC system.

☐ JOB SHEET 54–1

Identifying the Type of Air-Conditioning System in a Vehicle

Name _____ Station _____ Date _____

Objective

Upon completion of this job sheet, you will have demonstrated the ability to use a service manual and the labeling in the engine compartment to identify the type of air conditioning used in a vehicle. This task applies to ASE Education Foundation Tasks MLR 7.A.1 (P-1), and AST/MAST 7.A.2 (P-1).

Description of Vehicle

Year _____ Make _____ Model _____

VIN _____ Engine Type and Size _____

Mileage _____

PROCEDURE

1. Look at the instrument panel controls for the air-conditioning system. Describe the controls. _____

   _____

   _____

2. Look at all of the labeling in the engine compartment and locate any label that addresses the air-conditioning system. Summarize the contents of the labeling.

   _____

   _____

3. Look carefully at the fittings used at the compressor. What types of fittings are they?

   _____

   _____

   _____

   _____

4. What refrigerant is this system designed to use? _____

   What type of refrigerant oil must be used? _____

5. Locate any service precautions about this vehicle's air-conditioning system that are in the service manual. Record your findings.

   _____

   _____

   _____

6. Read through the material given in the service manual and determine what controls the flow of refrigerant through the evaporator. Summarize your findings here.

_____

_____

_____

7. Describe the air-conditioning cycle for this vehicle.

_____

_____

_____

_____

_____

_____

_____

Problems Encountered

_____

_____

_____

Instructor's Comments

_____

_____

_____

☐ JOB SHEET 54–2

Identifying the Components in an Air-Conditioning System

Name _____ Station _____ Date _____

Objective

Upon completion of this job sheet, you will have demonstrated the ability to locate and identify the main components of a vehicle's air-conditioning system. This task applies to ASE Education Foundation Task MLR/AST/MAST 7.A.2 (P-1).

Description of Vehicle

Year _____ Make _____ Model _____

VIN _____ Engine Type and Size _____

Mileage _____

Describe the type of air conditioning that this vehicle is equipped with.

_____

_____

PROCEDURE

1. Using the service information as a guide, locate the following components and describe their location.

   Note: *Not all of these components will be found in one particular air-conditioning system. Locate the ones that are in the system being studied.*

   a. Compressor

   _____

   _____

   _____

      1. What type is it? _____

   b. Evaporator

   _____

   _____

   _____

   c. Condenser

   _____

   _____

   _____

d. Receiver/dryer

_____

_____

_____

e. Thermostatic expansion valve/orifice tube

_____

_____

_____

f. High-pressure line

_____

_____

_____

g. Low-pressure line

_____

_____

_____

h. Sight glass

_____

_____

_____

i. Accumulator

_____

_____

_____

j. Blower motor/fan

_____

_____

_____

k. Ambient temperature switch

_____

_____

_____

l. Thermostatic switch

_____

_____

_____

m. Pressure cycling switch

_____

_____

_____

n. Low-pressure cutoff or discharge pressure switch

_____

_____

_____

o. High-pressure cutout switch

_____

_____

_____

p. High-pressure relief valve

_____

_____

_____

q. Compressor control valve

_____

_____

_____

r. Electronic cycling clutch switch

_____

_____

_____

Problems Encountered

_____

_____

_____

Instructor's Comments

_____

_____

_____

☐ JOB SHEET 54–3

Conducting a Visual Inspection of an Air-Conditioning System

Name _____ Station _____ Date _____

Objective

Upon completion of this job sheet, you will have demonstrated the ability to conduct a thorough visual inspection of an air-conditioning system before beginning detailed testing of the system. This task applies to ASE Education Foundation Tasks MLR 7.A.1 and AST/MAST 7.A.2.

Tool and Material

Feeler gauge set

Description of Vehicle

Year _____ Make _____ Model _____

VIN _____ Engine Type and Size _____

Mileage _____

Describe the type of air conditioning that this vehicle is equipped with.

_____

PROCEDURE

1. Inspect the condition of the compressor's drive belt. Describe its condition.

_____

_____

_____

2. Inspect the refrigerant hose and fittings from each of the following.

   a. Compressor to the condenser
   b. Receiver/dryer to accumulator
   c. Condenser to the evaporator
   d. Evaporator to compressor

   Describe your findings.

_____

_____

_____

3. Inspect electrical connections for the system. Describe your findings.

_____

_____

_____

4. Inspect the condenser by looking for any dirt or debris buildup that could cause decreased airflow through the condenser. Describe your findings.

_____

_____

_____

5. Locate the compressor clutch air gap specifications.

   The specified gap is _____.

   Measure the gap with a feeler gauge. The measured gap is _____.

   If the gap is outside of specifications, what should you do?

   _____

   _____

   _____

6. Start the engine and turn on the air conditioning system.

7. After the system has been on for a few minutes, check the condenser by feeling up and down the face or along the return bends for a temperature change. Describe your findings.

   _____

   _____

   _____

   _____

   There should be a gradual change from hot to warm as you go from the top to the bottom. Any abrupt change indicates a restriction and that the condenser has to be flushed or replaced.

8. If the system has a receiver/dryer, check it. The inlet and outlet lines should be the same temperature. Any temperature difference or frost on the lines or receiver tank is a sign of a restriction. Describe your findings.

   _____

   _____

   _____

9. If the system has a sight glass, check it. Describe your findings.

   _____

   _____

   _____

10. Feel the liquid line from the receiver/dryer to the expansion valve. The line should be warm for its entire length. Describe your findings.

   _____

   _____

   _____

11. The expansion valve should be free of frost, and there should be a sharp temperature difference between its inlet and outlet. Inspect it and describe your findings.

_____

_____

_____

12. The suction line to the compressor should be cool to the touch from the evaporator to the compressor. If it is covered with thick frost, that might indicate that the expansion valve is flooding the evaporator. Inspect it and describe your findings.

_____

_____

_____

13. Inspect the entire system for the presence of frost. Typically the formation of frost on the outside of a line or component means there is a restriction to the flow of refrigerant. Describe your findings.

_____

_____

_____

14. On vehicles equipped with the orifice tube system, feel the liquid line from the condenser outlet to the evaporator inlet. A restriction is indicated by any temperature change in the liquid line before the crimp dimples the orifice tube in the evaporator inlet. Describe your findings.

_____

_____

_____

15. Based on the visual inspection, what are your conclusions about this air-conditioning system?

_____

_____

_____

Problems Encountered

_____

_____

_____

Instructor's Comments

_____

_____

_____

☐ JOB SHEET 54–4

Recycling, Labeling, and Storing Refrigerant

Name _____ Station _____ Date _____

Objective

Upon completion of this job sheet, you will have demonstrated the ability to recycle, label, and store refrigerant safely and properly. This task applies to ASE Education Foundation Task AST/MAST 7.E.3 (P-1).

Tool and Material

Recycling machine

Description of Vehicle

Year _____ Make _____ Model _____

VIN _____ Engine Type and Size _____

Mileage _____

Describe the type of air conditioning that this vehicle is equipped with.

_____

PROCEDURE

1. Refer to the manual for your specific recycling machine to determine the procedure for recycling refrigerant. Summarize this procedure.

   _____

2. Isolate the recycling machine from the vehicle system. Explain how this is done. _____

3. Recycle the refrigerant.

4. Label the container as "empty" or "empty, evacuated, and ready for disposal" or "Freon" or "contains recycled Freon." Label Used _____

5. After the container has been identified, store or dispose of it accordingly. Describe how refrigerant is properly stored at your location. _____

Problems Encountered

_____

_____

_____

Instructor's Comments

_____

_____

_____

REVIEW QUESTIONS

Review Chapter 54 of the textbook to answer these questions:

1. List five components of the ventilation system. _____

   _____

   _____

2. Define latent heat. _____

   _____

   _____

3. Explain the benefits of using CO_2 as a refrigerant. _____

   _____

   _____

4. Explain why special refrigerant oil is required with electrically driven A/C compressors. _____

   _____

   _____

5. Describe the basic refrigerant cycle. _____

   _____

   _____

AIR-CONDITIONING DIAGNOSIS AND SERVICE

CHAPTER 55

OVERVIEW

The heating, ventilation, and air-conditioning (HVAC) systems share many components, such as the interior blower fan, but often have their own electrical control circuits. Many aspects of the operation of the HVAC system are monitored and controlled by the on-board computer system.

Safety is of primary importance whenever servicing the HVAC system. Care should be taken when working on components of the heating system due to the pressure and high temperature of the coolant and components of the system. Air-conditioning lines carry refrigerant under high pressure and at high temperatures. Refrigerant lines should never be disconnected while under pressure. Refrigerant that contacts skin can cause frostbite and blindness if it gets into your eyes. Refrigerants should never be exposed to a heat source because poisonous gas can be created.

Concept Activity 55-1
Air-Conditioning Service Precautions

1. Locate the air-conditioning service decals for a vehicle assigned to you by your instructor. Describe any safety precautions they contain. _____

2. Refer to the manufacture's service information for any system service precautions or safety warnings. Describe them here. _____

3. Explain what precautions should be taken when handling refrigerants. _____

4. Explain how exposure to refrigerant can cause frostbite. _____

Air-conditioning diagnosis begins with verifying the customer complaint. The most common A/C complaints are no cooling and/or poor cooling. As with diagnosing other systems, a thorough visual inspection should be performed, as well as a check for diagnostic trouble codes (DTCs), if applicable.

Concept Activity 55-2
Air-Conditioning Diagnosis

Match the following faults with the symptoms given below: no cooling, poor cooling, excessive noise, and water in passenger compartment.

1. Open compressor clutch coil   _____

2. Open low-/high-pressure switch   _____

3. Leaking seals   _____

4. Loose drive belt   _____

5. Restricted condenser air flow   _____

6. Plugged evaporator case drain   _____

7. Improper blend door adjustment   _____

_____

■

Tables 55-1, **55-2**, and **55-3** will aid you in diagnosing and servicing the HVAC system.

TABLE 55-1 Heating System Troubleshooting.

Symptom	Possible Cause
Insufficient, erratic, or no heat or defrost	Low radiator coolant level due to coolant leaks
	Engine overheating
	Loose fan belt
	Faulty thermostat
	Plugged or partially plugged heater core
	Loose or improperly adjusted control cables
	Kinked, clogged, collapsed, soft, swollen, or decomposed engine cooling system or heater system hoses
	Blocked air inlet
Improper blower operation	Defective blower motor
	Defective blower resistor
	Defective blower wire harness
	Defective blower switches
Too much or too little heat (air comes out of incorrect outlet for any function control lever position)	Loose or improperly adjusted control cables

TABLE 55-2 Typical Interpretation of Pressure Gauge Readings.

Low Side	High Side	Aux. Gauge	Sight Glass	Air Temp.	Diagnosis of Problem
N	N	N	B	W	Moisture in the system
H	H	H	B	C-W	Air in the system, overcharge
L	N	L	CL	C-W	Thermostatic switch stuck closed
L	L	L	B	W	Low charge
H	H	H	CL	W	Overcharge
L	L	L	CL	W	Expansion valve stuck closed
L	L	—	—	W	Orifice tube clogged
H	H	H	CL	W	Expansion valve stuck open
L	H	L	B	W	High-side restriction
H	L	L	CL	C-W	Suction throttling valve (STV) or evaporator pressure regulator (EPR) stuck closed
L	H	L	CL	C-W	STV or EPR stuck open
H	L	H	CL	W	Compressor malfunction

Key: B = bubbles, C = cool, C-W = cool-warm, CL = clear, H = high, L = low, N = normal, W = warm.

TABLE 55-3 General Air-Conditioner Diagnostic Procedures.

Symptom	Possible Cause
Little or no airflow	Open circuit indicates short in electrical system. Test blower motor switch.
	Inspect wiring to blower motor.
	Check blower motor for operation.
	Check resistor block.
	CAUTION: Never try to bypass fuse in resistor block. To do so could cause the blower motor to overheat, resulting in serious damage to the air-conditioning system.
	Restrictions or leaks in air ducts
	Ice on evaporator coil
Warm airflow when the air conditioner is on.	No refrigerant charge in system
	Moisture in system
	Refrigerant compressor not operating
	Air-conditioner microswitch not working
	Ice on evaporator coil
Foaming in receiver/dryer sight glass	Insufficient refrigerant charge in system
	Leaks in system
	Restriction in line between the condenser and receiver/dryer slowing refrigerant flow
Low evaporator coil outlet pressure (low compressor suction pressure)	Expansion valve not working
	Restrictions in line to expansion valve
	Insufficient refrigerant charge in the system

© Cengage Learning 2015

(Continued)

TABLE 55-3 *(Continued)*

Symptom	Possible Cause
High compressor discharge pressure	Shutters are not opening.
	Airflow through condenser is restricted.
	Internal restriction in condenser (return tube ends should all be same temperature)
	Air present in system
	System has too much refrigerant.
	Heavy frosting on suction line suggesting that evaporator coil flooded
	Engine overheated
	Restriction in compressor discharge line
Evaporator outlet air temperature increases as compressor discharge pressure drops.	Leaks in system
	Expansion valve setting too low
	Too much oil is in system (indication is clutch or belt slippage at governed engine speed).
Compressor operates too often or continuously.	Too little refrigerant in system
	Ice formed on evaporator coil
	Restriction in refrigeration system
	Dirt and debris clogging condenser fins
	Thermostatic switch not working
Quick or delayed cycling of compressor	Thermostatic switch operates but out of adjustment
	Loss of refrigerant causing delayed cycling of compressor
Temperature in cab too low or no heat	Water-regulating valve not open
	Water-regulating valve not opening completely
	Water-regulating valve not working
	Heater hose pinched or twisted
	Coolant leaking from system
	Dust or dirt clogging heater core fins
Condensed water leaking from the air conditioner	Drain tubes plugged

The following job sheets will prepare you for diagnosing and servicing the HVAC system.

☐ JOB SHEET 55–1

Conducting a System Performance Test on an Air-Conditioning System

Name _____ Station _____ Date _____

Objective

Upon completion of this job sheet, you will have demonstrated the ability to conduct a thorough system performance test on an air-conditioning system. This task applies to ASE Education Foundation Task AST/MAST 7.A.3 (P-1).

Tools and Materials

Thermometer

Manifold gauge set

Assortment of hose adapters and fittings

Description of Vehicle

Year _____ Make _____ Model _____

VIN _____ Engine Type and Size _____

Mileage _____

Describe the type of air-conditioning that this vehicle is equipped with:

_____

PROCEDURE

Note: *Because R-134a is not interchangeable with R-12, separate sets of hoses, gauges, and other equipment are required to service vehicles. All equipment used to service R-134a and R-12 systems must meet Society of Automotive Engineers (SAE) standard J1991. The service hoses on the manifold gauge set must have manual or automatic backflow valves at the service port connector ends. This prevents the refrigerant from being released into the atmosphere during connection and disconnection. Manifold gauge sets for R-134a can be identified by one or all of the following: labeled FOR USE WITH R-134a, labeled HFC-134 or R-134a, and/or have a light blue color on the face of the gauges.*

For identification purposes, R-134a service hoses must have a black stripe along their length and be clearly labeled SAE J2196/R-134a. The low-pressure hose is blue with a black stripe. The high-pressure hose is red with a black stripe, and the center service hose is yellow with a black stripe. Service hoses for one type of refrigerant will not easily connect into the wrong system, as the fittings for an R-134a system differ from those used in an R-12 system.

1. What refrigerant is used in this system? _____

2. Select the correct manifold gauge set.

3. Connect the manifold gauge set to the respective high- and low-pressure fittings. Where are these fittings found on this vehicle?

   High side _____

   Low side _____

4. Close the hood and all of the doors and windows of the vehicle. Explain why this is done.

_____

5. Adjust the air-conditioning controls to maximum cooling and high blower position.

6. Idle the engine in neutral or park with the brake on. For the best results, place a high-volume fan in front of the radiator grille to insure an adequate supply of airflow across the condenser.

7. Increase engine speed to 1,500 to 2,000 rpm.

8. Measure the temperature at the evaporator air outlet grille or air duct nozzle (35° to 40°F [1.6° to 4.4°C]). The temperature was: _____

9. Read the high and low pressures, and compare them to the normal range of the operating pressure given in the service manual. What were the pressure readings?

_____

What are the normal pressure ranges?

_____

10. What can you conclude from the results of this test? What service do you recommend?

_____

_____

_____

_____

_____

Problems Encountered

_____

_____

_____

Instructor's Comments

_____

_____

_____

☐ JOB SHEET 55–2

Checking the Clearance (Air Gap) of an Air-Conditioning Compressor's Clutch

Name _____ Station _____ Date _____

Objective

Upon completion of this job sheet, you will have demonstrated the ability to determine and measure the clearance of an air-conditioning compressor's clutch. This task applies to ASE Education Foundation Task AST/MAST 7.B.2 (P-1).

Tools and Materials

Measuring tool (nonmagnetic feeler gauge set)

Description of Vehicle

Year _____ Make _____ Model _____

VIN _____ Engine Type and Size _____

Mileage _____

PROCEDURE

1. Place fender covers over the front fenders to protect the paint.

2. Visually inspect the compressor clutch for damage. Note findings. _____

3. Loosen the drive belt to allow for rotation of the compressor pulley.

4. Using a nonmagnetic feeler, inspect the air gap between the compressor pulley and the clutch assembly's pressure plate. Rotate the compressor pulley one-half turn and again inspect the air gap. What are the two measurement values?

   _____

5. Compare the measurement values to shop manual specifications. Is the air gap within specifications? _____ YES _____ NO

6. How is the air gap adjusted on this model of A/C compressor?

   _____

   _____

7. Adjust the compressor drive belt.

   Belt Tension Specification _____

Problems Encountered

_____

_____

_____

Instructor's Comments

_____

_____

_____

☐ JOB SHEET 55–3

Inspecting, Testing, and Servicing a Compressor Clutch

Name _____ Station _____ Date _____

Objective

Upon completion of this job sheet, you will have demonstrated the ability to inspect, test, and service a compressor clutch. This task applies to ASE Education Foundation Task AST/MAST 7.B.2 (P-1).

Tools and Materials

Necessary pullers and installers

Test light

Digital multimeter (DMM)

Description of Vehicle

Year _____ Make _____ Model _____

VIN _____ Engine Type and Size _____

Mileage _____

Describe the type of air conditioning that this vehicle is equipped with.

_____   _____

PROCEDURE

1. Place fender covers over the front fenders for protection.

2. Visually inspect the clutch for damage. Note findings. _____

3. Inspect the electrical connections to the clutch. Note findings. _____

4. With key on engine running (KOER), use a test light or DMM with the A/C system on, check the electrical plug at the clutch for power.

   Results _____

5. If there is no power, locate and repair the problem. Retest.

   Location of Problem _____

6. When power is present, test the ground circuit with a DMM.

7. If it is determined that both power and ground are present at the compressor clutch, the clutch is defective and must be replaced. On some vehicles, the clutch can be serviced on the vehicle. On others, the compressor assembly must be removed to service the clutch. Refer to the service manual on your specific vehicle to determine what must be done.

   On vehicle _____ Off vehicle _____

8. Using the proper pullers for your specific application and following the procedure in the service manual, remove the clutch.

9. Reverse the procedure and install the new clutch.

10. Test clutch operation after repairs are completed.

Problems Encountered

_____

_____

_____

Instructor's Comments

_____

_____

_____

☐ JOB SHEET 55–4

Identifying and Recovering Refrigerant

Name _____ Station _____ Date _____

Objective

Upon completion of this job sheet, you will have demonstrated the ability to identify and recover refrigerant. This task applies to ASE Education Foundation Task AST/MAST 7.E.2 (P-1).

Tools and Materials

Manifold gauge set   Refrigerant identifier

Description of Vehicle

Year _____ Make _____ Model _____

VIN _____ Engine Type and Size _____

Mileage _____

Describe the type of air conditioning that this vehicle is equipped with.

_____

PROCEDURE

1. Place fender covers over the front fenders for protection.

2. Check the decal under the hood on the air-conditioning unit for the type of refrigerant used in your specific system. If this decal is not present, refer to the service manual for your specific vehicle. Find the type of Freon used in your system.

   Refrigerant Type _____

3. One other way to determine if the system uses R-12 or R-134a is to look at the service valves. If the valves are smaller, and threaded on the outside, the system uses R-12. If the valves are considerably larger, the system would most likely use R-134a.

   Note: *It is very important to determine the type of refrigerant in your system before servicing. These two types of refrigerant cannot be mixed. Before connecting the manifold set or refrigerant recovery equipment, the contents of the air-conditioning system should be identified. Systems with improper refrigerants can contaminate and destroy the air-conditioning equipment.*

4. Attach a refrigerant identifier to the system. Obtain a sample and record the findings of the tester. _____

5. If the system passes the identifier test, disconnect the identifier and connect the recovery equipment. _____

6. Begin the recovery procedure following the instructions for your specific vehicle. Describe the recovery procedure. _____

7. When the recovery is complete, close the valves on the manifold gauge set and turn off the recovery unit. You are now prepared to do necessary repairs on the air-conditioning system.

8. Describe the type of refrigerant oil used in the system.

9. If a major component is replaced, a specific amount of il must be added to the system. Record the amount of oil that is recommended after replacing the condenser.

10. After the repairs are made, the system must be recharged. Describe the procedure for recharging the system of the vehicle you are working on.

Problems Encountered

_____

_____

_____

Instructor's Comments

_____

_____

_____

REVIEW QUESTIONS

Review Chapter 55 of the textbook to answer these questions:

1. Explain why special safety precautions must be taken when working with refrigerants.

   _____

   _____

   _____

2. What customer complaint may be caused by a plugged evaporator drain?

   _____

   _____

3. An undercharged A/C system will display _____ pressure gauge readings.

4. Excessive compressor clutch clearance can cause what problems?

   _____

   _____

   _____

5. What is the purpose of pulling a vacuum on the A/C system prior to recharging?

   _____

   _____

   _____

ASE PREP TEST

1. The compressor clutch coil does not energize when the A/C system is turned on. However, when a jumper wire is used to connect the battery to the clutch coil, the clutch activates. All but which of the following could cause the problem?

   a. A blown fuse

   b. A burnt-out clutch coil

   c. A defective pressure switch

   d. A wiring problem

2. The compressor clutch does not engage, but the A/C pressures are normal. Technician A says that a defective pressure cycling switch could be the cause. Technician B says an open clutch coil could be the cause. Who is correct?

   a. Technician A

   b. Technician B

   c. Both A and B

   d. Neither A nor B

3. While conducting a pressure test on an A/C system, the ambient temperature is 85°F (29°C), the low-side gauge reads very low (10 psi [69 kPa]), and the high-side gauge also reads low (90 psi [621 kPa]). Which of the following could cause these readings?

   a. A low refrigerant level

   b. Normal operation

   c. A bad compressor

   d. A high-side restriction

4. What should you do to remove an orifice tube?

   a. Use a special tool to pull it out of the accumulator.

   b. Pull it from the evaporator inlet with a pair of needle-nosed pliers or a special puller.

   c. Unbolt it from the top of the accumulator.

   d. Remove it from a fully charged system.

5. There is a clicking noise when the A/C compressor clutch is engaged. Technician A says that low system pressure could be the cause. Technician B says that defective compressor bearings could be the cause. Who is correct?

   a. Technician A

   b. Technician B

   c. Both A and B

   d. Neither A nor B

6. In discussing the retrofitting of an R-12 system to an R-134A system, Technician A says to evacuate the A/C system for 30 minutes. Technician B says before starting the retrofit use a refrigerant identifier to make sure the system only contains R-12. Who is correct?

   a. Technician A

   b. Technician B

   c. Both A and B

   d. Neither A nor B

7. The high-side pressure of an A/C system is higher than normal. All except which of the following could be the cause?

   a. Overcharging of the refrigerant

   b. A bad compressor

   c. Air in the refrigerant system

   d. Restricted airflow through the condenser

8. In discussing a fluorescent leak detector, Technician A says to first introduce the dye into the system using a special infuser. Technician B says tracer protection goggles are used and the system is scanned with a black light. Who is correct?

   a. Technician A
   b. Technician B
   c. Both A and B
   d. Neither A nor B

9. Technician A says it is permitted to vent blended refrigerants to atmosphere. Technician B says using refrigerants other than R-134a may void the part manufacturer's warranty. Who is correct?

   a. Technician A
   b. Technician B
   c. Both A and B
   d. Neither A nor B

10. The heater of a car puts out very little heat. Which of the following could cause the problem?

   a. Low coolant level
   b. A thermostat that is stuck open
   c. A heater control valve stuck closed
   d. All of the above

11. When checking an A/C system, the evaporator's inlet and outlet tubes feel cold and about the same temperature. Which of the following could cause this?

   a. A plugged evaporator
   b. Normal system operation
   c. A restricted orifice tube
   d. Restricted expansion valve

12. When conducting an A/C system performance test, Technician A says to start then engine and set the A/C to MAX cooling. Technician B says to run the engine at 3,500 rpm during a performance test. Who is correct?

   a. Technician A
   b. Technician B
   c. Both A and B
   d. Neither A nor B

13. While visually inspecting an A/C system, the technician notices a buildup of frost around an A/C line. Which of the following could cause this?

   a. An overcharged system
   b. An undercharged system
   c. A refrigerant leak
   d. A restriction in the line

14. All except which of the following are reasons to replace an accumulator/dryer unit?

   a. The accumulator/dryer is punctured.
   b. The accumulator/dryer is saturated with water.
   c. The A/C system was open to the atmosphere for two or more hours.
   d. The outer shell of the accumulator/dryer is dented.

15. The filter screen of an orifice tube is covered with aluminum particles. Which of the following could be the cause?

   a. A leaking desiccant bag
   b. A plugged condenser
   c. A damaged compressor
   d. A damaged evaporator

16. Which of the following can be removed from most A/C systems without discharging the refrigerant from the system?

   a. A pressure cycling switch
   b. An accumulator/dryer
   c. A low-pressure cutoff switch
   d. A condenser

17. A special disconnect tool is required to separate the connection of a spring-lock coupling. What is this tool used for?

    a. To release the O-rings

    b. To reduce the size of the flare

    c. To collapse the cage

    d. To expand the garter spring

18. Technician A says that a sight glass is often used to determine correct charge on R-134a systems. Technician B says that a milky appearance in a sight glass indicates that there is air in the system. Who is correct?

    a. Technician A

    b. Technician B

    c. Both A and B

    d. Neither A nor B

19. An A/C system is being tested with pressure gauges. The high-side pressure is high and the low-side pressure is high. Which of the following is the *least likely* cause of these pressures?

    a. The system is overcharged.

    b. A restriction in the high side of the system

    c. Poor airflow over the condenser

    d. Inoperative cooling fan(s)

20. While diagnosing higher than normal system pressures, Technician A checks the condenser for dirt buildup in the condenser's fins. Technician B tests the engine for indications of an overheating condition. Who is correct?

    a. Technician A

    b. Technician B

    c. Both A and B

    d. Neither A nor B